TRANSPORTATION PLANNING FOR A BETTER ENVIRONMENT

NATO CONFERENCE SERIES

I Ecology
II Systems Science
III Human Factors
IV Marine Sciences
V Air–Sea Interactions

II SYSTEMS SCIENCE

Volume 1 Transportation Planning for a Better Environment
Edited by Peter Stringer and H. Wenzel

TRANSPORTATION PLANNING FOR A BETTER ENVIRONMENT

Edited by

Peter Stringer

University of Surrey
Guildford, England

and

H. Wenzel

Messerschmitt-Bölkow-Blohm
Munich, West Germany

Published in coordination with NATO Scientific Affairs Division by

PLENUM PRESS · NEW YORK AND LONDON

Library of Congress Cataloging in Publication Data

Main entry under title:

Transportation planning for a better environment.

(NATO conference series: II, Systems science; v. 1)

Proceedings of a conference on Transportation and Urban Life held in Munich, West Germany, Sept. 15-19, 1975, which was sponsored by the Special Program Panels in Systems Science and Human Factors.

1. Transportation planning—Congresses. 2. Underdeveloped areas—Transportation—Congresses. 3. Transportation—Environmental aspects—Congresses. I. Stringer, Peter. II. Wenzel, H. III. NATO Special Program Panel on Human Factors. IV. North Atlantic Treaty Organization. Science Committee. Special Program Panel on Systems Science. V. Series.

HE193.T67 380.5 76-46276

ISBN 0-306-32841-0

Proceedings of a conference on Transportation and Urban Life held in Munich, West Germany, September 15-19, 1975, which was sponsored by the Special Program Panels in Systems Science and Human Factors

A Division of Plenum Publishing Corporation
227 West 17th Street, New York, N.Y. 10011

Printed in the United States of America

EDITORIAL ACKNOWLEDGMENTS

The number of people to whom we are indebted for their help in preparing and running the conference at which the papers in this volume were presented is considerable.

Professor Dr.H.Schmidtke (Vice-Chancellor of the Technical University of Munich), Mr.E.Muller-Heydenreich (Mayor of the City of Munich), Professor Dr.K.Huber (Member of the NATO Special Programme Panel on Systems Science) and Dr.Schweikl (Bavarian State Ministry of Regional Planning and Environmental Control) severally made the conference participants welcome to the city of Munich.

We were fortunate to have advice on the selection of papers for the conference and their service as chairmen of the conference sessions, variously from: Professor Donald Appleyard (University of California, Berkeley), Professor Dr.P.J.D.Drenth (Free University of Amsterdam and Chairman of the NATO Special Programme Panel on Human Factors), Professor Dr.Girnau (Federation of Public Transportation Organisers, Cologne), Professor N.Stuart Kirk (University of Technology, Loughborough), Mr.H.G.Lange (Association of German Cities, Cologne), Mr.J.H.Léon (International Federation of Housing and Planning, The Hague), Professor Dr.John A.Michon (State University of Groningen), and Professor T.Sieverts (Technical University of Darmstadt).

Much of the credit for the organisation of the conference is shared by Dr.Kilian and Dr.Brill (Messerschmitt-Bölkow-Blohm, Munich). Considerable secretarial contributions were made by Mrs.Edda Brodersen, Mrs.Christine Bute, Mrs.Nicolette Gunn, and Mrs.Lynda Nelson.

Finally our thanks would be incomplete without acknowledging the kind assistance of Dr.Bayraktar, the NATO Scientific Affairs officer responsible for co-ordinating NATO Systems Science and Human Factors activities.

Three papers which were presented at the conference cannot be included in this volume for technical and copyright reasons. They were:

Activity Linkages Associated with the Journey to Work
Susan Hanson (Department of Geography, State University of New York at Buffalo)

Transport as an Instrument of Urban Development
D.Marx (Urban Development Office of the Bavarian Capital, Munich)

Transportation Strategies in Large Urban Areas
J.M.Thomson (Directorate for Science, Technology and Industry, OECD, Paris)

GENERAL INTRODUCTION

All the papers in this volume were presented at a conference on Transportation and Urban Life, held in Munich during the third week of September, 1975. The conference was sponsored by the Special Programme Panels on Systems Science and Human Factors of the Science Committee of the North Atlantic Treaty Organisation.

The distinguishing characteristic of the conference and of this volume lies in the combination of systems science and human factors contributions in the field of urban transportation. The initiative for attempting such a synthesis came from the sponsors. It is increasingly realised that the complexity of contemporary problems which applied scientists are being asked to solve is such that the coordinated efforts of several disciplines are needed to solve them.

The brief which we formulated for the conference and distributed in our international call for papers was as follows:

"The conference is intended to highlight significant psychological, sociological and economic aspects of transportation and urban life and to present new techniques which can be applied to these aspects and their interfaces".

Papers were invited in four topic areas: Communities (transportation problems in relation to social needs, residential planning, industry and commerce); Cities and large urban areas (the economic environment, material flow, resources); Regional development and transportation (administration and management at the project and multiprogramme level); and Quality of life and transportation (noise, visual intrusion, severance, air pollution).

It was suggested that these topics might be approached from one of the following orientations:

Technical development and advanced concepts, their assessment and evaluation; Current experience, practical situations, case study, projection; Reports of recent research; Research and

and administrative methodology, forecasting, modelling, monitoring, cost-benefit, survey, experiment, public participation.

The arrangement and content of the papers in this volume demonstrate that the response to the call for papers did not reflect these categories in any balanced way. However, we believe that all the topic areas and orientations are represented at some point. Moreover, we were particularly fortunate that our widespread call for papers attracted submissions from a considerable range of academic disciplines. Amongst the papers published here are contributions from architecture, civil engineering, economics, ergonomics, geography, industrial engineering, management, psychology, social sciences, traffic studies, transport technology and urban and regional planning.

Two other important sources of variation among our contributors are their professional responsibility and their nationality. More than one third of the papers given at the conference were by non-academics, including officers of OECD, of national and state governments, and environmental and engineering consultants. Authors come from Australia, Canada, Germany, the Netherlands, Turkey, the United Kingdom and the United States of America. Conference participants came also from Austria, Belgium, Denmark, France, Greece, Israel, Italy, Spain and Sweden.

The substance of this volume is readily apparent from the list of contents. However, we would draw particular attention to certain themes that have emerged - and have emerged more or less spontaneously in response to the call for papers. The general conceptual background to recent urban and transportation planning, its problems, methods, and directions, is clearly of concern to many practitioners, as shown by the papers of Leibbrand, Richrath and Vohlwahsen. Some of the more significant areas of innovation and development are dealt with in detail in the sections on transportation planning in developing countries, environmental assessment, and citizen participation and open planning. The move to environmental assessment studies is perhaps a major facet of recent planning methodology. The high technology that one has come to expect from quantitative work and modelling approaches takes on distinctive educational overtones in the papers by Ensminger and DiCesare and by Truelove.

The study of traffic accidents might well be the subject of a whole conference. Its inclusion here is particularly justified by the examples offered by Shaoul and by Drury of how systems science and human factors approaches can be integrated. In other sections, Warman, Wicht and Uzzell each make useful demonstrations of the same point.

In some areas one might have hoped for more contributions. The papers on policy by Appleyard, on theory by Michon, and the case studies by Wicht and Navin are relatively isolated. But they are of a standard and significance that makes one hope that more attention will be devoted to these approaches in the near future.

The value of this set of papers, we believe, lies in its recency and in the cross-section of concerns with which it deals. However, because work was submitted and never commissioned, we cannot claim that it is fully representative of the field. It does not necessarily give our view of what the field ought to contain, but is a selection of current work which seemed to us to be of interest and of reasonable scientific merit. The only influence we had on these proceedings, apart from our initial selection among submitted abstracts and minor editorial changes, was to encourage each author to take into account the double focus of the conference, on human factors and systems science.

CONTENTS

SECTION 1. REGIONAL AND URBAN TRANSPORTATION PLANNING

Throughout this volume the simultaneously positive and negative effects of transportation are seen to present one of the major dilemmas of the transport professions and of the public. In Kuhnemann's view the servicing of increased mobility has recently led to such a decline in the environment that it has become urgent to balance the dilemma. In our future, largely urban environment satisfactory environmental conditions will depend on the management of the car and of public transport. The survey by OECD of 300 cities in member countries has shown that it is now generally accepted that cities will be improved if they have less traffic. Measures for achieving car-limitation and alternative modes of transport - buses, trams, pedestrianisation, bicycles, para-transit - are beginning to be widely implemented. The kind of comprehensive planning strategy in which these measures must be embedded is indicated in several of the papers which follow.

The last twenty years have seen considerable changes in the procedures of transportation and urban planning. Some of the major trends are summarised in Leibbrand's paper. There is a closer co-operation today between the various professions involved - engineers, architects and others. Cities are changing their shape under the impact of the motor car and expanding to become low-density, polycentric agglomerations. New areas of interest highlighted by Leibbrand are sophisticated models of traffic forecast, traffic assignment and modal split; issues of traffic restraint; a revival in public transport; environmental concern; and 'democratic' planning which encourages the co-operation of citizens. The last two of these areas define the concluding sections of this volume.

These changes in part reflect responses to the problems which Richrath sees as characterising the recent history of urban planning. He analyses planning issues in terms of a series of conflicts between objectives and means. Techniques for locating mutually accessible residential areas, workplaces, retail sources and recreational facilities, at the same time as achieving a reduced intensity of land-use and less traffic congestion, seem to elude us. If intended solutions such as subsidies, traffic restraint or improved vehicle design do not work, it is because we are clinging too long to the form of the historic city. Rather than continuing to design our cities in response to immediate needs, Richrath believes that we should determine the environment we would like to have, and in a flexible and adaptable form deliberately create it.

Four of the papers in this section describe viewpoints on transportation planning in Germany. The next two draw more directly on the practical experience of planning exercises in that country. Vohlwahsen's focus on analytical techniques forms a link to Section 3, which deals with the high technology of planning methodology. He describes two projects which attempted to reconcile the goals of quality of urban life with the constraints imposed by transportation systems offering necessarily limited accessibilities. A significant aspect of his approach, which anticipates much of the discussion in the final section of this volume, is the formal determination of the quality of urban life in terms of the expectations of the population. Value systems are transformed into land-use patterns.

While Vohlwahsen believes that comprehensive planning efforts at the regional level are necessary to improve urban life, Wicht offers a fascinating case-study of an unusual town which actually contains within itself the ingredients of a solution. In Kaufbeuren the familiar problem of integrating residential and working areas is defined anew by the particular employment characteristics of many of the population. They work in an imitation jewellery industry which consists of some 500 small-scale, independent units. Although considerable inter-unit traffic is generated and although the road network is generally poor, few instances of critical congestion occur. In spite of similarly poor networks for pedestrians and cyclists, less than 50 per cent of journeys to work are undertaken by car. Kaufbeuren offers an encouraging demonstration of how undeveloped a community's transport system need be, if residential areas and workplaces are closely integrated spatially.

BETTER TOWNS WITH LESS TRAFFIC

Jorg K. Kuhnemann

Urban Environment and Land Use Division

O.E.C.D.

The automobile, invented in the last century more or less as a toy or a sports facility to serve a minority of people who believed in economic and technical progress, met with hesitation and even opposition on the part of the wider public, because of its adverse effects on the existing environment. As the advantages of this new means of transport became more and more evident, most of the objections faded away. An object originally designed for pleasure and prestige has managed to become so widespread in the course of a few decades - in our Western society, at least - that it has completely changed our life style, the form of our environment and even some objectives of the entire economy. This development took place with very little serious consideration of the negative impacts associated with the automobile.

At present, as we know only too well since the rise in oil prices alerted us to the question, it is quite likely that we will leave behind us a body of unsolved problems. It is doubtful whether the next generation will be able to cope with them. Our generation has been providing education, techniques and skills that are inadequate because they are based on the applicability of a philosophy of growth which has developed gradually over the centuries. What is called for now is a more comprehensive and balanced approach towards the solution of present and future issues.
The interrelations between environment, economy and ecology need to be viewed from a new angle. Transportation will play an important role in achieving satisfactory environmental conditions through better management of the use of the private automobile and public transport facilities.

The future environment will be an urban one, with approximately

70% of the world's population living in urbanised areas by the turn of the century. What the cities will look like will largely depend on the demands they make and how the decision-making mechanisms respond to these demands.

Although precise investigation has not yet taken place, there is an increasing conviction among decision-makers, planners and economists that it would be both undesirable and unlikely that a total world population of 8 to 11.2 billion - depending on when and at what level zero population growth can be obtained - could be provided with automobiles for private transport at rates achieved in the most developed countries today (U.S.A.: 430 vehicles per 1,000 inhabitants). Theoretically this would mean, bearing in mind the objective of greater equality for the world's population, that world automobile distribution must have room for about a ten-fold growth over the 1985 estimated world automobile figure of 372.5 million units.

The highest figures for car distribution appear at present in the Western countries, who in general are members of the OECD. In 1970, for instance, these 24 countries accounted for 9/10 of all automobiles, whereas they represented only approximately 1/8 of the world's total population.

It will not be too difficult, in the opinion of raw material experts, to make enough metal available; especially when recycling of scrap becomes more common, which requires between 8 to 20 times less energy (aluminium 5-9%, steel 10-15% for recovery). The question of energy for propulsion is more difficult to assess, since no economically feasible substitute for oil is in sight. Depending on the price ratio, the exploitable resources of crude oil, oil sands and shales will be depleted as early as the middle of the next century. This means that severe shortage problems will appear several decades before that date.

Year	OECD countries	World
1960	88.8 millions	97.3 millions
1970	169.1	186.8
1985	313.2	372.5
2000	406	524

Source: OECD Long-Term Energy Assessment, 1974

TABLE 1.

On the assumption that our present generation feels responsible for the living conditions of future generations, we would have to gear the development of our urban areas to a point where people were less dependent on the demand for travel. This process has to be initiated now, since urban investments are long-term investments, designed to serve for 50 to 100 years.

This, in a few brief remarks, is the side of the coin which emphasizes a global concern for a prudent future transport policy. Applying better husbandry to our resources, which include space, time, raw materials, manpower, finance and the environment's capacity for bearing pollution, is more than a nostalgic way of overcoming present constraints.

There is only one world and, as was pointed out by the Australian Minister of the Environment at the OECD Ministerial Conference on the Environment last November, we have not inherited the earth from our fathers and are hence entitled to use it according to our wishes. We have rather borrowed it from our children and have to maintain it properly until they can take over.

It can be assumed that in the future cities throughout the world will be more acceptable if they can be managed with less traffic. But there is also reasonable evidence that our cities today could be better with less traffic.

Our civilisation developed when wandering nomads gave up mobility, between six and ten thousand years ago. The city became a product of culture, an artifact, a homestead for a society, a focal area for human communication and activity. With the development of greater mobility, through more widespread use of the automobile during the last decades, the inherited structure of the city and pattern of human behaviour have changed in such a way that it remains an open question whether planners and decision-makers have achieved environmental improvement by servicing these changes and whether the individual has enhanced the quality of his life, whatever it may include. This is not the time to argue whether or not the positive effects outweigh the negative ones. They probable do, but nevertheless the adverse effects have reached such a magnitude as to compel us to pay serious attention to them if we are to assure that, on the one hand, the advantages of an optimal city pattern with good accessibility to goods and facilities will be maintained and, on the other hand, the deteriorating effect on the quality of the environment will be avoided.

Since the establishment of an Environment Committee at OECD in 1970, the Organisation has been engaged in investigating the effects of the automobile on the environment and the demand of the population for urban transport. In November 1974, based upon the Organisation's previous work, the Environment Committee at

Ministerial level adopted a recommendation on Traffic Limitation and Low-Cost Improvement of the Urban Environment. Considering that "air pollution, noise, congestion and accidents which result from motorised traffic severely hinder attempts to improve the quality of urban life", and considering that "the need to lessen dependence on private automobiles is reinforced by the necessity to conserve energy resources", the OECD Council recommended that Member countries:

"1. seek to strike a better balance between private and public transportation by encouraging local authorities and other responsible bodies, particularly in congested cities, to expand and improve the quality of transportation services, and to encourage the use of means of transport other than private automobiles whenever these alternative means are or can be made available;

"2. include in the above efforts, traffic management and other measures which can be implemented at relatively low cost - such as bus priority measures, provision of bicycle-ways, car pooling, traffic signal systems, parking control and establishment of car-free areas;

"3. complement the above policies with low-cost measures to improve the quality of the local environment, for example through the creation of small parks and open spaces, pedestrian areas, and conservation of the urban landscape;

"4. support and encourage further experimental projects aimed at demonstrating the feasibility and financial, environmental and urban development implications of the measures referred to above and co-operate in the international dissemination of this experience;

"5. in the longer run, seek to reduce the need for transportation through comprehensive land use and transportation planning and through other measures affecting the patterns of human activities".

Taking into account this Council recommendation, an OECD Conference was held in Paris on 14th-16th April, 1975. The Conference was devoted to investigating and discussing the practicality and consequences of limiting automobile traffic in towns and cities.

How great a limitation can be put on the use of cars in cities without hindering mobility and causing economic disruption? What other kinds of transport are likely to be attractive to motorists and used by them, at least for some of their trips? What can be done quickly and inexpensively to relieve cities - their central areas as well as their residential quarters - from the burden of traffic?

Even three years ago the answers to these questions would have shown, with few exceptions, much less emphasis on the importance of environmental integrity and quality. It was a common belief that even the severest implications of private motorisation would have to be accepted as a form of tribute to the progress of our civilisation. This appeared in a survey carried out by OECD in 1971-72 on the impact of pedestrian zones on cities, their economy and environment. None of the cities admitted that pedestrian zones were created to improve the quality of life for their residents and visitors.

It is perhaps because the car-free areas have been such a success in all the cities concerned - Munich is an outstanding example of such success - that cities are now less hesitant to base planning decisions on requirements for a better environment. This emerged from a questionnaire circulated by the Urban Division of OECD to 450 cities, with more than 100,000 inhabitants, in 19 countries. The 300 cities responding to this questionnaire reported that they tended to move on from measures such as parking restrictions and pedestrian streets to devices such as bus lanes, bus-activated traffic signals, and planned congestion for individual vehicles in order to give priority to public transport, cyclists and pedestrians.

A growing interest in all forms of semi-collective transport - also called para-transit - such as car-pooling, dial-a-bus, shared taxis, is apparent, particularly in cities in North America, where in general there are fewer public transport facilities than in Europe and therefore the danger of competition with subsidized public transport does not have to be watched as carefully. Tram services seem to be undergoing a renaissance as an efficient, less polluting system. However, from the standpoint of safety, a reasonable argument can be made against the reappearance of trams in the streets, especially in the innovatory form of light rapid-transit vehicles. For short-distance travel, as a feeder service and also to a large extent for leisure excursions, two-wheelers - bicycles and mopeds - are enjoying increasing attention where the topography and the climate are suitable and the authorities have improved safety for two-wheeler traffic by the creation of bikeways and separate lanes.

Case studies from seven cities selected because of their advanced approach to the creation of a better town with less traffic - Besançon, Bologna, Munich, Nagoya, Nottingham, Uppsala and Singapore - showed that a comprehensive and coordinated policy is more likely to be successful in improving both accessibility for people and goods and the environment as well.

The case studies from the seven cities and expert panels showed that, given the limited financial resources of most of our municipalities, significant improvements can be obtained, generally on a low-cost basis, by better utilisation of existing fleets and equipment. Bus lanes, bus-activated traffic signals and complementary measures, such as para-transit, have the same potential servicing effect as an increased fleet. In addition, faster service by buses makes buses more attractive to the user, with the result that part of the investment for the installation of a costly underground service might be directed to the expansion of public transport into areas as yet unserved. Many cities have had good experiences when diversifying their fleet in keeping with actual demand - by using small units or mini-buses. Areas formerly neglected because of insufficient demand to fill an 80-person bus, and hence totally dependent on the private car, can be connected to the public transport network. Although it is a longer-term process, improvements can also be made by adapting the form of buses more closely to the demands of urban traffic.

Traffic limitation combined with short-term, low-cost improvements to existing public transport potential will not, as is occasionally feared, have a significant influence on automobile production figures; and thus will not represent a serious threat of lay-offs in the industry and balance-of-payment rate changes in national economies. Traffic limitation measures are implemented on a gradual, step-by-step basis, and may affect the annual mileage of individual cars only in cities of a certain size. Other trends and changes in human behaviour, on the one hand, and misinterpreted market demands of the industry, on the other, have a more far-ranging effect than endeavours of public decision-makers for improving the urban environment. Market opportunities for buses and other transport equipment more than compensate the minor effects of traffic limitation.

This side-issue should be borne in mind, since it is evident that there is an inclination on the part of some lower-level representatives of interest groups to question or even hinder the attempts of municipalities to arrive at better solutions.

Unfortunately, as we must say today, overall energy-demand curves are not decreasing as dramatically by means of traffic limitation schemes as one might have hoped. But national energy saving policy justifies good husbandry in this area also, even if the results can be considered minimal in the total framework. These comparatively minimal savings, however, can be an important contribution to a better budget balance for transport authorities. At any rate, it should be borne in mind that these savings are made in limited geographical areas where they have the most welcome side effect of reducing air pollution and noise levels.

The Conference has shown that a more pleasant environment can be achieved without negative implications for economy and trade, so that any reasonable step in this direction by decision-makers deserves support.

The delegates, from 23 OECD countries, agreed on a number of final conclusions which will form part of OECD's contribution to the United Nations Conference on Human Settlements-HABITAT - to take place in Vancouver in June 1976.

I quote briefly from the major findings:

> "Towns are better with less traffic, so long as adequate provision is made for tbe mobility of workers and residents and the distribution of goods. It is of urgent necessity that national and local governments in OECD Member countries develop their efforts to reduce the adverse effects of motor traffic in urban areas. In so doing they should make provision for the needs of people who by choice or necessity do not have access to a car..."

> "Where and when traffic congestion and its cost are severe and frequent, measures to reduce the use of private cars and goods vehicles and to improve alternative forms of transport should be introduced".

The kinds of possible measures to limit traffic and to improve the efficiency and quality of service of buses and trams are then mentioned in detail.

> "Pedestrian schemes should be introduced whenever possible to improve the environment and safety for pedestrians. The use of bicycles should be encouraged wherever practical".

As regards the use of para-transit, it is pointed out that "treating all forms of urban passenger transport as a system and improving interchange between them can also in itself bring about major improvements in service".

> "Programmes of the kind discussed above should be designed as integrated packages and planned and progressively implemented with full public consultation. The potential social, economic, distributional and environmental effects should be evaluated quantitatively and qualitatively at all stages".

In view of the fact that traffic and transport problems cannot be solved with optimal effectiveness by improved urban planning and better coordinated traffic management alone, the Conference concluded that "these policies must also be part of a comprehensive urban strategy, including or taking account of

(i) measures which have effects in the short term and reinforce the impact of traffic limitation on accessibility, environmental and energy conservation aims - appropriate staggering of work hours, technical improvements to vehicles (e.g. improved fuel efficiency, activation of traffic signals by buses), and enforcement of emission and noise standards;

(ii) measures having effects in the longer term to reduce the need for travel through comprehansive land use and transport planning, and other measures affecting the pattern of human activity;

(iii) measures requiring more costly investment in renovation, new infrastructure or, where appropriate, the development of new technologies;

(iv) national matters - economic instruments, and environment standards; considerations such as energy conservation, trade, and the production of the necessary transport equipment and vehicles".

The science of cities provides a wealth of information and expertise on most aspects of urban activity and functions. But it is generally based on hypotheses and assumptions about economic development that are no longer taken for granted as a precondition for urban management. In the recent past, environmental constraints and foreseeable limitations of raw material resources (primarily oil supply), together with reduced growth rates in all Western economies, have resulted in a significant shift of emphasis in values and goals. This new morale or value judgement forms a new background for science and must be taken into account in the future. It requires new research in areas which may formerly have been of lesser concern or had only secondary or tertiary priority in time and budget schedules.

Bearing in mind the urgent demand for more expertise, the Conference finalised its conclusions in a paragraph I should like to quote:

"The practical experiences and experiments reviewed at this Conference show that policies combining selected traffic limitation measures and public transport improvement can achieve a better urban environment, enhance accessibility for people and goods and conserve energy. Therefore the Conference concludes that the national and local governments should actively support and encourage the design, implementation and evaluation of such programmes, through their own political processes. The Conference emphasizes the need for continuing experimentation and research in this area; to promote cooperation between various international organisations on environmental problems, and the dissemination through OECD and other international bodies of the experience and knowledge on these new approaches to the improvement of the urban environment".

The Environment Committee of OECD has therefore decided that, in cooperation with other international bodies, methods being used to evaluate and monitor schemes of this kind should be examined and the effectiveness of both the evaluation methods and the schemes themselves should be studied.

The general objective of our future work at OECD will be to ensure, through a broader approach, that the quality of life will be maximised and that waste of resources will be minimised to such an extent that an optimal urban and natural environment can be achieved.

RECENT CHANGES IN TRANSPORTATION AND URBAN PLANNING

K. Leibbrand

Consulting Engineer

Frankfurt

COMPREHENSIVE PLANNING

In the Fifties town planners and transportation engineers prepared their plans independently. Sometimes they made different proposals concerning the goals, regulations, priorities, distribution of residential or commercial areas, or certain other details. There were misunderstandings, delays and losses. In some cases planning was even more divided: into architecture, land use, landscape, monuments, public transit, roads, etc.

Experience showed that a close cooperation from the very beginning is indispensable. Several cities began to enforce the exchange of views between the experts. In some cases the planners were incorporated into teams. A uniform philosophy has not yet been commonly accepted, but the exchange of views between engineers and architects has been greatly improved.

In a planning team the number of responsible persons should be limited to two or three experts. The best solution seems to be a comprehensive Town Development Plan, including the former independent Master Plan and Master Transportation Plan.

The planners have to work in close contact, use the same basic data, check the effects of every proposal on the whole, avoid contradictory statements, and arrive at a well-balanced result. A mutual understanding is essential. Therefore they need a common philosophy. They have to be aware of the interdependence of urban and transportation planning.

It is surprising that knowledge of the impact of transport on urban structures - the development of polycentric agglomerations, density of population, land use - is still scarce. Very often there are no data available. Experience will help to overcome this handicap.

DEVELOPMENT OF TOWNS

Before modern means of transport were introduced there were only

> pedestrian cities covering up to 14 km^2 of surface, with a density of up to 600 inhabitants/ha, like the old cities of Rome, Byzantium, Alexandria, or Babylon.

The railways brought a complete change. Due to the railways, within just one century the population of most European cities increased tenfold. At the same time a migration from the core areas to the outskirts began. New suburbs were built around the railway stations. From 1851 to 1911, 100,000 people out of 120,000 left the City of London to settle outside. The so-called

> railway cities covered up to 100 km^2 with a density of up to 300 inhabitants/ha

In the motor-age an even more dramatic change began. The transformation of towns into polycentric agglomerations was initiated. Urban sprawl is a result of growing motorisation. Within ten years, 100,000 people out of 700,000 left the city of Dusseldorf to settle in neighbouring communities. The migration is still going on, much faster than the exodus in the railway age. The so-called

> motor cities may cover more than 1,000 km^2 at a very low density.

This in turn means more traffic, for the distances between home and work are growing. Many citizens are not yet fully aware of this basic change of urban development which has not come to an end and which will not be reversed.

In 1960 we prepared a Master Transportation Plan for the City of Munster/Westfalia which was unanimously adopted by the City Council (See Figure 1). Seven years later due to the changing situation a new plan had to be established. We have just finished a third version. Table 1 shows the changes which occurred.

In 1972 there were 10% more cars than first anticipated for 1990. The number of cars entering and leaving the Central Business District during the rush hour and the number of bus passengers all

TABLE 1. CHANGES IN POPULATION AND CAR OWNERSHIP IN MUNSTER, 1959-1972

	1959	1966	1972
actual population	175 000	199 000	200 000
estimated population	225 000(1990)	230 000(1995)	220 000(2000)
cars per 1000 inhab.	78	178	265
cars per 1000 inhab.estimat.	240(1990)	355(1995)	425(2000)

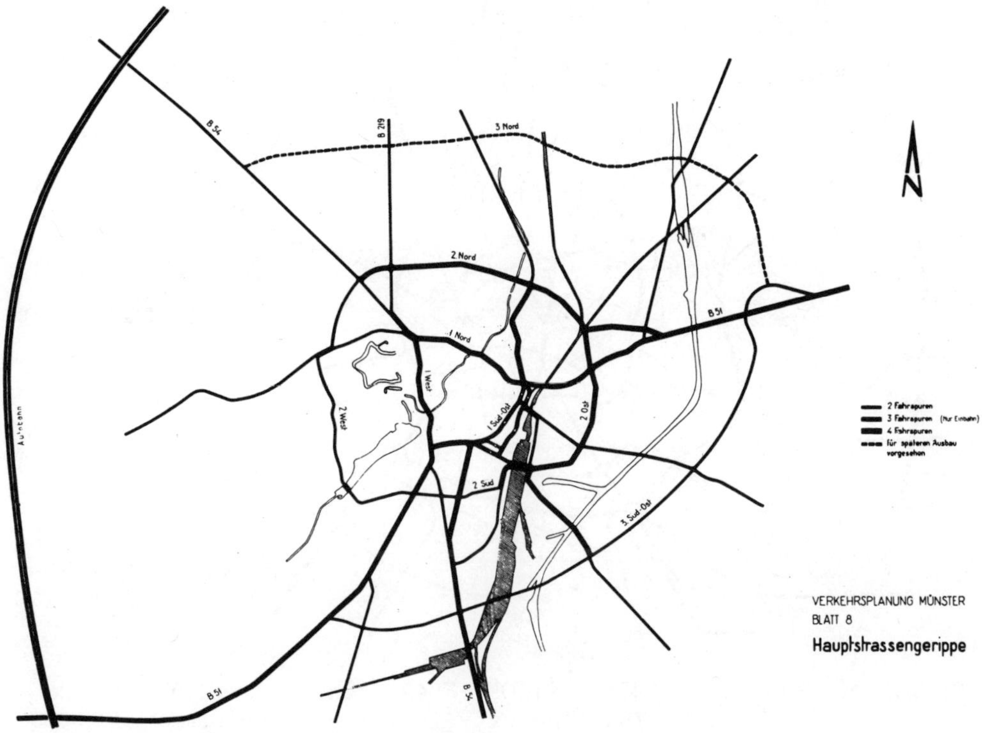

FIGURE 1. MASTER TRANSPORTATION PLAN FOR THE CITY OF MUNSTER, 1960. Most improvements in the core area. Legend: 2 lanes, 3 lanes (one-way only), 4 lanes, to be built later.

over the city were:

	cars/hour	passengers/year
1960	20 100	24,3 million
1966	21 800	25,8 million
1972	23 000	22,9 million

This means that the importance of the core area is decreasing. The number of entering cars should have been 20 100, 52 000 and 78 000 respectively. Or there should have been a marked increase in the number of transit riders. People are using their cars. But a good deal of working, shopping and other activities is now displaced from the old centre to other parts of the city, which is being transformed into a polycentric region. Consequently the Master Transportation Plan had to be adapted.

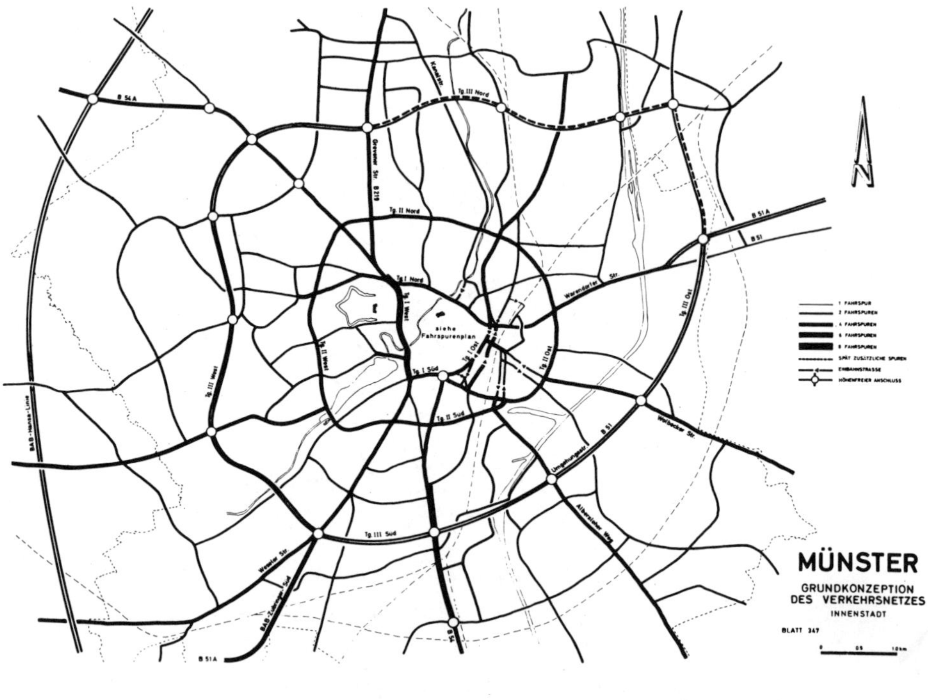

FIGURE 2. MASTER TRANSPORTATION PLAN FOR THE CITY OF MUNSTER, 1975. The city is being transformed into a polycentric agglomeration. The network of arterial roads is to be extended to the newly incorporated communities; the third ring is to be completed. Legend: 1, 2, 4, 6, 8 lanes; lanes to be added later; one-way road; two-level intersection

In the 1960 plan all projects still centred on the inner part of the city. The 1975 plan mainly recommends future investments in the road network to be made in the outskirts.

TRAFFIC RESTRAINT

Congestion by private cars is a common concern. Many cities try to fight it by traffic restraint, by parking restrictions, and by giving preference to public transit. At first the situation may be eased to a certain extent. However, in the long run there seems to be no great difference between the effects of congestion and restraint on urban life. The economic activity in the affected areas will slowly decline. The number of people who enter the core area will diminish, since there are new centres outside the old business district offering easy access and convenient parking, such as shopping centres, commercial and industrial centres, and even school and hospital centres. The transformation of the city into a polycentric agglomeration will be accelerated. The new equilibrium between urban development and transport will be reached earlier.

PEDESTRIAN AREAS

Pedestrian areas are in vogue. There are really wonderful malls. They are highly successful as long as they are small and have convenient access from parking facilities and transit stations. Today there is a trend to extend these areas. In a city of 230 000 inhabitants we were asked to establish a car-free district of 1 km diameter, closed even to public transit. Walking distances would have been up to 500 m. They would not be accepted by workers or customers. A great number of cars would have to be deviated around the central area. It would be encircled by a highly congested ring rendering the access to the core very difficult.

NEW SUBURBS

In many regions new centres, suburbs or towns are being planned and built, regardless of traffic connections. The land is cheap, but transportation added later is expensive and takes a long time to implement. The distance to the "mother town" is very important. It enforces changes in living and riding habits and in the modal split. A dividing green belt is beautiful; but it means more person-kilometers and ton-kilometers, more oil consumption, noise and fumes every day. This has to be taken into account from the very beginning.

ENVIRONMENT

'Environment' is a new key word. People say that planning by engineers and architects is no longer sufficient. Twenty years ago they were happy to have a new road built. Today there is strong opposition everywhere. In a public ceremony a car was buried. Sometimes the word 'environment' stands for a certain nostalgic feeling which is opposed to any technical progress at all. Every new building reduces the green space, every new road produces only noise and fumes. The art of planning becomes even more difficult. There may be a new road project to deviate all through-traffic from the heart of a town. The citizens of the outer districts insist the road be moved further and further outside. The final project is much longer and therefore more expensive. And many drivers do not use the new road because it takes less time to pass through the centre of the town than before. This is but one example. Sooner or later the one-sided exaggerations of this kind will be overcome, and a balanced judgement will develop.

STATE OF THE ART

The classical form of transportation planning consisted of

traffic survey
traffic forecast
project of future network
calculation of capacity.

Three of the four steps applied more or less mathematical formulae; for instance, for the evaluation of field observations and for traffic assignment. A great number of highly sophisticated rules were developed; but none of them was fully convincing. Last year Research Group T 7 of the OECD Road Research Programme on "Urban Traffic Models: Possibilities for Simplification" stated that a new approach was necessary to avoid the great losses of time and money involved in the classical method. The group asked for better ways and means of planning to shorten the process. Several years ago the City of Basel, Switzerland, spent about one million francs on a traffic survey. In another metropolitan area in Western Germany the results of a survey were not yet available three years after it was taken. Meanwhile some of the premises already had changed.

In the computer era many experiments are being made to computerize the urban and transportation planning process. Some people expect that one day it will be possible to feed all relevant data into a computer and to have the complete plan of a new town or of a transport network plotted electronically, all environmental needs taken into account. The only handicap is that the right software has not yet been developed. We fear that even painstaking experiments will not succeed. The behaviour of man - habits of dwelling, working, leisure, and riding, his standard of living - will never follow mathematical rules. It will always be submitted to unforeseeable changes. A common denominator for all activities and influences in human lives does not exist. Transportation forms a part of the economy which can never be calculated in advance. Therefore all these experiments are likely to be in vain. The computer is an excellent tool, but it will never replace the planner and his experience.

The outlook for urban traffic and transport is highly controversial. Early in 1975 the Deutsches Institut für Wirtschaftsforschung, Berlin, and Prognos, Basel, published forecasts for public transit in Germany up to 1985. The first one predicted a 20% decrease, the other one a 100% increase. This is disappointing, since the decision-makers in parliament or in the city councils do not know who is right. They need a reasonable judgment. Who will be responsible if billions of marks are wrongly invested? Reliable forecasts are urgently needed.

MODELS

Maybe it will be possible to replace most of the surveys and forecasts by models. This seems to be very promising. Different types of models are being used. Gravity models are often preferred. However, once again we point to the fact that transportation is not just applied mathematics. Therefore the results of gravity models very often differ from reality. We prefer more sophisticated models which include data on social and economic activities. For instance, we have used such models for:

- a traffic forecast for Zurich Airport in 1955 for a period of 15 years; in 1970 the results proved to be

passengers	4% wrong
air freight	10% too low
air mail	10% too high

- an EEC study of traffic in the frontier regions of Belgium, Netherlands, and Germany giving exact data for the future road and rail networks without any field survey. A comparison with data from neighbouring regions showed the results to be reliable

- a study of the Brussels agglomeration limiting

the zones to be served mainly by the Metro, and fixing the network of arterial roads using earlier surveys for calibration only.

A complex problem is a model of modal split. Up to now no rule has been found as to how to assign traffic demands to the different modes of transport. When preparing forecasts it would be a mistake to replace one car passenger by one transit rider, or vice versa. As an example, Figure 3 shows that in Brussels for many years transit was decreasing while private traffic increased. However, there is no regular relation.

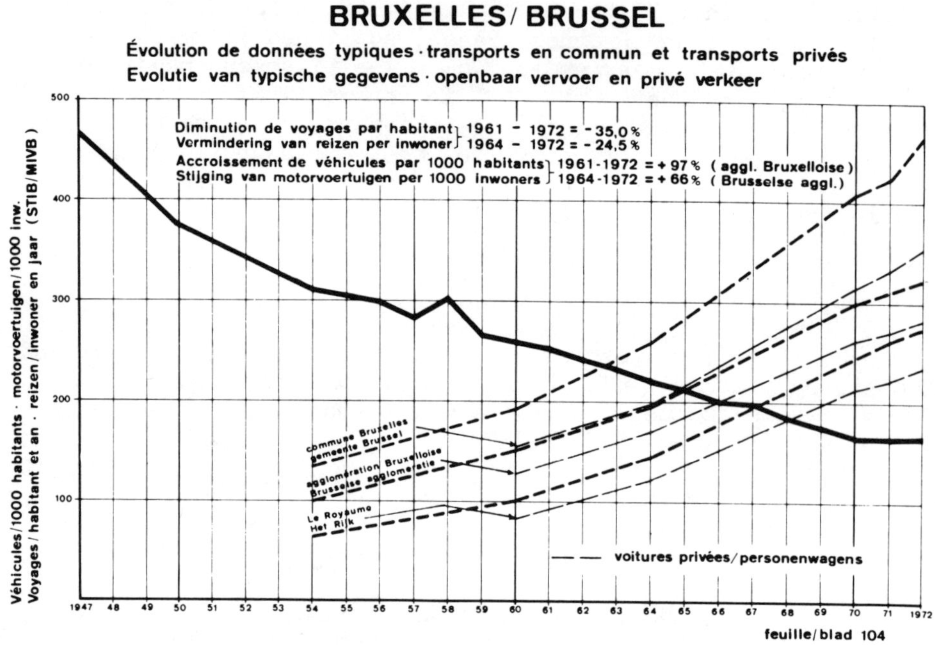

FIGURE 3. DECREASE OF PUBLIC TRANSIT RIDES PER CAPITA, AND INCREASE OF THE NUMBER OF CARS PER INHABITANT. There is no mathematical relation between these two important data. From 1970 to 1972 public transit remained stable due mainly to new pre-Metro lines, while car ownership went up even more than before.

The thick hatched lines refer to all vehicles in the City of Brussels, the agglomeration, and Belgium respectively, while the thin lines refer to private cars only.

In several cities both modes of transport increased at the same time. Once again, the computer will not be able to tackle the problem for there is no mathematical base. For many years public transit has been neglected, sometimes for the simple reason that the planners were not familiar with the needs of bus and rail transport. Or they believed that the trend shown in Figure 3 would continue unchanged. Recent years have brought a revival of public transit and its ability to solve all urban transport problems. This led to many proposals of "new" means of transport. Many of them revived proposals made even 100 years ago. There was a wave of more than 300 different new types. However, most constructions face the same problems as bus and train: time-table, headway, distance between stations, walking distance, changing between lines, layout of the network and, last but not least, economy. Many experiments could have been spared if these common problems of public transport had been studied more carefully before starting work on the construction of cabins, tracks, engines, etc. For the foreseeable future we will rely on conventional road and rail transit.

Many cities - even towns of less than 500 000 inhabitants - began to build underground lines, or to add new lines to existing networks. However, due to the financial situation work on these projects is slowing down. Construction and operating are too expensive. Therefore, the chances for long tunnels seem to be less promising. Short underground connections are more advantageous. Anyhow, planning of public transit should always form an integral part of urban planning.

SOCIOLOGY

Social factors are becoming more and more important. For example, the declining birth-rate in many European countries will influence all planning activities. All statistical data related to the population will change. The sharpest decrease of the population will occur in the inner districts of the cities. It will not be possible to revitalize these quarters by just renovating the old houses.

Most immigrant workers do not own cars for their first years, but depend on public transit. Their riding habits differ from those of the resident population.

Social aspects should be taken into account more intensively.

Sociologists should give more attention to urban and transport problems.

DEMOCRATIC PLANNING

In recent years "democratic" planning has become another new issue. Continually City Councils had to decide on Master Plans and new projects. Now the immediate cooperation of the citizens is encouraged. They are asked for their assistance in preparing decisions and in sharing responsibility. This is an excellent idea. However, in some cases there are pressure groups working in favour of private or financial interests. Better rules and means for the democratic way of decision-making are needed.

TRANSPORTATION AND LAND-USE: PROBLEMS IN RECENT URBAN PLANNING CONCEPTS

Klaus Richrath

Institute for Local, Community and Regional Planning

University of Karlsruhe

COMMUNITY, COMMUNICATION AND URBAN LIFE

Urban life happens amidst a structure of adapted spaces and flow systems. Mainly defined by building structures, the spaces for various activities are linked by transportation and service channels according to certain rules and by common consent.

Part of the common consent is to achieve as low communication costs as possible in respect of the economic and technical conditions involved. Low communication costs are an outstanding advantage of urban communities. Physically this means the ability to reach the many exchanges where goods and information are handled, not only as easily, fast and cheaply as possible, but also comfortably and safe. "Communities comprise people with common interest who communicate with each other." (Webber, 1964).

It can be doubted whether common interest is any longer a sufficient feature of urban communities. But if communication is to be an important feature of urban life, it means living in an interaction space where artefacts and the channels and flow of communication have accumulated to a high level of standard and density. Physical means of communication help people to approach other people and artefacts. And they affect the choice of site, size, density and gestalt of cities and urban areas. The history of the city and of city planning is about improving human interaction. Spatial patterns are the result of choice and compromise between the locations which are considered best for a particular land-use and the transportation systems which are economically and technically available.

URBAN PLANNING CONCEPTS

This latter point is particularly easy to appreciate in the programmatic concepts of ideal cities and other city models which have been developed since the end of the nineteenth century. Irrespective of varying social or economic objectives, transportation systems and land-use are the central subjects dealt with in the technical and spatial determination of these models.

When division of labour and locational differentiation increase, point-concentric models try to reduce the increasing need for transportation by their scale and by their allocation of the main land-uses, e.g. Howard's Garden-City. The linear models try to accomodate to the increasing need for transportation by more efficient transportation devices, e.g. Soria y Mata's or Miliutin's linear cities. In fact, most of these models combine both strategies, as also do most of the recent concepts of spatial organisation in large urban areas.

In the German Federal Republic, as in many other developed countries, most of the concepts applied concentrate the main land-use areas in distinct places of urban development which are connected radially to one or several main centres by development axes of 'bundled' transportation and service channels. Land-uses which are considered relevant are usually the origins and destinations of daily and weekly trips, which are the main influences on traffic conditions. These are residential and recreational areas, places of employment, schools and stores. Places of urban development are mostly identical with the traditional 'central places'. The development axes usually extend alongside mass transit systems and suburban corridors developed in the last one hundred years (see Figure 1).

The general planning objective is to improve the conditions of urban life. The detrimental consequences, which arise from too intense a land-use and traffic congestion in the core, are to be counteracted by the extension of new subcentres along efficient lines of local traffic, especially lines of public transit. This should also improve the service standard in suburban areas in regard to quantity, quality, possibility of choice and accessibility.

This spatial pattern is accepted uncritically and applied without modifications in quite different urban areas. Nevertheless, the critical articulation of a city has seldom been as intense as it is now. Successful planning with this concept is very rarely found. The supposed design influence of mass transit systems is as much antithetical to changes in modal split as is the spatial pattern based on it to the actual changes in urban patterns. Urban sprawl and "loss of urbanity" are better fitted

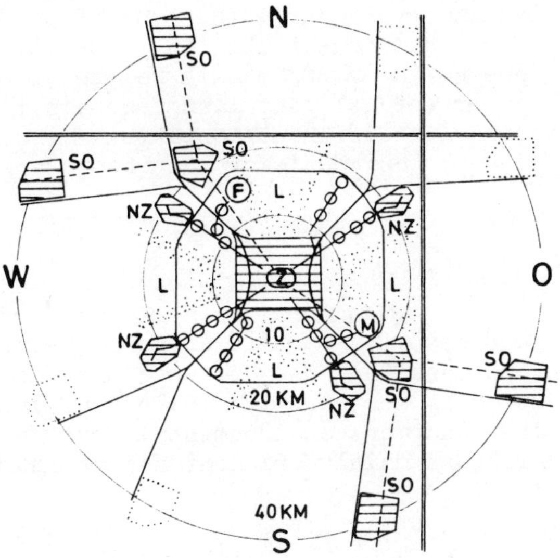

FIGURE 1. PATTERN OF A REGIONAL CITY (from Hillebrecht, 1962)

to the third model of Broad Acre City which Frank Lloyd Wright once promoted, some kind of "nowhere city".

SOME CONFLICTS IN URBAN PLANNING CONCEPTS

The divergencies between urban planning concepts and urban reality are based on conflicts between objectives and means of planning. They result, for example, from the contrast between a rather limited number of locations which can be reached by mass transit systems and the desire to have more locations and opportunities to choose from. Conflicts also result from the chosen transportation system itself. That is, a basically concentric spatial organisation of urban areas prevents the peak load in the core from being corrected.

Mass transit systems allow a far more intense land-use in the core than would be possible otherwise. As land costs increase with improved transit facilities, the land is used even more intensively, thus creating more traffic and diverting less profitable institutions to the periphery. These effects make it difficult to develop a net of polycentric and polyfunctional centres, or economic housing programmes for those people who have to rely on easy access to public transit stations and who represent the economic basis of public passenger transportation.

A remarkable number of conflicts result from the call for high densities near mass transit stations. Relative to those determining factors which are primarily related to transportation, the locational factors of relevant land-uses are more manifold and of varying importance in different contexts. If communication costs are going down, other decision components are entering the picture.

Residential Areas

This point applies with particular consequences in the case of locations for residential areas because they determine the spatial distribution of most trips. Increasing demand for more floor area per unit and more environmental quality is met by offering locations which are mainly orientated in terms of accessibility by public transit.

The economic constraint of public transit demands high population densities within easy reach of the station in order to minimize access distances. The increasing demand for more floor space per person and the diminishing size of the family, make it necessary to increase built-up densities, thus contradicting the tendencies of the housing market.

Granting certain assumptions and simplifications the built-up density (GFZ) of 1955 should be more than doubled in order to keep the number of inhabitants and users of public transit at the same level. The numbers of households and dwelling units have increased as well as the space needed for transportation facilities, public services, etc. (Table 1).

But there is still another problem concerning recent changes in population growth and housing programmes. Even if we assume invalid and optimistic projections of the annual increase of housing, and a rather high ratio of high-density housing, the realisation of these concepts will be difficult, except maybe in some fast-growing urban areas (Table 2).

If we assume the more probable slight decrease of population as well as different regional growth-rates, and estimate to what extent the location of high-density housing alongside public transit lines by means of zoning can be managed in reality, these planning concepts prove to be huge growth models. They cover far longer a period than their own infrastructure will last, and exceed the horizon of forseeable technical and economic development.

1955	persons per dwelling unit		av. 4.0 EW/WE
	percentage of residential estates		av. 0.7
	floor area per dwelling unit		av. 75 m^2
Dwelling units (WE)	$= \frac{18{,}000\ EW}{4.0\ EW/WE}$		= 4,500 WE
Floor area ratio (GFZ)	$= \frac{4{,}500\ WE \cdot 75\ m^2}{0.7 \cdot (\pi \cdot 600\ m^2)}$		= 0.4
1975	persons per dwelling unit		av. 2.75 EW/WE
	percentage of residential estates		av. 0.6
	floor space area per dwelling unit		av. 75 m^2
Dwelling units (WE)	$= \frac{18{,}000\ WEW}{2.75\ EW/WE}$		= 6,500 WE
Floor area ratio (GFZ)	$= \frac{6{,}500\ WE \cdot 95\ m^2}{0.6 \cdot (\pi \cdot 600\ m^2)}$		= 0.9

TABLE 1. AVERAGE NUMBER OF DWELLING UNITS AND FLOOR AREA RATIO at rapid transit stations in 1955 and 1975 (GFR), 20,000 inhabitants, 90% within a radius of 600 metres.

(without week-end houses and rebuilding)	p.a.	400,000 WE
about 35% WE in multi-storied houses to be erected in urban areas	p.a.	140,000 WE
Persons in multi-storied houses to be erected in urban areas (2.5 EW/WE)	p.a.	350,000 EW

TABLE 2. INCREASE OF DWELLING UNITS IN GFR 1972-1985

If plans are to be realised in a conceivable time-span, and following the completion of an intense housing recovery programme, the other main condition would be a revolutionary new distribution of population in urban areas. This is because only a diminishing group of people is still relying on public transit services, and therefore lives or has to live in high-density areas in easy reach of transit stations. But it is precisely low cost housing which is affected by the consequences of investing in public transportation (i.e. increasing land costs) and the costs of high built-up density.

These conflicts between planning and the real demand for housing affect the location of public and private institutions. Trips to and from places of employment, schools, shopping and recreational areas mostly begin or end at home. Close to stations there arise conflicts between the increasing need for space for these land-uses and the demand for high densities.

Workplaces

Looking back to the early sixties, when these urban planning concepts came into use, the main problem was how to connect the fast-growing residential areas at the periphery with the increasing number of workplaces concentrated in the central city by mass transit systems. In the meantime modal split has changed, as well as employment patterns. Car ownership has increased and urban sprawl makes it difficult to run attractive public transit services.

The spatial concentration of employment facilities in the core, for historical reasons, has not continued in the years following the fifties. While spatial deconcentration of industry is clearly visible, it is still only beginning in traditional downtown employment. Locations at the periphery often become favoured because they can offer better transport conditions as well as an increasing number of employees living there. Central locations lose their attractiveness as access gets increasingly difficult and commuting distances grow.

Of course, the historical concentration of employment will be reduced only in the long term. But the realisation of urban planning concepts will also need a long time. Decisions to locate large enterprises in the central city are scarcely the way to improve its "urbanity". On the contrary, the demand for land and for more streets and parking lots, is in opposition to conserving historical settings and buildings. This problem is unlikely to be solved.

By changes in employment patterns the traditional rush-hour

flows from the suburban corridors to the core are superimposed by 'cross-commuting' and 'reverse-commuting'. Given the presumably stagnating rate of commuter trips the economic basis of mass transit is unlikely to be improved. Nevertheless, the improvement of public transit systems is motivated particularly by an increasing number of downtown commuter trips. Excluding trips to and from schools, these trips will increase in number. The private car is nearly perfect for them. It is extremely mobile for recreational use and it is a basket on wheels (Figure 2).

Retail Services

Due to new methods of preserving food and altered shopping habits, the importance of proximity to shops is diminishing. Easy access, sufficient parking lots and a good assortment of shops for one-stop shopping is more important. The necessary space for new shopping centres and their complementary facilities is usually not available in the core, because of their scale and the cost of land.

Nevertheless, the planning choice between a shopping centre at the periphery or increased downtown retailing is often decided

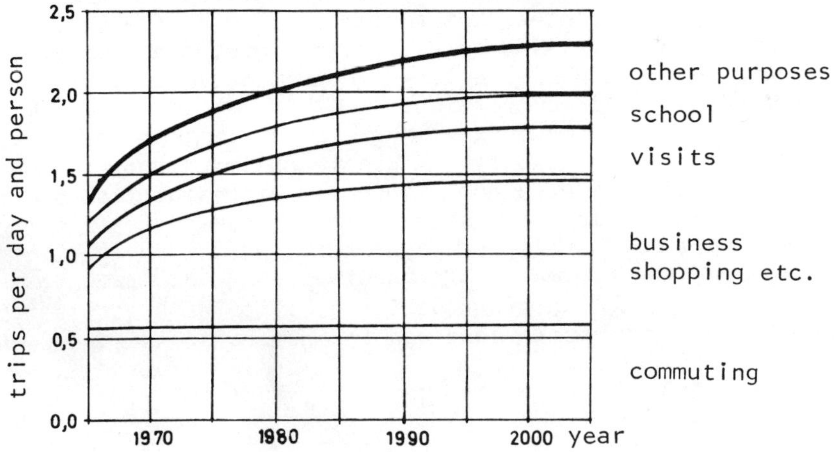

FIGURE 2. PROJECTION OF DAILY TRIPS PER PERSON AND TRIP PURPOSE 1970-2000 (from Kessel, 1972)

in favour of the latter, although the main objective should be to diminish problems in the central-city. Anyway, this restrictive location policy could not stop the development of new retail techniques and locations. The recent tendency of shopping centres to come back and re-integrate into the city will probably cause additional traffic instead of "urbanity".

The decentralisation of retailing, and at the same time its concentration in huge units, means that a hierarchy of centres has to be achieved by a dispersed and often monofunctional structure. The integration of retail capacities close to public transit stations involves high building costs, if it is not to affect the necessary residential capacity. Further, it means giving up a principle of transportation which gives priority to public transit in general.

Recreational Areas

To a similar extent, changes have been caused by private mobility in recreational activities. The car is no longer the only means of transportation, but the means of recreation are the same. When transporting several people at the same time the car is in fact cheaper than public transit.

Planning concepts restrict recreational areas to the spaces between the development axes, and if possible within a short walking distance from home. As far as weekend traffic problems are concerned, this concept is only partly helpful. Because the location of recreational areas is tied to distinct natural qualities, there is no guarantee that people will go to the nearest areas rather than those they like best. Furthermore, the fact that people have to live in high built-up densities, will probably increase weekend traffic congestion.

As environmental quality has become an important component in residential location, a good synthesis of residential and recreational areas would presumably diminish weekend traffic problems. But that would raise new contradictions in the public transit orientated arguments of planning concepts. Anyway, the one-sided interpretation of outdoor recreation as a contrast between city and countryside conceals the neglected leisure potentials of the city.

CONCLUSION

These considerations about current urban planning concepts as a sectoral view on transportation and land-use are limited and incomplete. But they disclose some of the conflicts between objectives and means in the long-term planning concepts which aim to determine the environment of urban life. With regard to economy in traffic, the expectation that, if energy gets expensive individual passenger transportation could be restricted, has proved unrealistic until now. Apparently the public budget is more affected by this phenomenon than the private. But even without regard to recent economic changes, the technical and economic facilities would prove unequal to the task of effectively altering urban reality in favour of the planning concepts. The same applies to the legal instruments involving planning and land-use.

Neither increased subsidies for public transit nor road-pricing seems to be able to solve peak-load problems without affecting locational structure. Economic limits come mainly from the indivisibility and long-range commitment of capital in transportation facilities built up over recent decades. The sophisticated dual-mode vehicles which may serve tomorrow's traffic, and which are equally suitable for passenger transportation in low and high-density areas, would strengthen the trend towards urban dispersal and central-city congestion.

In principle, these concepts enlarge the radial suburban corridor growth of the city, although in a much more sophisticated way. They presuppose that order and design have to be identical with the principles of the historic city. And they assume that society will defend these principles, and by technical means. The pattern of today's cities since mass transit systems have come into use is only one hundred years old. There is no reason to regard patterns which have grown in this way as final and valid for the future. In the long-term affiliation of urban patterns to recently available transportation techniques, the flexibility and adaptability of these concepts proves to be as shallow as the megastructures of modern architecture.

Finally, economic postulations of public transit serve as a legitimation for urban planning concepts achieving a certain pattern and defined densities; although there is no proof that high densities are related to urban life in a causal or correlative way. If the diminishing influence of public transit on urban design is not to be followed by other comparable determinations, society will have to define precisely in what kind of environment it would like to live. Cities will not be formed by fulfilling needs, but by discipline and design. Or, as Bloch (1973) puts it, by "phantasy and horizons".

REFERENCES

Bloch,E. (1973) Funken im Pulverfass. Bauwelt 64, 44

Hillebrecht,R. (1962) Stadtbau und Stadtentwicklung. Archiv fur Kommunalwissenschaften, 1.

Kessel,P. (1972) Verhaltensweisen im werktaglichen Personenverkehr. Bonn.

Webber,M.M. (1964) The urban place and the nonplace urban realm. In Explorations into Urban Structure. Philadelphia.

THE QUALITY OF URBAN LIFE AND OF TRANSPORTATION SYSTEMS AS DETERMINANTS OF REGIONAL DEVELOPMENT PLANS

Andreas Volwahsen

Department of City Planning

Technical University of Darmstadt

In the 1950's and 1960's, attempts to improve the quality of life in residential areas led to their separation from industrial areas and commercial districts in numerous German towns. The improvement of the local and regional road systems and rapid transit systems was one aspect of this development. But another was an increasing monotony in the segregated parts of urban life. Planners very soon started to feel uncomfortable with the emerging settlement patterns, particularly in the major urban municipalities, where - as in many other countries - commuting habits brought about an unexpected increase in vehicular traffic. In Germany, regional planning authorities were established to cope with the resulting planning problems on a regional level.

The main thesis of this paper is that in order to balance our goals for the quality of urban life with the constraints imposed upon this quality by transportation systems, comprehensive regional planning efforts are necessary.

The role of city and regional planning as a mechanism for synchronising goals and constraints - or as the link between efforts to improve urban life and efforts to improve transportation systems - can be demonstrated in the context of two very down-to-earth planning tasks.

In the first example, one of the oldest European planning authorities - the Rhine-Ruhr-Planning Authority (Siedlungsverband Ruhrkohlenbezirk) - tried to link all the relevant goals and restrictions of urban life, transportation and environment in a comprehensive modelling effort.

This model - or rather, this series of interconnected models - served as a means to analyse interdependencies between the transportation system, the intra-regional migration pattern, environmental quality (i.e. what residents consider as the essential features of urban life), the government-funded new towns policy, and the location of a major new employment centre. This example clearly shows the role that regional large-scale land-use policy plays as a necessary link between transportation systems and the quality of urban life.

In the second example - a government-funded research project on new methods of land-use planning - a similar approach was chosen in order to define on a much smaller scale the optimal location for a few housing projects, for new sites for industrial firms and for a net of local recreation centres. The area chosen for testing these methods was the fast-developing axis in the Rhine-Main region (Regionale Planungsgemeinschaft Untermain), the axis linking the cities of Frankfurt and Darmstadt. Figure 1 shows the Rhine-Ruhr industrial area, where land-use and transportation models were applied on a regional scale, and the axis between Frankfurt and Darmstadt, where comparable models were used on a local scale.

In both projects it has been assumed that a substantial improvement in the quality of urban life can only be achieved by concentrating all public investments in so-called developement centres, or a development axis, normally one or several medium-

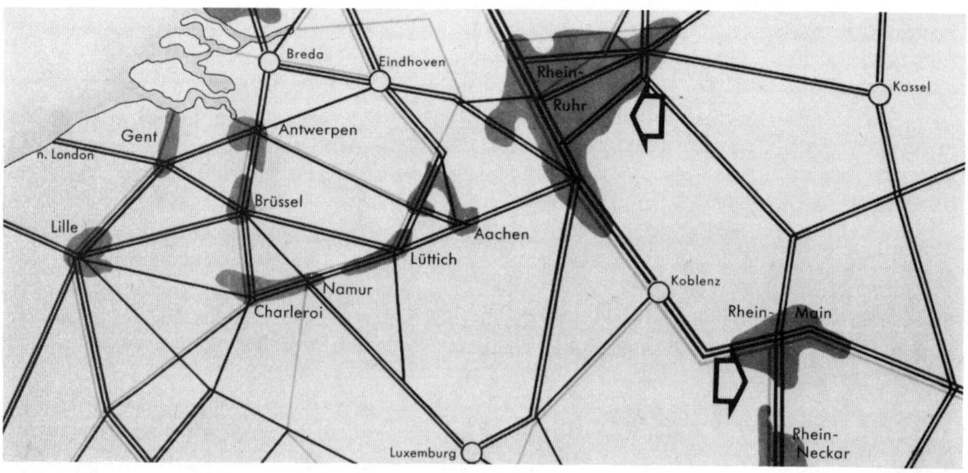

FIGURE 1. LOCATION OF THE RHINE-RUHR INDUSTRIAL REGION AND OF THE RHINE-MAIN REGION

size towns. The reasoning behind this uneven distribution of government subsidies and public investments is that the quality of life can be more effectively improved by providing all citizens with a highly developed social infrastructure within a reasonable distance (in the development centres) instead of evenly distributing funds among all communities.

The official government policy, which I believe also reflects the opinion of the public, helped the planners to narrow down the question of how to improve the quality of urban life in general, of how to find the optimal locations for the new development centres to be built, and of how to choose between competing potential development centes.

In both projects we started these tasks by listing the criteria that different socio-economic groups would apply to evaluate the relative suitability of each statistical unit as a residential development centre, industrial area, commercial centre, recreation area, etc. In the process two problems emerged; one regarding the method, and one the content of the list of criteria, i.e. the data. The methodological problem arose from the need to weight the criteria against each other before including them in the specific suitability analysis. The data problem became apparent when we realised that most of the criteria were accessibility indicators and not simple census data. The first problem was solved by the formation of goal hierarchies. This allowed for a step-by-step transformation of some generally accepted planning goals into weighted sets and subsets of specific goals and evaluation criteria. The second problem was solved by the inclusion of a number of transportation and accessibility models. They performed the necessary transformation of practically all statistical data into accessibility data.

At this point it should be apparent why I chose as the title of this paper "The quality of urban life and of transportation systems as determinants of regional development plans". Apart from some methodological suggestions on how to handle the problem I have described, I want to draw attention to the necessity of explicitly including the topic of land-use policy in our discussions.

The quality of urban life, described in a set of goal hierarchies, and the quality of transportation systems determining the accessibility of activities considered essential for an evaluation of this quality, seem to be the two critical issues in land-use planning. In my opinion, there is no direct causal relation between certain types of transportation systems and different qualities of urban life. Both topics can only be discussed within the framework of an analysis of specific settlement patterns,

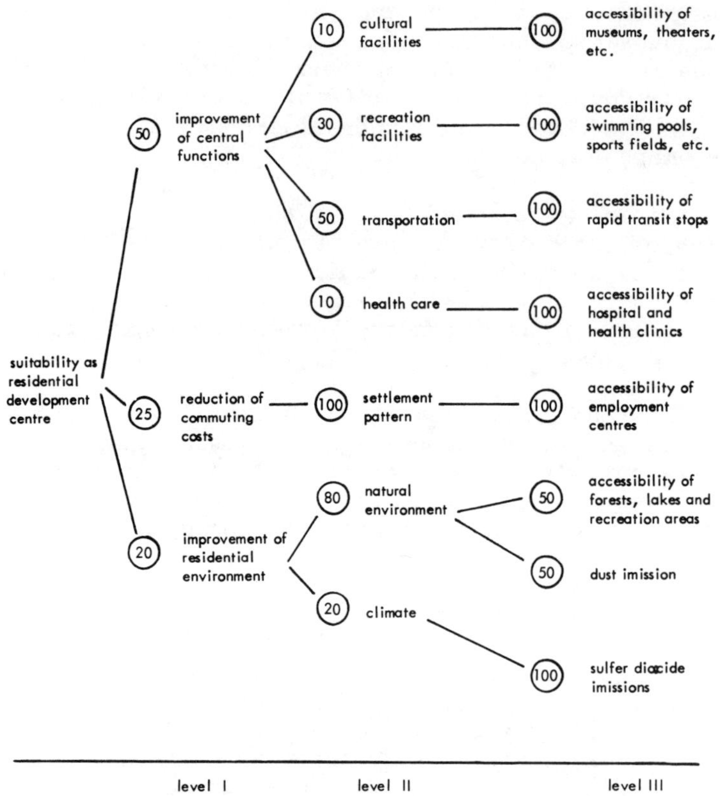

FIGURE 2. GOAL HIERARCHY USED FOR THE COMPUTER-AIDED SEARCH FOR POTENTIAL HIGH DENSITY RESIDENTIAL CENTRES

commuting habits, shopping habits, etc.

After these general remarks, let me show by the two examples mentioned before how we included goal hierarchies and accessibility models in the planning process.

Our search for potential sites for a limited number of development centres in the Rhine-Ruhr industrial region had to be based on different goal hierarchies, depending on the type of development centre to be planned - high density residential centre, commercial centre, recreation centre or industrial centre. For each category a separate hierarchy was designed and weighted in close collaboration with local planning officials. Figure 2 shows the goal structure used for the computer-aided search for potential high-density residential centres. Six out of eight indicators finally used in this suitability analysis are accessibility variables. The remaining two indicators refer to environmental quality.

Weights were distributed stepwise from the top to the bottom of the hierarchy. The main purpose of the whole planning effort is an improved provision of social and cultural facilities. This goal appears on the first level of the hierarchy and gets a much higher weight than two other goals which ought to be achieved at the same time: provision of improved environmental quality; and reduction of commuting time. On the second and third level similar decisions regarding planning priorities have to be taken in order to arrive at the relative importance of each evaluation criterion.

Having dealt with the problem of the quality of urban life, I will now turn to the transportation aspect of this project. During the last two decades a dense grid of highways had been built in this region. Only in the last few years had a shift in government policy towards stronger support for municipal and regional mass-transit systems started to change the picture. The detrimental effect of vehicular traffic on densely populated urban areas, and the economic need to concentrate new public investment in certain development centres led to a series of government regulations, according to which all development centres have to be situated on one of the major links of existing mass-transit systems. Communities off these routes will not get support, even if they have direct access to the highway system.

The high priority put on mass transit logically narrows down all accessibility analysis to calculations for one transportation mode only. This undoubtedly distorts the actual accessibility under modal-split conditions; but it is one of the major policy inputs in the planning of development centres. A conventional traffic model was used to compute the matrix of travel times for the morning peak-hour traffic between about 350 traffic cells, taking all relevant data from the present time-tables. To check the results, isochrones around a large number of traffic cells were calculated (Figure 3), displayed and compared, and some travel times were recalculated by hand on the basis of the time-tables.

Next, a gravity model was added and for each statistical area, accessibility values for a large number of variables were calculated. Figure 4 shows mapped values taken from the resulting data sets, - here, the relative accessibility of cultural facilities. Having computed all relevant accessibility values, it is a trivial task to weight and overlay the standardised indicators of the third level of our hierarchy of goals and thus to arrive at the suitability values mapped in Figure 5.

Parallel to this search for optimal residential centres, other analyses covering the problems of industrial, recreational

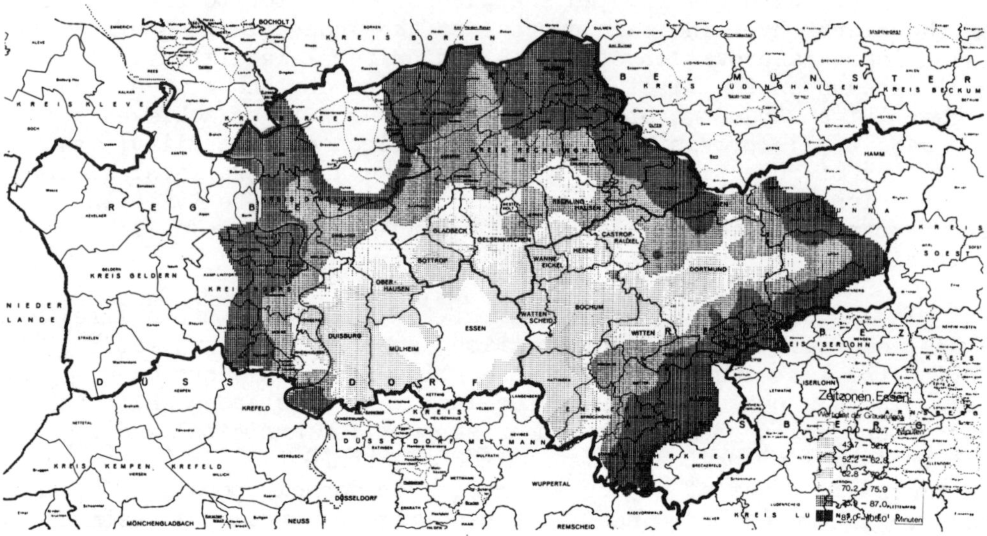

FIGURE 3. ISOCHRONES COMPUTED AROUND THE RAILWAY STATION OF THE CITY OF ESSEN

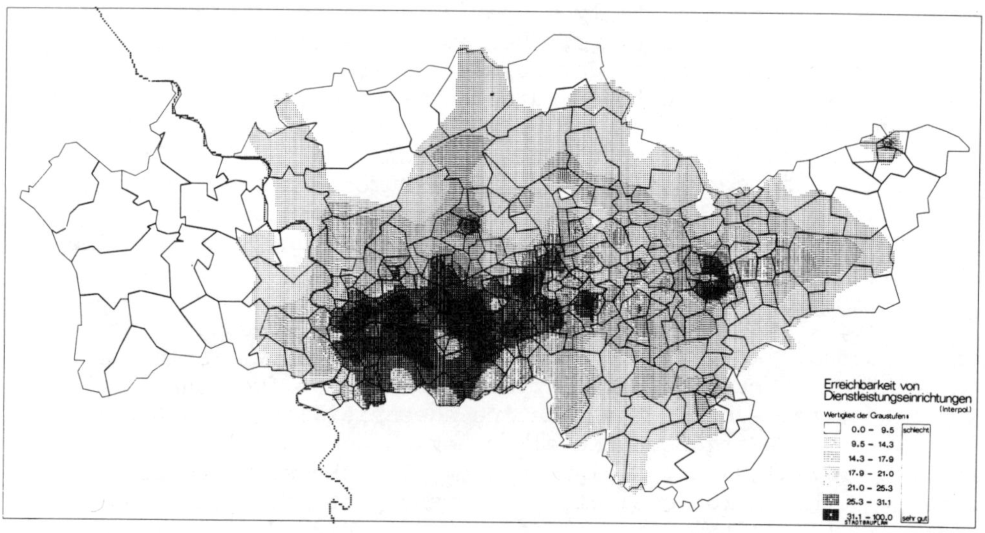

FIGURE 4. ACCESSIBILITY OF CULTURAL FACILITIES

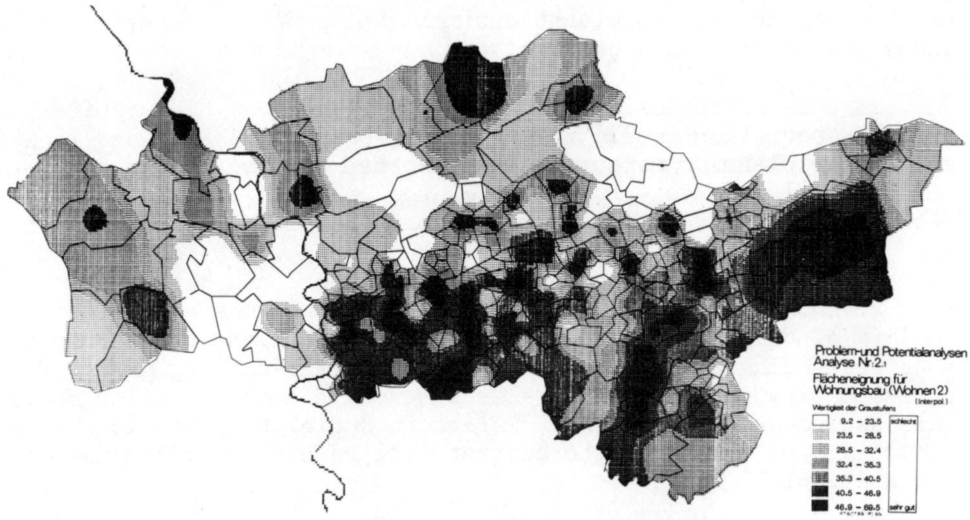

FIGURE 5. RESULT OF THE SUITABILITY ANALYSIS FOR HIGH DENSITY RESIDENTIAL CENTRES

and commercial location were carried out, and step by step the regional pattern of development centres emerged from a simultaneous display of all the results and their comparison with sites actually available for future development. Figure 6 shows the highest suitability levels extracted from all the analyses. The

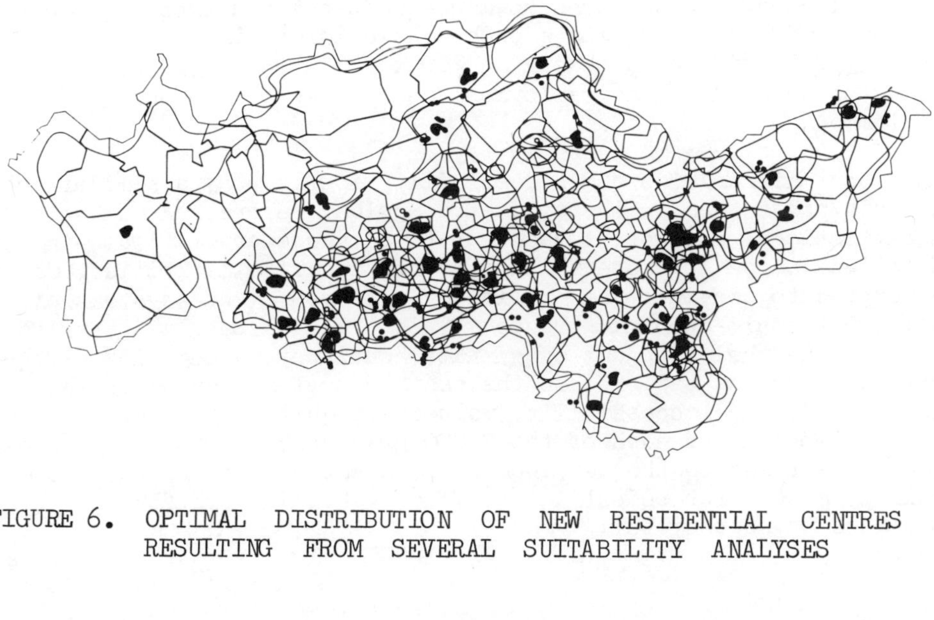

FIGURE 6. OPTIMAL DISTRIBUTION OF NEW RESIDENTIAL CENTRES RESULTING FROM SEVERAL SUITABILITY ANALYSES

black areas are the resultant optimal choice for residential and industrial development centres.

In this short description of how we used goal hierarchies and transportation models in a planning project, I have had to skip many other planning aspects which also had a considerable influence on the final results - for example, the regional demographic projection or detailed analyses of economic development.

In the context of the other project mentioned before, I now want to show how the same method of expressing some aspects of the quality of urban life in a set of goal hierarchies and of computing a corresponding distribution of different land-uses by means of accessibility indicators can also be used on a much smaller scale. This second example not only differs in scale, but also in the type of model applied and in the geographical reference system used for the statistical data.

Changing administrative boundaries and street patterns, and thus the changing boundaries of census tracts, are a great hindrance at present to the introduction of efficient methods of small-scale statistical analysis in the planning process. The German census data of 1970 is hardly comparable to the data of 1961; and this situation will not change very much in the near future.

Therefore the administration is considering the introduction of the UTM-grid as a general geographical reference system for all statistical data. The grid system was tried for the first time in Germany in the Rhine-Main planning project. In order to get the census data for this study we actually had to go back to the original census data stored on magnetic tapes by street and house number, and had to associate each household or firm with one of about 7,000 grid cells measuring 200 x 200 m in size. Figure 7 shows as an example the employment density mapped in the grid system.

Goal hierarchies and accessibility models play a similar role in this project as in the project described before. A number of modifications, however, turned out to be necessary. First, in our detailed search for potential sites for specific activities, we attempted to consider the explicitly different value-systems of different socio-economic groups. This policy aspect was carried through the whole project by simultaneously operating with different goal-systems tailored to the assumed preferences of these groups. Second, accessibility values for quite a few variables had to be computed for each of the 7,000 grid cells. This methodological aspect led to the development of a set of interrelated accessibility models for pedestrian traffic, vehicular traffic and for the mass transit system.

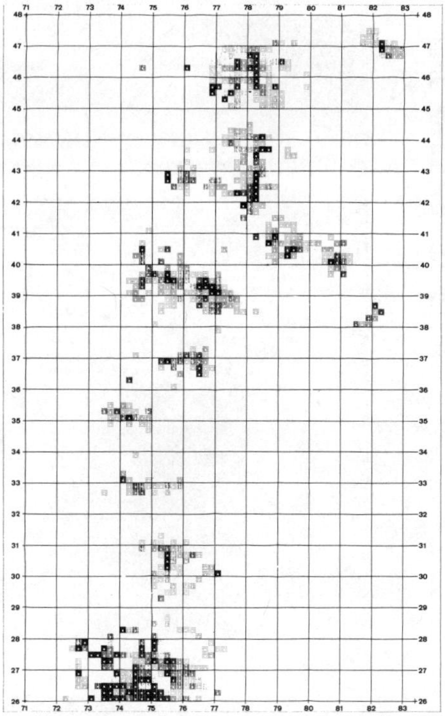

FIGURE 7. DATA MAPPED IN THE UTM GRID, RHINE-MAIN REGION, SIZE OF CELLS: 200 x 200 m

Which are the main features of these accessibility calculations? The relatively small scale and the fine grain of our planning proposals necessitated, in addition to the analysis of private and public transportation, an analysis of the considerable local pedestrian traffic. Pedestrian accessibilities were computed by a 'moving-circle' model. This model automatically jumps from one cell to another, drawing each time a circle with predefined radius around the respective cell and adding up each time what is found within this radius. Figure 8 shows some mapped results, with the accessibility of rivers, lakes and brooks on the left. The darker a cell the better it is in relation to these water sources. In the middle, the relative pedestrian accessibility of places of work is mapped; and on the left, the accessibility of the rail system.

Private and public transportation is taken into account by a separate model which we called a 'moving isochrone' model. This model first computes isochrones for each transportation mode around

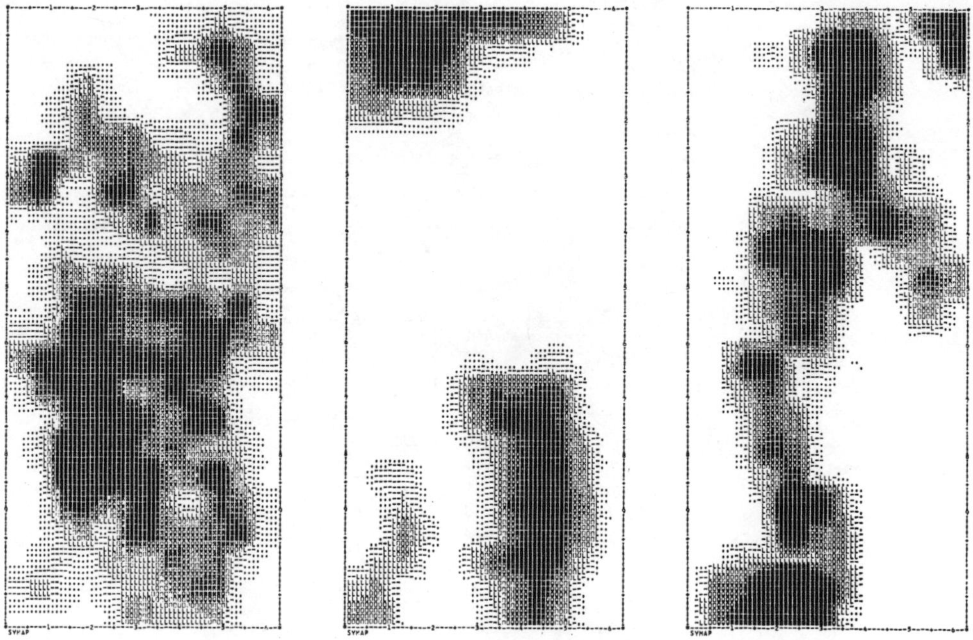

FIGURE 8. RESULTS FROM A PEDESTRIAN ACCESSIBILITY MODEL. FROM LEFT TO RIGHT: ACCESSIBILITY OF RIVERS, OF PLACES OF WORK, AND OF THE RAIL NETWORK

each grid cell, and then computes accessibility-values according to the gravity principle. Figure 9 shows one set of isochrones. Remember that 7,000 separate isochrones had to be taken into account for this calculation. Both models, the moving-circle model evaluating the immediate surroundings of each grid cell from the view of a pedestrian, and the moving-isochrone model evaluating the accessibility of different activities by car or train, produce results which cover most accessibility goals referred to in the goal structure.

This leads us to the question of how the quality of urban life, that is the expectation of different social groups, is included in this planning project for the Rhine-Main region. Let me start by giving you an example of how different value-systems have been analysed with regard to their potential for conflict. We will then see how the value-systems were transformed into land-use patterns - occasionally, of course, into conflicting land-use patterns - and how these group-specific intentions to change the present land-uses have been put together into a logically consistent development concept.

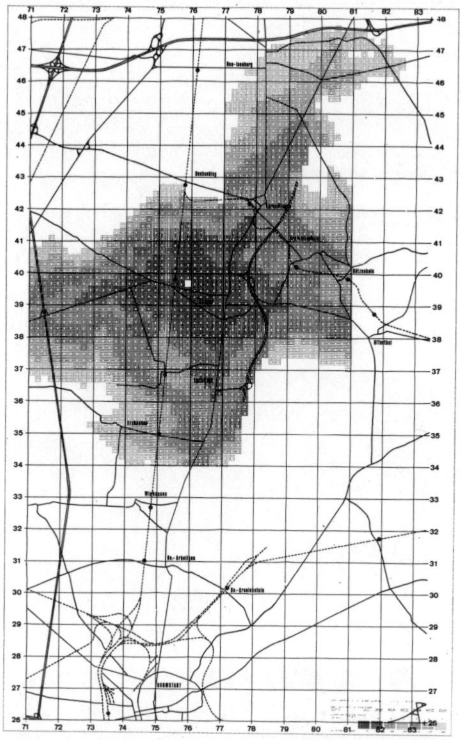

FIGURE 9. ISOCHRONES COMPUTED BY MEANS OF A TRANSPORTATION MODEL CONSIDERING ABOUT 7000 SQUARE TRAFFIC ZONES

The nine socio-economic population groups that we had defined - considering age, household size and education - have different preferences in their choice of residential locations. They put different weights on the evaluation criteria of the respective goal hierarchies. Similarly, four different commercial groups which we defined have different locational preferences expressed by different weights in their respective goal hierarchies.

Even before using all the hierarchies for specific spatial suitability analyses, we tried to compare some of them by using a special type of cluster analysis in order to determine the major conflicts. Figure 10 shows the graphic interpretation of such an analysis. Each goal of two hierarchies is here represented by a circle. The radius of the circle corresponds to the weight put on an evaluation criterion. The distance between the circles indicates the compatibility of the goals. Large distances denote serious incompatibilities. One of the major conflicts in this comparison of preference systems turned out to be between one of the population groups and one of the commercial groups, who were in competition

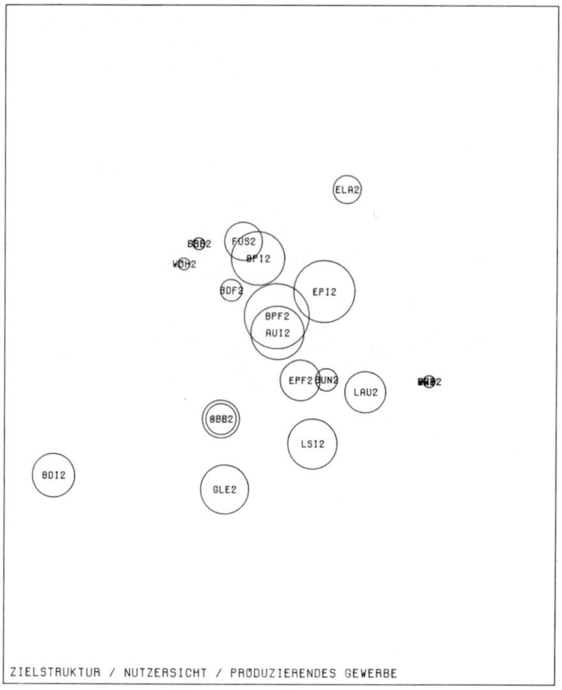

FIGURE 10. GRAPHIC INTERPRETATION OF AN ANALYSIS OF PLANNING GOALS (COMPROGRAPH)

for location near the mass transit stations. But apart from this somewhat trivial result, other conflicts of a more subtle nature also became apparent.

The spatial suitability analyses based on these hierarchies (Figure 11) were interpreted individually (Figure 12), and were finally put together in order to arrive at a comprehensive development concept for the region.

Let me summarise my main points. In the first project we were looking for an optimal distribution of development centres or a development axis on a regional scale. In the second, we used similar methods for the selection of optimal individual sites within a development axis. On both levels of spatial planning the main determinants were first, the expectations of the population regarding the quality of urban life, structured as a series of goal hierarchies; and second, the transportation systems, on which all accessibility calculations (i.e. most evaluation criteria) were based. I hope that the two examples I have given will help to define the role that these two topics play in our efforts to improve the present techniques and results of land-use planning.

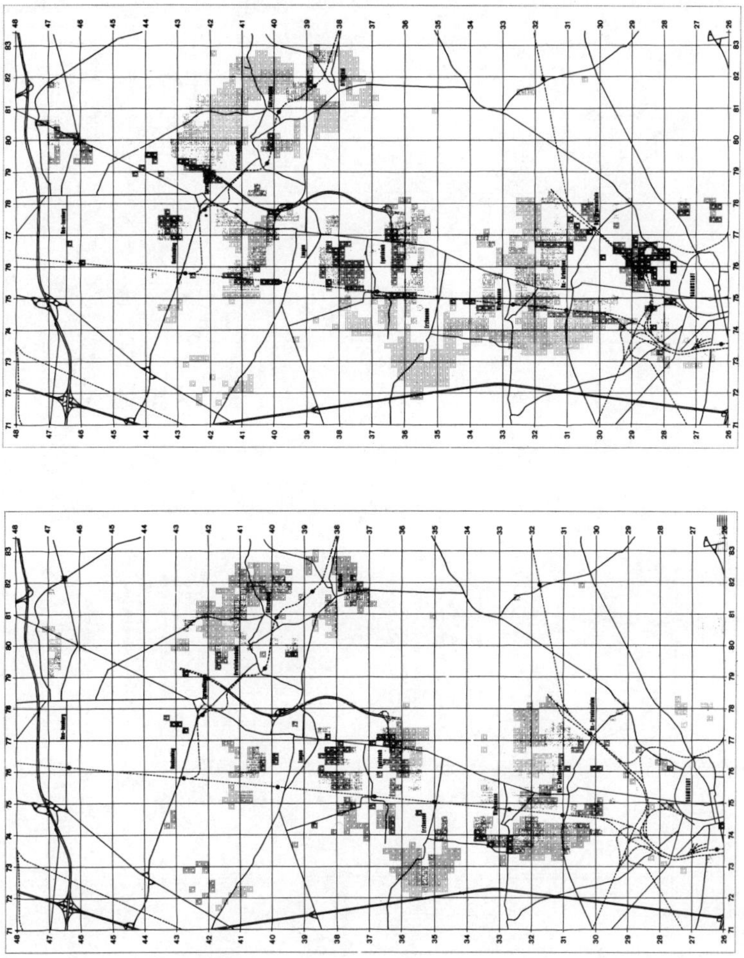

FIGURE 11. RESULTS OF COMPUTER-AIDED SEARCH FOR POTENTIAL INDUSTRIAL SITES (RIGHT) AND POTENTIAL RESIDENTIAL SITES (LEFT)

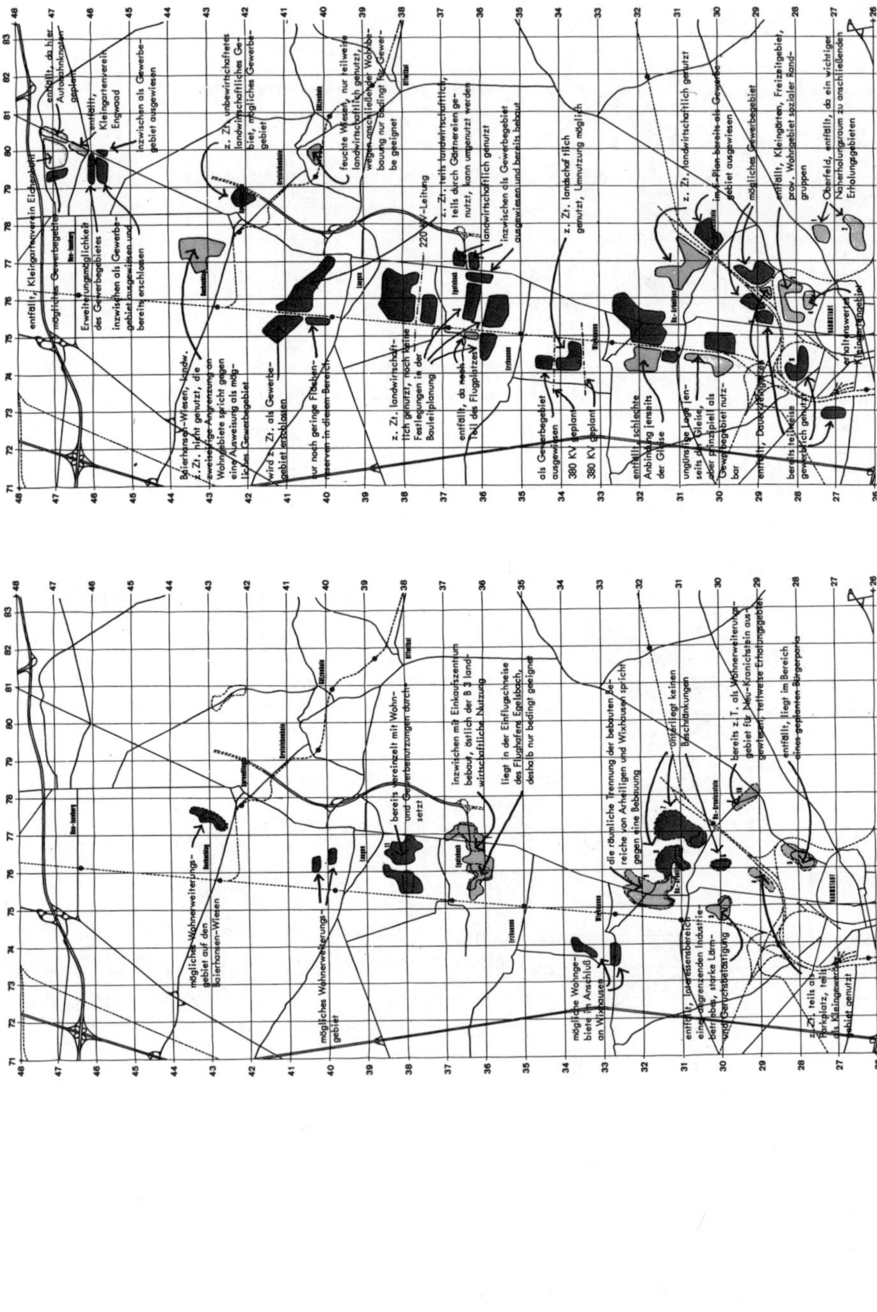

FIGURE 12. INTERPRETATION OF THE SUITABILITY ANALYSES. RIGHT: INDUSTRIAL SITES. LEFT: RESIDENTIAL SITES.

LAND-USE AS AN ELEMENT OF COMMUNITY STRUCTURE AND TRANSPORT

Hans-Jürgen Wicht

Dornier System GmbH

Friedrichshafen

THE REINTEGRATION OF RESIDENTIAL AND WORKING AREAS; OR URBAN TRANSPORT POLICY STARTS WITH TRAFFIC GENERATION

The familiar four-step transport planning process has as phases:

1. Traffic generation
2. Traffic distribution
3. Selection of modes of transport (modal split)
4. Routing and dimensioning

Present short distance passenger transport policies (see for example, the transport concept of the Federal Republic of Germany for 1970; also the objectives of its municipal transportation finance law)*, focus largely on phases 3 and 4 of the transport planning process. This means, for instance, that modal split is to be controlled by a preponderant use of indirect rather than restrictive measures. These measures are to be taken according to a cost benefit relation which is not an optimal investment strategy, but satisfactory enough to solve our economy-wide transport problems.

In 1971, West Germany pinned its hopes on the realisation of the GVFG. A solution to the short distance transport problem seemed more possible, at least in the metropolitan areas.

*Under this law (abbreviated as GVFG) a large part of the investment costs of public transport are financed by federal funds.

This hope was only partly fulfilled, because:

1. Conurbations are unconstitutionally given preference, although the oil tax is collected in the whole Federal Republic. Thus the other regions - particularly rural areas - are discriminated against.

2. Those public transport systems are subsidised that require high route investments. Only by spending federal funds in this way, can the unconventional transport systems (like CAT) and the conventional investment systems with maximum transport capacity, such as the rapid transit system (S-Bahn) and the subway, have a market chance at all. A disadvantage of this subsidising practice is that it was only recently that transportation systems with low route investment costs (e.g. buses) finally became a part of the government research-and development programme for intra-urban traffic. Consequently, the cities did not develop those less expansive systems.

3. Financing public transport investments was so easy for communities and traffic enterprises that they opted for systems which were too expensive in their subsequent costs. With a completely immovable price policy, these costs have led today to the transport services showing growing deficits. These deficits have reached such an amount that the public will no longer bear them, and transport facilities have had paradoxically to be restricted (see the urban railway/bus network in the Munich area).

From the situation described above, the following thesis has been derived:

Short distance transport policy in the Federal Republic of Germany, as well as in many other developed and developing countries, tries to intervene almost exclusively in the modal split of private and public short distance transport. It does not intervene in the field of traffic generation and distribution.

The success of measures which might be taken on traffic generation can be demonstrated by the example of the town of Kaufbeuren.

SITUATIONAL ANALYSIS OF KAUFBEUREN

Geographic Location

Kaufbeuren is located at a distance of about 90 km. west of Munich (Figure 1). Kaufbeuren has 43,000 inhabitants; and, in our forecast to 1990, will just about keep this number.

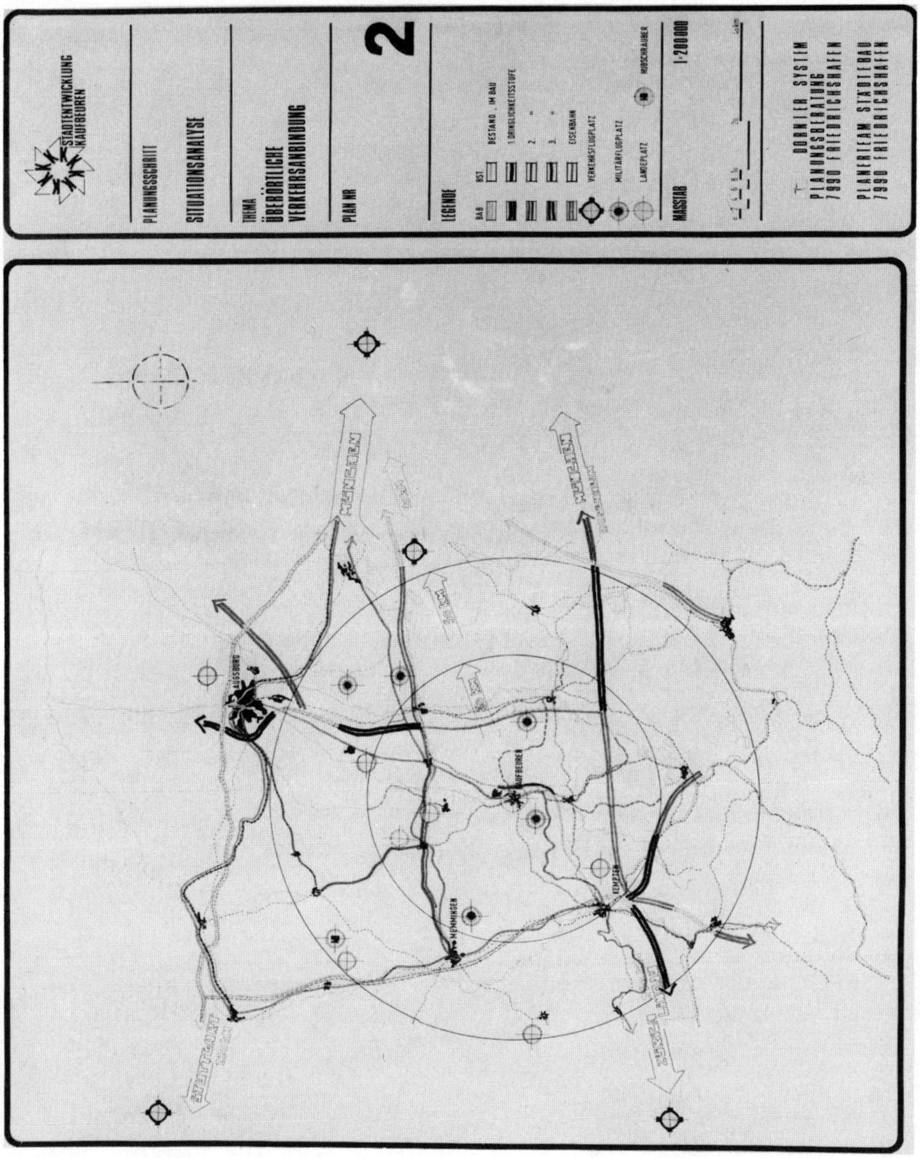

FIGURE 1. REGIONAL TRANSPORT CONNECTIONS

Natural and Man-made Restrictions for Planning

Figure 2 shows some of the natural landmarks and zones which characterise Kaufbeuren and restrict planning.

The most important structural element in the topography of the urban area is the valley of the river Wertach, running from south to north (600-800 m above sea level). At its widest point it is about 1000 m. The valley is limited to the west and east by steep terraces up to 20 m in height, which pass in a broad ridge to the west. This ridge is 840 m high and represents a natural border to the urban area. In the east, however, there are wide, slowly rising plains with many moraine hills.

The topography of the Wertach Terraces forms one of the biggest obstacles for the road network to the east and west of the town. For this reason, there are almost no important traffic connections between one quarter and another, which means that the transportation system is predominantly urban.

The topographical restrictions also determine the location of the airport near the town centre, which has a strong influence on transportation in the southern part of the town.

Economic Structure and its Effect on Housing, Urban Patterns and Urban Development

The economic structure of Kaufbeuren is largely characterised by the Gablonz industry, which specialises in imitation jewellery. Two-thirds of all employees in the manufacturing sector work for it. The Gablonz jewellery industry is not - as its name suggests - an industry in the usual sense of the word. It is rather a 'production pool', as was found in the pre-industrial centuries in Europe, and as still exists in many developing countries nowadays.

The Gablonz industry consists of about five hundred small-scale enterprises which are legally independent and are organised in a so-called 'industrial association'. As the holding company, the Gablonz industrial association has no legal influence on the enterprises' decisions, but is based on voluntary co-operation, like a chamber of commerce.

The production programme of the Gablonz industry has traditionally concentrated on the manufacture of fashion jewellery in glass, metal and plastics. Semi-finished products for other

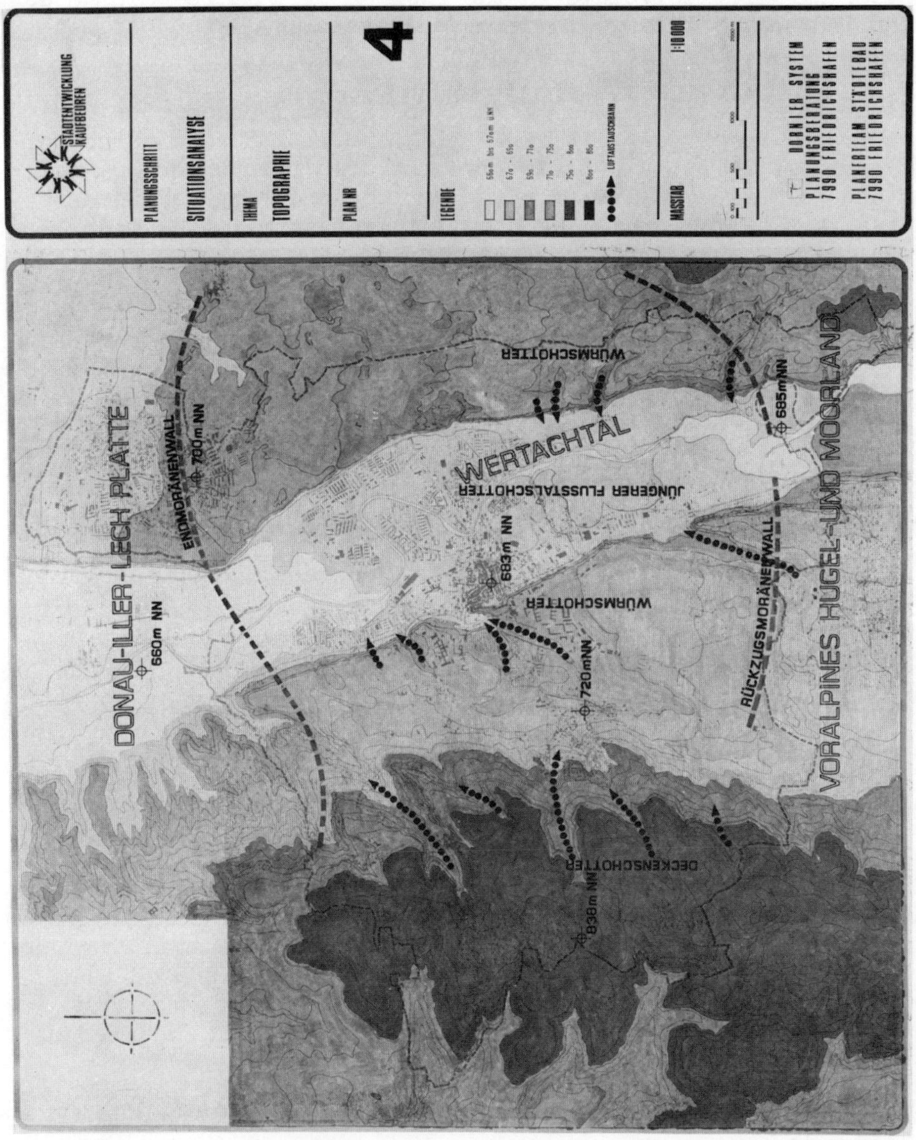

FIGURE 2. TOPOGRAPHY

industries are, however, increasingly gaining importance, above all in the few large-scale companies. Each product and each service is generally offered and manufactured by a legally independent company. The daily supply of the individual companies produces a considerable traffic volume in the town and the surrounding areas. In other industries this would form an intra-plant transport network.

Since this traffic often consists of skilled production workers travelling by car, it is very expensive. For this reason, one of our first proposals to improve the town's transport situation and to make economies for the Gablonz industry has been to suggest a kind of forwarding agency, to service the pick-up and delivery trips. The larger-scale companies made use of a similar facility before World War Two, when the Gablonz industry was still quartered in the town of Gablonz, which today is in Czechoslovakia.

The characteristics of work-places in the Gablonz industry are low investment per place of work due to relatively simple machines, low space per work place, low noise emission and low pollutant and heat emission; but also an increased demand for skilled labour, technical know-how, product design skills, a feeling for the market and commercial training.

The characteristics of the place of work enable 2000 employees out of 6000 to work at home. A third of all workers in the Gablonz industry do not need to commute. On the other hand, there is a relatively large volume of goods transport, albeit without marked peaks.

The number of workers in a company determines the distance and kind of the commuters' daily journey. Generally, with decreasing company size, the length of the journey to work decreases. The 6000 employees of the Gablonz industry are distributed among 500 companies. Company size ranges from a maximum staff of 400 to 500 persons for one company, to a large number of one-man companies. The greater part of all employees work in companies of less than 10 persons. Short trips correspond to a small company size. A household survey carried out in Spring 1974 showed that 70 to 80% of the workers do not use their cars, while 50 to 60% go by foot or bicycle, in those areas whose inhabitants are mainly employed in the Gablonz industry (e.g. Neugablonz). The time spent travelling is correspondingly short (see Tables 1 and 2). About 50% of all employees in the whole town need less than 10 minutes for their trip to work. A typical Gablonz area has as many as 60% of inhabitants with trips to work of less than 10 minutes.

A characteristic feature of Kaufbeuren is the 'bipolar' urban structure: the main pole developed around the medieval centre;

Means of Transport	by Foot	Bicycle Motorcycle	Car	Bus	Other
Whole town	30.6	14.4	45.5	6.2	3.3
Typical Gablonz quarter	54.8	9.0	18.2	9.0	9.0

TABLE 1. MODES OF TRANSPORT USED ON THE JOURNEY TO WORK HOUSEHOLD SURVEY - SPRING 1974 (PERCENTAGES)

Minutes	up to 10	10-20	20-30	30-40	40-60	more than 60
Whole town	51.2	29.8	9.8	3.6	1.5	3.8
Typical Gablonz area	60.0	25.0	0.0	5.0	0.0	10.0

TABLE 2. TIME SPENT ON THE JOURNEY TO WORK (PERCENTAGES)

the other is formed by Neugablonz which was newly established after World War Two (Figure 3).

The quarter of Neugablonz began its development in the years after 1945, on the northeast Wertach Terrace 3.5 km away from the centre of Kaufbeuren, amidst a dense forest zone. The new location of the Gablonz industry does not really represent the result of comprehensive planning. Rather the quarter arose out of the situations and necessities of the post-war period.

At first, housing was restricted by the need to use the remaining parts of the former ammunition stores from World War Two. Ammunition roads and bunkers were also adapted to the new requirements. The inadequate urban development which resulted can still be seen in the housing to the east.

The planned construction which started in 1952 was based essentially on the land-use plan produced by the city surveyor, Mr. Gunther. The population was to be 12,000 inhabitants in the first development stage, and 20,000 inhabitants at the second stage.

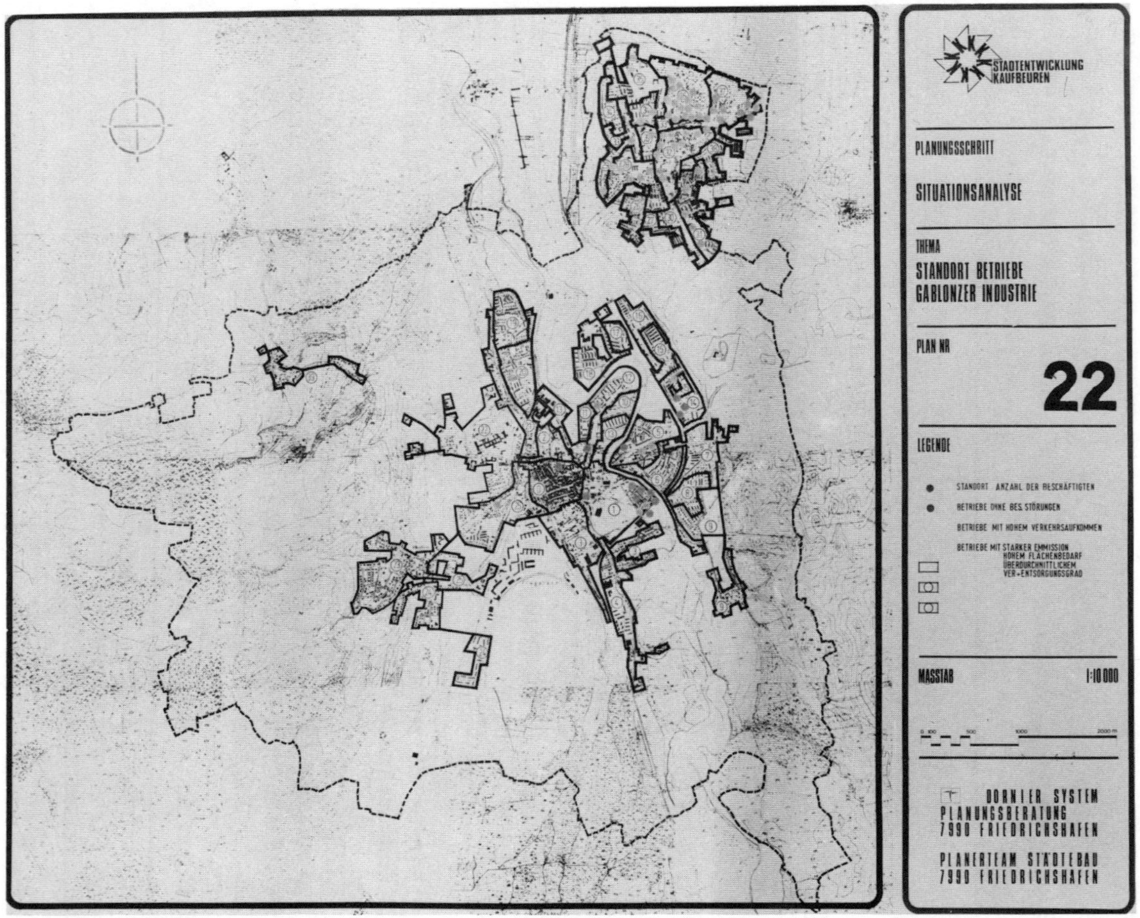

FIGURE 3. LOCATION OF THE GABLONZ INDUSTRY

Because of its proposed economic and social structure, but also in appreciation of its special landscape value, Neugablonz had to be developed as an industrial and garden town with an average density of 100 inhabitants per hectare. The spatial concept emphasised the main north-south axis and the centre "Neuer Markt" at the mid-point of the axis.

Due to the influx of refugees, almost 70% of today's buildings were started in the years 1948 to 1964. By 1964 housing areas had reached the restricting forest zones surrounding Neugablonz. The forest was partly cleared, despite a conflict between housing and landscape values. The basic idea behind urban development in the south of Neugablonz can be seen in the attempt to link together the two poles of Neugablonz and the medieval centre of Kaufbeuren. At the moment, the housing areas of Neugablonz and Kaufbeuren are only 500 m apart.

Compared with old Kaufbeuren, Neugablonz appears as a largely compact settlement, above all in the centre. The areas situated east and west of the main street have quite distinct urban patterns. The eastern part still reflects the restrictions of the 'pioneer' period, and the characteristics of the Neugablonz industry; while the western area appears as a planned urban structure.

The extensive industrial areas in Neugablonz are situated at the periphery, above all in the east. There is some conflict with recreational land-use, though this is tempered by prevailing wind directions. The glass factory is located in the north-western area, well-screened by the woods. The industrial areas near the centre conflict with the desired high population density.

The eastern part of Neugablonz represents almost entirely a residential area intermingled with industrial sites (in future referred to as a 'mixed area'). In many respects this is an ideal kind of land-use and structural texture. It was generated by the individualistic work-places that are characteristic of the Gablonz industry. This structure is, however, contrary to the required density towards the centre of Neugablonz, and some production processes in this quarter are no longer compatible with a residential area. Since extensive urban development is restricted by forest and administrative boundaries, the problems which arise from mixed areas have to be solved by a resettlement and renewal programme.

The mixed areas amount to about 43% of the complete housing area in Neugablonz, and in Kaufbeuren altogether to about 25%. Other towns comparable in size show proportions in the range of 15-20%.

Traffic Situation

Figure 4 shows the traffic situation. The medieval radial road network still exists, and housing development, except in Neugablonz, runs fan-wise according to the existing network.

A forecast of traffic volume to 1990 shows an increase from about 10000 cars/16 hours to 20000 cars/16 hours, for the maximally loaded section to the north of the old town.

The deficiencies of this traffic situation are illustrated in Figure 5. The concentration at the crossing of the road network (Spittelmuhl intersection) is striking. Here, in the town centre, two federal roads, east/west (B 12) and north/south (B 16), meet, together with the heavily loaded urban artery to Neugablonz, and an access to the old town. Here, as well as at other points of the road network, the car-driver can orientate himself only with difficulty.

It is not possible for long-distance traffic to by-pass the city of Kaufbeuren, nor can traffic within the city by-pass the centre. Moreover, there are no linkages between the quarters and the old town centre.

Despite the high percentage of bicycle and pedestrian trips to and from the work-place, there is no network of routes for them. Car separated paths are not readily apparent and are unattractive; although there are excellent natural conditions for combining the green axis with a pedestrian network.

Public transport services are provided by a private bus company. This company benefits by the well-balanced relation between Neugablonz and the old town centre. But the bus network does not cover the whole of the town because of the incomplete road system (Figure 6).

Despite all these important deficiencies in the road system in Kaufbeuren, it is remarkable that the traffic maintains just tolerable conditions short of collapse. Moreover, it is remarkable that the proportion of pedestrians and bicyclists on the journey to work reaches 50%, despite the bad pedestrian and bicycle route network.

This fact can only be explained by the peculiarity of the economic base of the town and by the housing structure.

Statistically, this peculiarity becomes obvious if you compare the land used for transport with the land used for housing. The figures suggest the very low investment costs of the transport system.

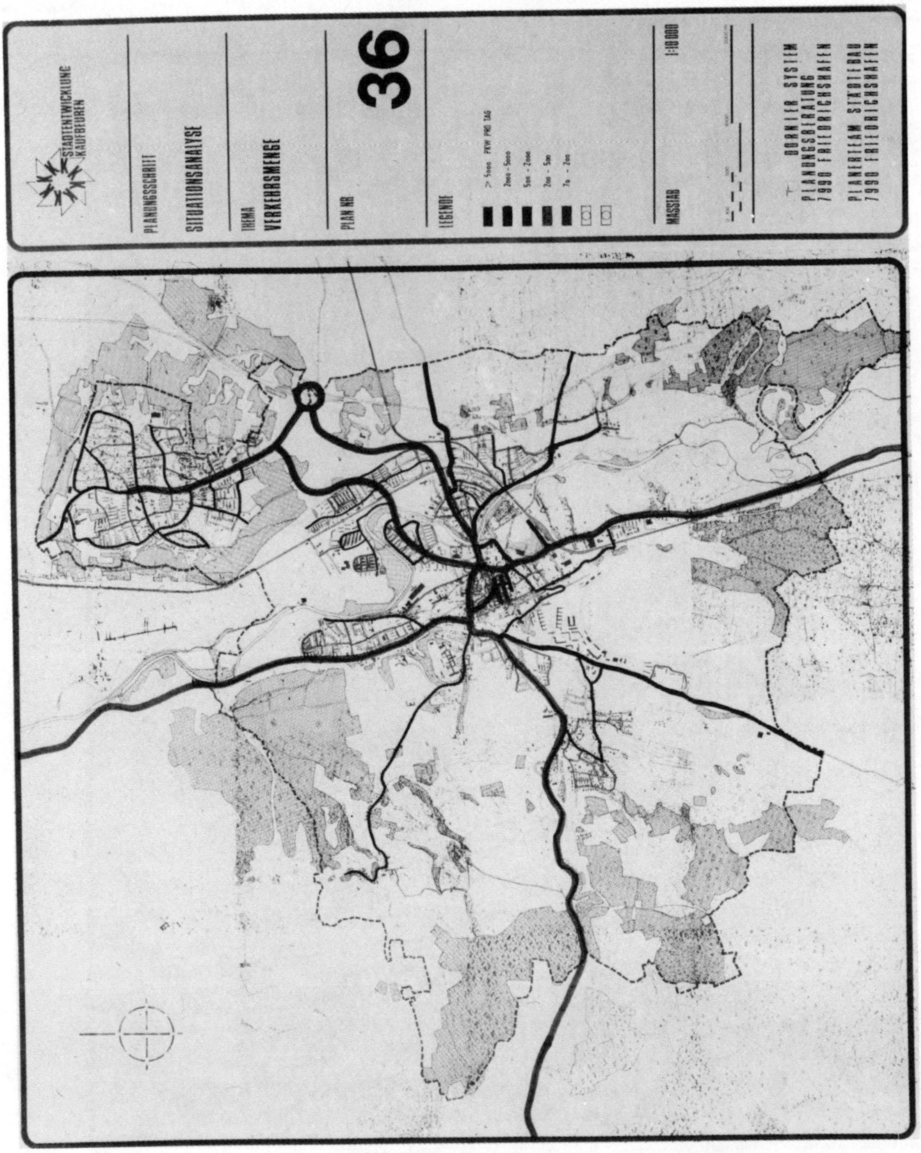

FIGURE 4. TRAFFIC VOLUMES

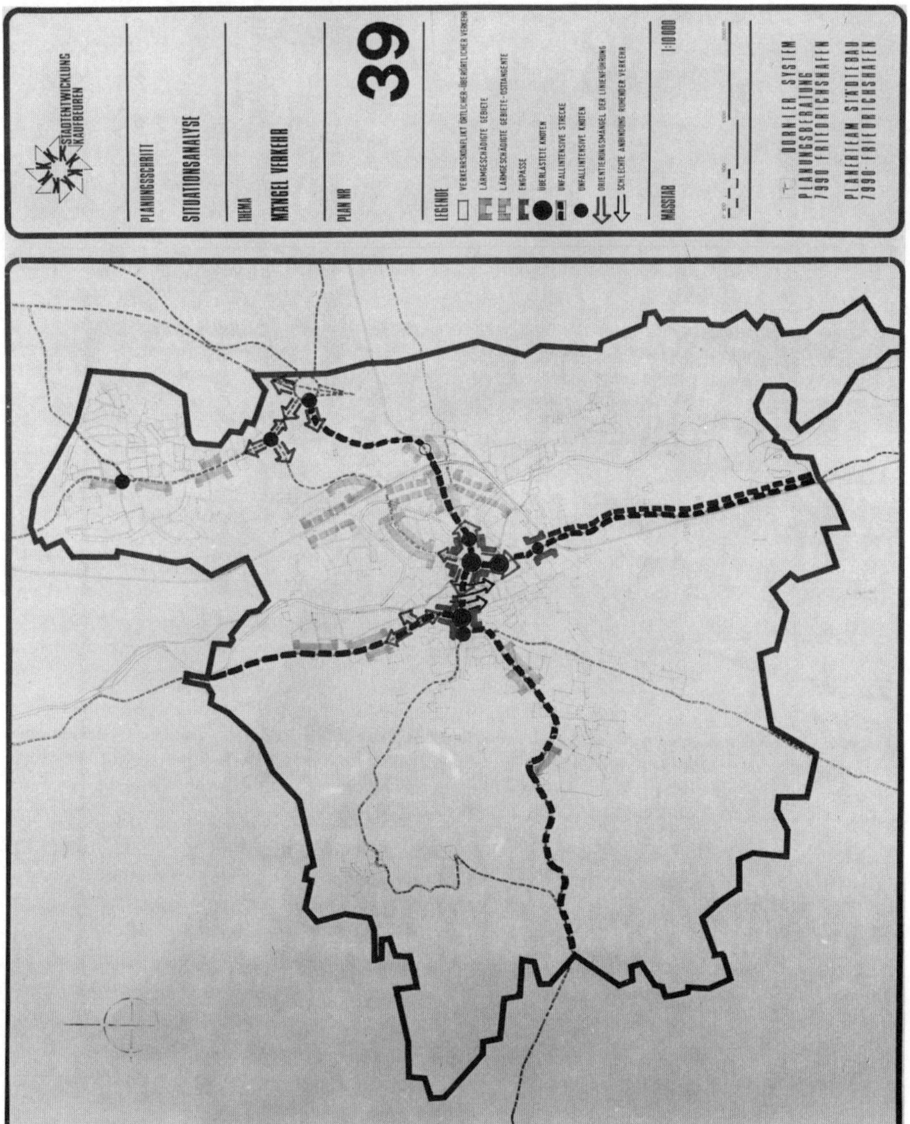

FIGURE 5. DEFICIENCIES OF THE ROAD NETWORK

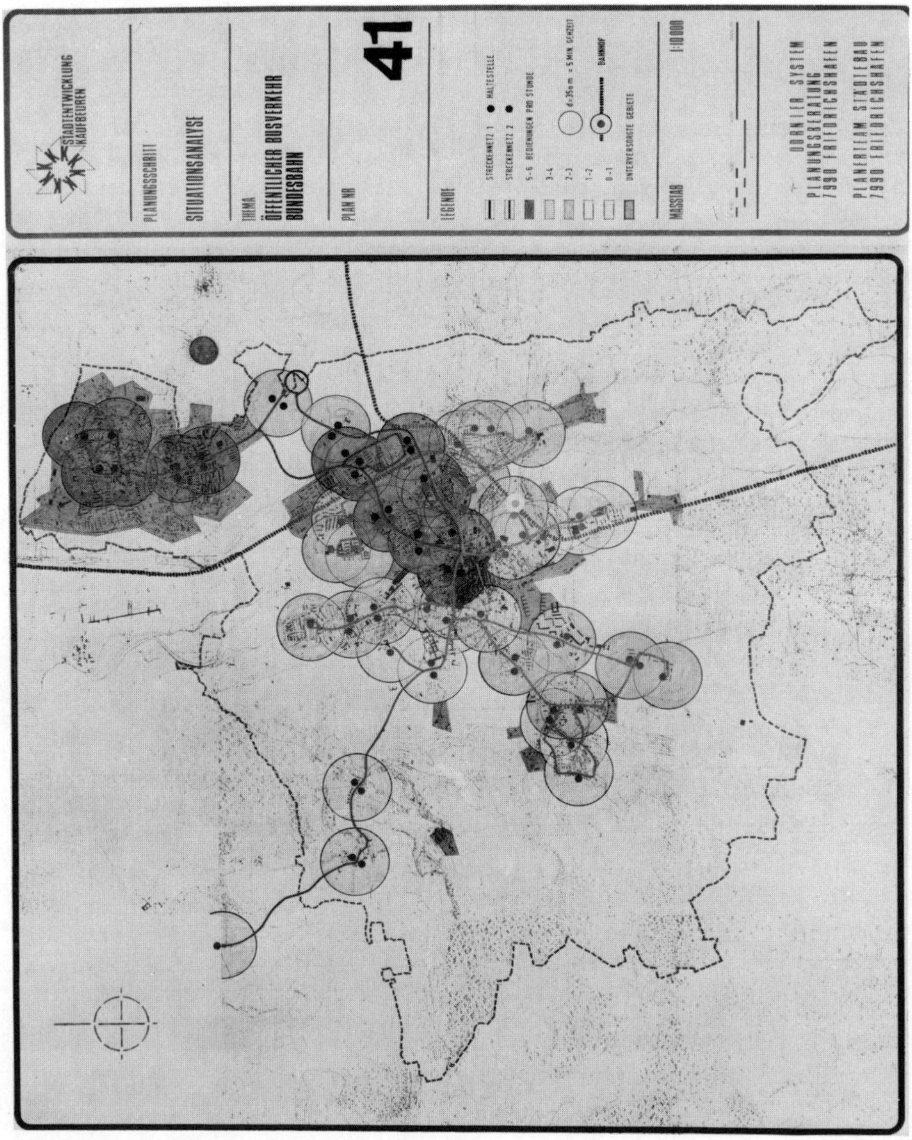

FIGURE 6. PUBLIC TRANSPORT NETWORK

WHAT FACTORS WILL AFFECT FUTURE TRAFFIC GENERATION IN KAUFBEUREN?

Besides the variables which generally affect the population's mobility, such as increasing leisure-time, cars per household and higher incomes, the economic development trends in Kaufbeuren may help to change decisively the mobility pattern of the population, especially in the journey to work.

The Gablonz industry is in stiff competition in the foreign markets with Eastern Bloc countries, Hongkong, Singapore and Taiwan. Therefore, change of internal structure has been begun in the Gablonz industry, demanding different processing methods and raw materials and leading also to bigger production units. Since the working force is at a standstill in the Gablonz industry, the city of Kaufbeuren is forced to look for other industrial employment. The internal structural changes in the Gablonz industry, as well as the need to offer industrial sites to companies willing to settle in the town from outside, put the planner on the horns of a dilemma if he is to integrate living and working. Separating residential areas from work-places, however, may cause heavy overloading on the transport infrastructure of the town. If the city continues to hold to its goal of maintaing population-size, considerable funds have to be spent not only on the construction of industrial areas but also on the road network.

URBAN DEVELOPMENT CONCEPT AND TRANSPORT PLANNING CONCEPT

When looking for the location of industrial areas we tried to put them in the vicinity of residential areas and to protect the skyline and landscape. Figure 7 illustrates which goals had to be considered in the development of the traffic concept(missing or inadequate arterials; and missing or inadequate bicycle and pedestrian linkages).

Figure 8 shows the same goals in relation to the landscape and recreation. A partial system of green areas and water-courses can be recognised. It is introduced into the bicycle and pedestrian route network and linked with the corresponding network within the city. In this way, as illustrated by Figure 9, inhabitants are able to use the nearby recreation facilities in the very beautiful landscape surroundings of Kaufbeuren without recourse to a car.

The land-use plan and the solution proposed to preserve town and landscape structure and image aim to ensure that journey-to-work distances are attractive to pedestrians and cyclers; and to define an industrial area which does not conflict with the

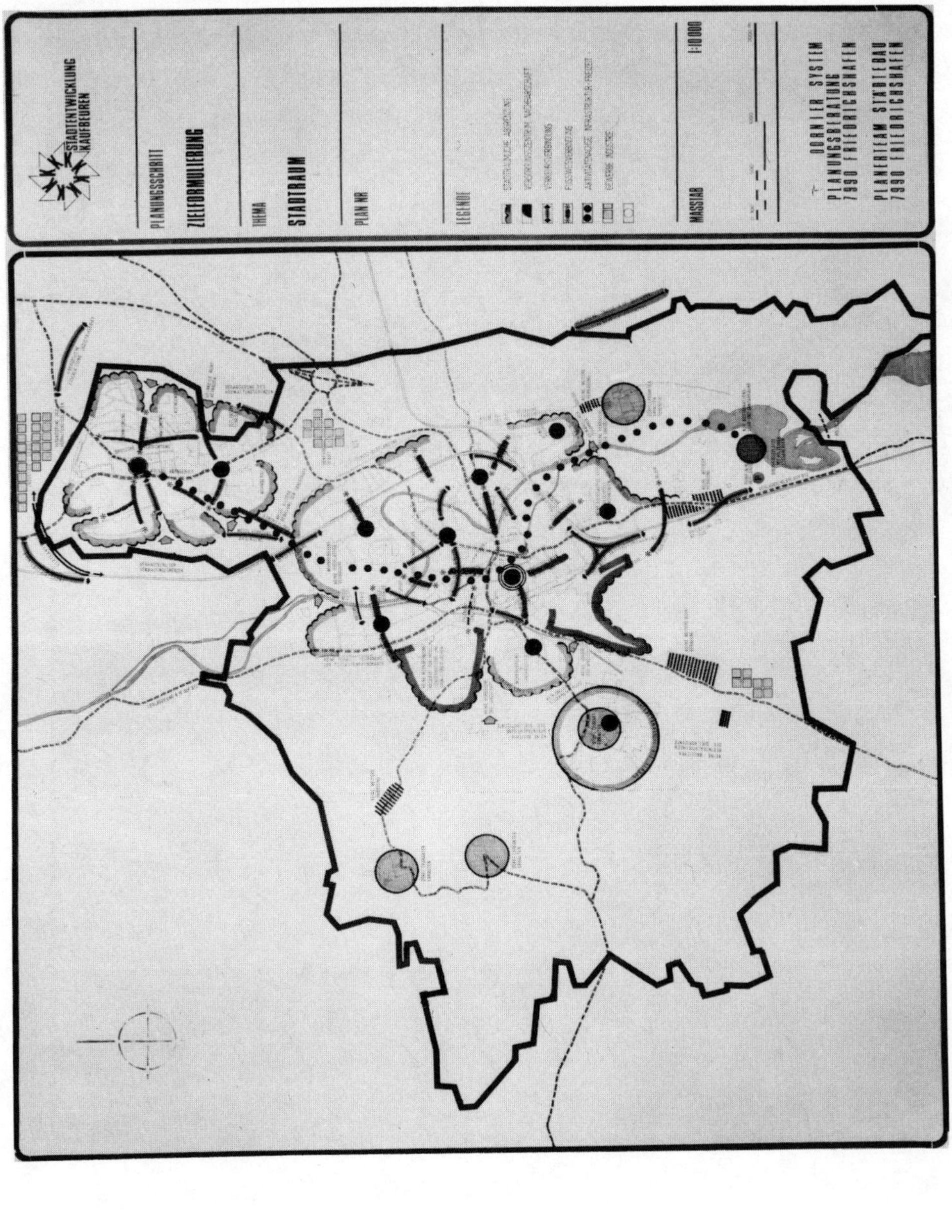

FIGURE 7. BUILT-UP AREAS

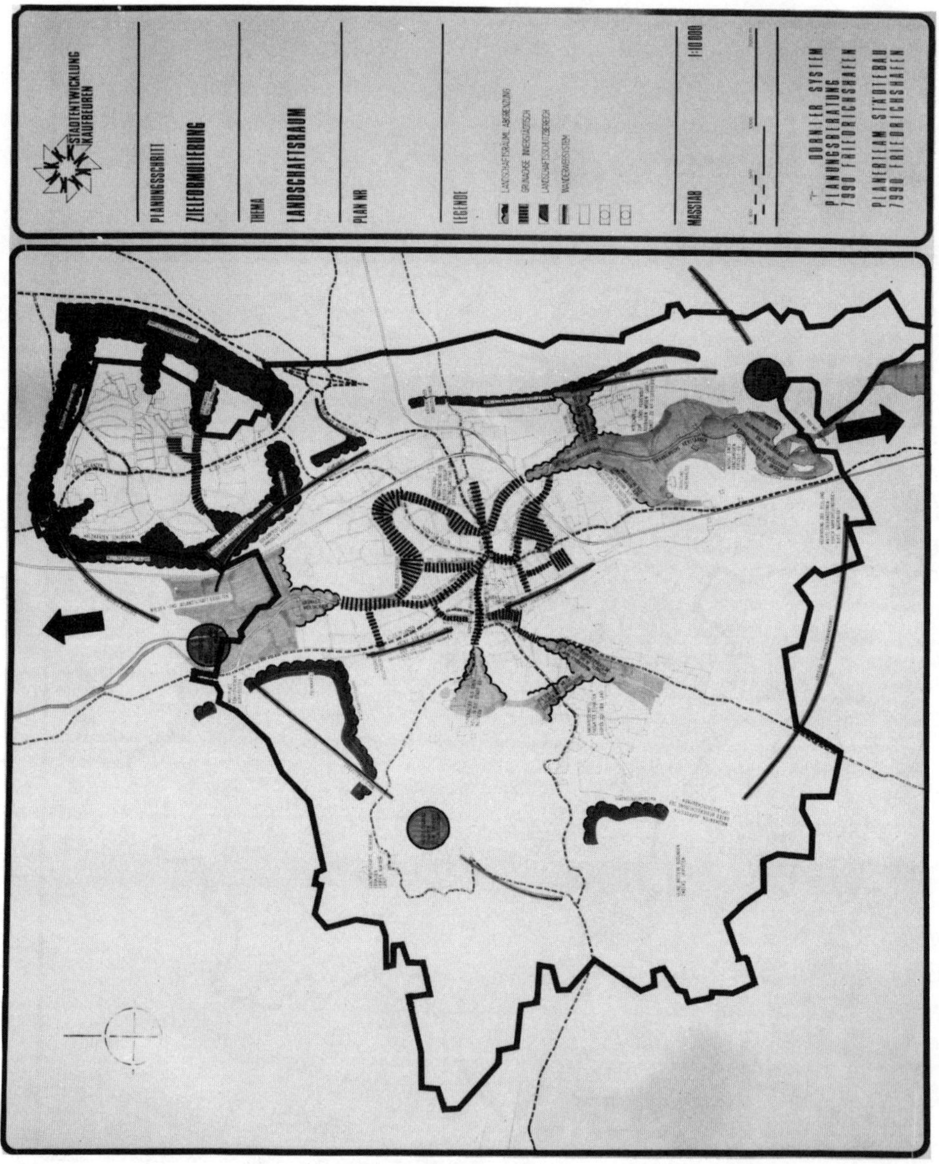

FIGURE 8. RECREATIONAL AREAS

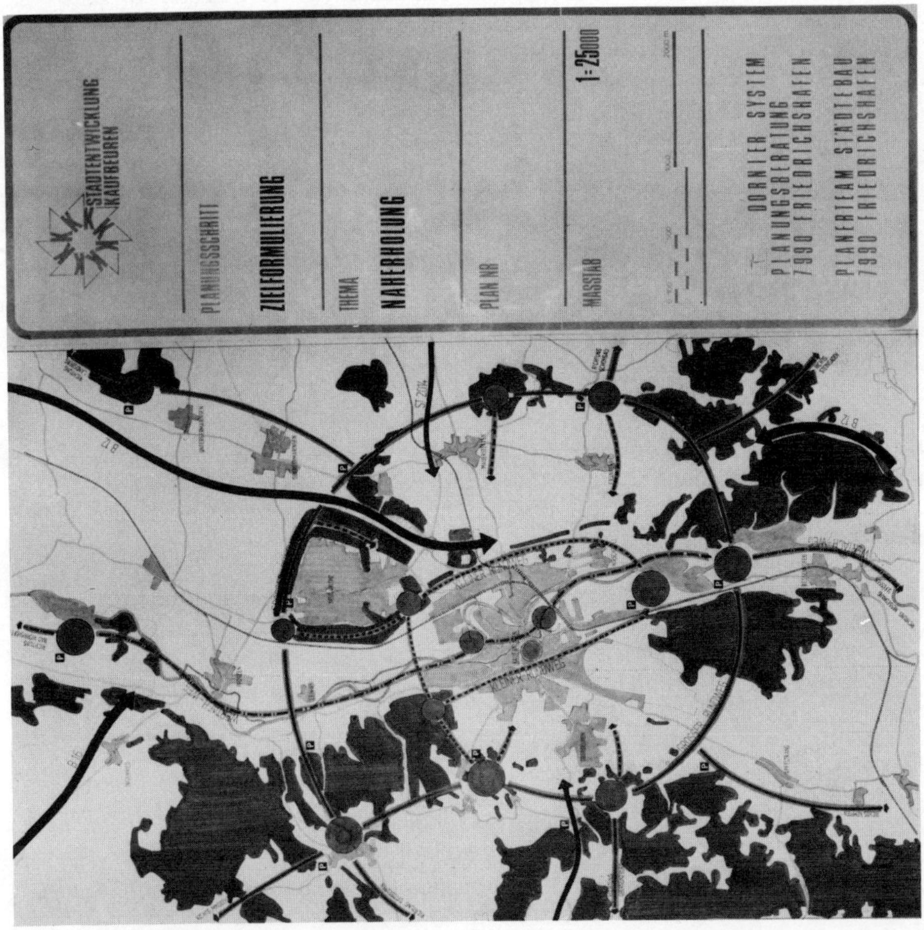

FIGURE 9. LOCAL RECREATION

residential zones. Industrial areas are located at a distance of 500 to 1000 m from the residential areas (particularly Neugablonz), with only one exception.

The internal renewal of Neugablonz is achieved by siting industrial areas for the less disruptive enterprises next to the existing mixed areas. The wish to protect the town and landscape structure prevented the location of an industrial area with high polluting rates within the municipal boundaries. A suitable area easily reached by foot was found north of Neugablonz.

To avoid the use of cars for the journey-to-work, we designed a bicycle and pedestrian network that links residential areas and industrial areas by the shortest route, while cars have to take detours. The bicycle and pedestrian network is linked to the green axis. Mixed areas were put next to those residential areas which still had potential for additional inhabitants.

The traffic concept is based on the thesis of Buchanan, that car traffic should not enter and disturb homogenous quarters, but pass around them. Only origin and destination trips are allowed to enter these calm cells. Cyclists and pedestrians have an attractive shortest-path network.

With this concept we hope to attain the same high percentage of non-car users on the journey-to-work as nowadays and in the past. However, we have not been able to integrate residential and industrial areas as successfully as before.

On the other hand, we hope to keep at a minimum the use of cars for recreation. We have linked the inner urban pedestrian routes with the footpaths outside the city. The design, of course, is favoured by the attractive natural elements of the Kaufbeuren landscape.

GENERAL REMARKS ON KAUFBEUREN AND THEIR CONSEQUENCES FOR A TRANSPORT POLICY

The economic structure of Kaufbeuren and the largely unplanned development of the housing, particularly in Neugablonz, is surely an exceptional case. This special case shows, however, how restricted the transport system of a community may be, provided the residential areas and work-places are strictly integrated.

Since in all developed countries the role of the tertiary sector and the number of non-distributing plants is increasing, the integration of housing and work-places is a pre-requisite. Pollution, visual intrusion, noise and other disturbing factors may be reduced by appropriate investments.

Another trend that favours the integration of housing and work-places is that the advantages of agglomeration are not as great as are usually claimed. Peripheral business and administration districts, as in Hamburg or Frankfurt, prove that modern tertiary sector work-places need not be located in central business districts. Often the only reason for locating in the central business district is prestige and image. My own investigations in the banker's quarter of Frankfurt found that contacts between bankers involve only a very small percentage of the personnel and nearly all are carried out by telephone.

However, these peripheral business centres also represent a negative example, as they are established without any connection to residential areas.

Conditions similar to those in the Gablonz industry are found in the many small-scale industries in developing countries. The bazaar, for example, represents one of the most successful models of the integration of living and working. Unfortunately, urban development practices in developing countries rarely adopt their own historic urban design concepts, but orientate themselves to stereotyped European and American styles.

So far, the planner still lacks the instruments he needs for this task. The gaps in our knowledge about the integration of living and working could be filled by a research programme that aims at finding out where the cost/benefit break-even point of integration is.

The results of such a research programme could be adopted by a short distance transport policy that dealt with traffic generation.

It is indeed not out of the question that such measures to integrate living and working should be encouraged such as

- bicycle and pedestrian route networks seperated from traffic
- support for the additional cost of integration, such as the establishment of lawns, noise-absorbing measures, additional anti-pollution facilities, etc.
- organisational measures intervening directly on traffic generation, e.g. no cars to be used on Sundays and holidays. (In Switzerland in Spring 1975, more than 50% of persons polled agreed to the introduction of a Sunday without cars every four weeks.)

Although the technical pre-requisites for integrating living and working are available in many cases, the realisation of such a policy may be hindered by transport politicians and the interests of the car-lobby.

ACKNOWLEDGEMENTS

I am grateful to the Council of Kaufbeuren and the Gablonz Industry for their material; and thank Dipl.-Ing. Hagen Binder, Dipl.-Ing. Kurt Schließmann, and Dipl.-Ing. Roland Fakler who elaborated with me the development plan of the city of Kaufbeuren.

SECTION 2. TRANSPORTATION PLANNING IN DEVELOPING COUNTRIES

The first two papers in this section see the forecasting of travel behaviour as the major task confronting transportation planners in developing countries. Particular problems occur with forecasts for the very rapidly expanding large urban areas which are characteristic of these countries. However, despite the difficulties introduced by such factors as the lack of reliable statistical information, or the greater role of public transport than in developed countries, Ueberschaer believes that classical transport planning models are generally applicable. His belief is illustrated with information from an urban transportation study carried out in South East Asia. He concludes that one may reasonably expect that the future demand for personal mobility in developing countries will take the same path as in most countries in the West. He emphasises the need for traffic models to reflect human behavioural characteristics.

Kozanoğlu sees the planner's main problem as being one of producing the most appropriate balance of services and of environmental and living standards. The congestion which is so common in urban areas of developing countries may best be tackled by planning whole communities from a systems' point of view. A combination of decentralised growth at the macro-level and intervention in land-use development at the micro-level should be adopted. For planning to be successful, a variety of incentives and restrictions will be needed to direct growth into the desired locations.

A concrete example of the place of transportation and communication in the socio-economic state of a developing country is given by the statistical study of Sorguç and Şener. They draw up eight indices of development for Turkey's 67 provinces and by multiple regression analysis show the relative contribution of each index to socio-economic development. 'Transportation and communication' emerges as one of the major factors; and is in turn shown to be particularly related to the indices of 'social and cultural' and 'commercial and financial' development. Clearly data analysis of this kind needs validation and further research to relate it more pragmatically to the underlying planning issues.

PROBLEMS OF LONG-RANGE TRAVEL BEHAVIOUR FORECASTS IN DEVELOPING COUNTRIES

Manfred Ueberschaer

Intertraffic Gmbh

Düsseldorf

CONDITIONS IN DEVELOPING COUNTRIES WITH RESPECT TO URBAN TRANSPORTATION

The traffic planner is expected to be in a position to foresee how traffic will develop in a certain area over the next ten, twenty or more years and what consequences will result, in order to promote the future economic, cultural and other development of the area in question in a favourable manner. The traffic planner, however, supports his forecasts on an estimate of the economic and demographic development, in a circular fashion. These facts are well known and everyone is aware that the statements of the traffic planner, at best, can only be as good as the inputs which he gets in order to make his predictions.

Now, it is a fact that such basic inputs for the forecast are relatively easy to obtain in the industrial countries of the Western world. There are now available a great many reliable statistics and observations of how tendencies are developing over decades. Information can be obtained under a variety of categories. In addition, people have been trying for a long time to describe the interrelationships discovered, either statistically or by way of models. This has been done in the field of transportation planning where models describe the behaviour of the population as road or transit users.

None of these prerequisites exists in the developing countries. There are only poor, and mainly unreliable, statistics available. Observation of travel behaviour over long periods of time, from which development trends could be fairly validly derived, do not exist, so that it may be advisable to apply travel behaviour models for traffic predictions only with considerable reservations.

It causes no surprise, therefore, that traffic forecasts made by American, Japanese and European consultants turn out to be simple trend prognoses, in which the growth of population, motorization, agricultural production, and of certain branches of industry or mining typical of the country under review, play their role. However, the travel behaviour of the resident population remains insufficiently covered in the forecasting process. This procedure may be adequate if a prediction about the future development of a corridor area or of a region has to be made, but it proves insufficient if it is applied to the forecasting of future traffic in urban centres with a concentration of several million residents, which is often the case with cities in developing countries.

Until now, there are real gaps and many questions in this field, but only a few satisfactory answers. In addition, the extremely rapid growth of residents in these metropolitan areas has shown a trend which can be analysed only with great difficulty. The enormous need to catch up in these countries in almost every sphere of economy and technology has led to rocketing developments during the last few years which have been sometimes so rapid that one scarcely dares to make further forecasts.

The problems arising from all these factors for traffic forecasting will be shown by taking the example of some traffic investigations in South East Asia. Reference will especially be made to data and experiences which were collected in Jakarta, the capital of Indonesia.

FACTORS INFLUENCING TRAFFIC BEHAVIOUR

First, in order to have a sufficient basis for transportation forecasts in developing countries, it is necessary to make comprehensive surveys of the land use and the travel pattern and behaviour of the population. That is, however, no simple matter because the most rudimentary statistical prerequisites for such surveys are lacking in the majority of cases: for example, official registration files of the population, lists of addresses, population statistics, etc. Such facilities would be desirable because of the vigourous migration of the population concerned; but are often not available for a wide variety of reasons. Where they do exist, they seldom present a reliable basis for surveys. The unlegal settlement of large tracts of land is, in many cases, not registered at all until the construction of a road or factory at a particular place becomes acute.

The increase of motorization is just as hard to follow in many cases. There are annual growth rates of 10%, and more, in almost every developing country despite the disproportionately high prices of vehicles resulting from import duties. The age of the vehicles is correspondingly old, and the resulting traffic safety on the road low. Scarcely any developing country has an automobile production of its own; though assembling factories are frequently found which, in some cases, show respectably high capacities.

Industrial plants are established only in large cities or in places where certain mineral resources are found. Growing industrialization again leads to further migration from the rural areas and increasing density of settlement in the towns.

Public modes of transport bear the greater portion of traffic demands in the majority of the developing countries, both in urban as well as cross-country travel. Buses, mainly operated by private bus companies, are the most important mode of transport. Trips are made according to demand so that the buses, in general, are filled to capacity and frequently overloaded. The public modes of transportation, often offered in great variety - from vehicles driven by manpower for one or two persons only to air-conditioned railcars - provide the sole means of transportation which can be afforded by the majority of population in developing countries.

This short review is to show that the following important external factors have to be taken into consideration as influences on the existing and future traffic behaviour of the population in these countries:

- rapidly increasing population growth
- rapidly increasing urbanization
- rapidly increasing motorization
- industrialization and technical modernization
- improvement of the transportation network, that is the street network and the public transportation facilities.

Added to these factors are several human behaviour characteristics which are typical of each particular country or each region. They are brought about by the situation and development of each individual country in its socio-political, social, cultural, historical, and other aspects. They express themselves in the population's social values and organization.

The aforementioned external factors, as well as human

behaviour characteristics, influence the traffic and travel attitudes of the population. To analyse these interrelationships is an essential prerequisite for a forecast of the future traffic situation.

PLANNING METHODOLOGY IN METROPOLITAN AREAS

Great efforts have been made in the last two decades to describe and simulate traffic patterns with the help of mathematical or numerical models. In order to be able to get an overall view of this quite complicated planning and simulation procedure, which can only be handled by the use of large computers, the process has been divided up into several stages. Feed-back loops take care of the fact that reciprocal effects and influences between traffic, the transport network, and the different land-use parameters of the planning area transform the simulation process into a uniform whole and, after a corresponding number of iterations, bring equilibrium into the whole system.

In the following, it is not intended to deal with the group of models whose forecasts are produced on the basis of trend extrapolations, although it is just these models that are favoured in the developing countries. They are useful where large-scale regional traffic prognoses and corridor investigations are concerned. But they cannot be used for traffic forecasting in urban metropolitan areas with high population densities and a differentiated traffic network which is supplied by a relatively dense road and mass transport network. In such cases the only way to get a solution is to use traffic models.

A variety of model concepts exist, of course. The decision as to what type of model should be applied depends on the planning goals and objectives in individual cases. Attention, however, will be directed in what follows to classical planning methodology, which has proved useful on so many occasions and which also, as the calibration of models with the data collected in Jakarta has shown, can be used in the urban centres of developing countries.

CHANGE OF TRANSPORTATION MODEL COMPONENTS AND PARAMETERS WITH RESPECT TO TRAVEL BEHAVIOUR CHANGES

The complete sequence of simulation usually occurs in four stages. The first step is the trip generation procedure, in which the total number of future person-trips is generated in each subarea of the planning area, using relationships existing between human activities and traffic.

As trips made by individuals always have a purpose, the generating factors causing traffic movements must be persons and not factors like industrial areas, sales areas, etc. The most significant factors are individuals who live in a zone and go outside their zone to pursue some kind of activity, such as employed people, schoolchildren, or housewives, and who thereby generate traffic. Schoolchildren or employees also generate traffic, either directly or indirectly, ending or originating at their schools or work-places.

The number of trips made by individuals in an area under examination depends demonstrably on a series of socio-economic factors. The interrelationships which reveal themselves in this respect have a certain general worldwide applicability. These interrelationships can be explained by some examples. Their influence on traffic generation, especially with regard to the forecast, will be shown.

A very clear interrelationship exists, for example, between vehicle ownership and daily trip frequency per person or per household in public and private traffic. The specific trip frequency grows with increasing motorization. This can be derived from the analysis of statistics collected in German and US-American cities and could be demonstrated in the analysis of the Jakarta data (Figure 1).

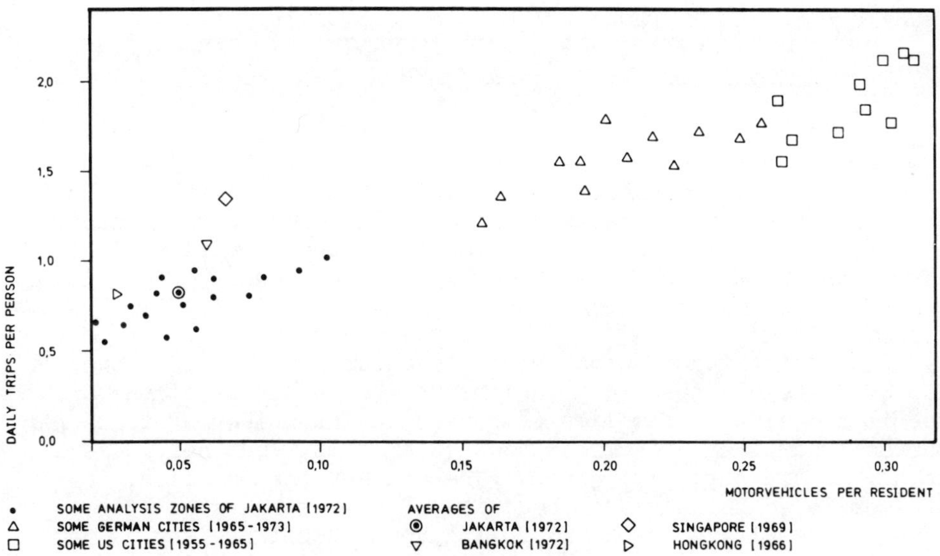

FIGURE 1. AVERAGE NUMBER OF DAILY TRIPS PER PERSON BY MOTOR VEHICLE OWNERSHIP

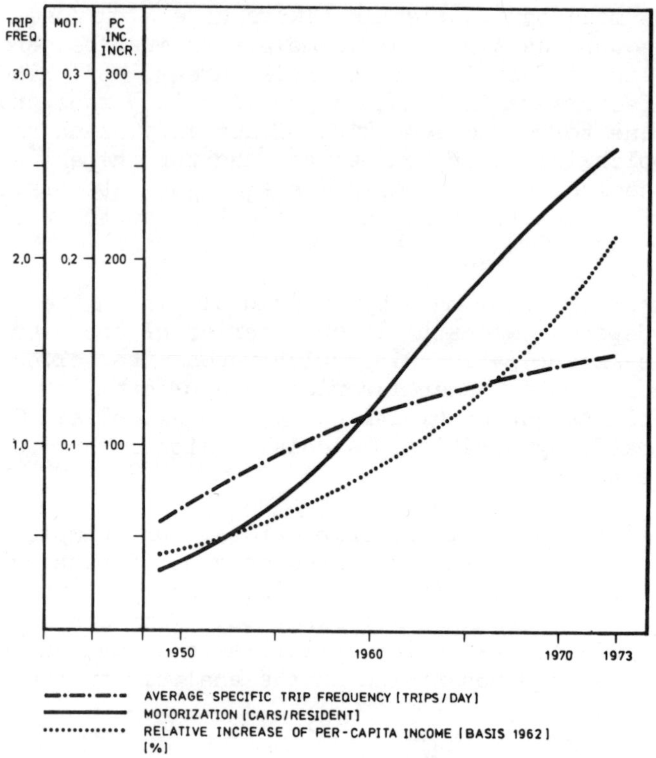

FIGURE 2. AVERAGE SPECIFIC TRIP FREQUENCY PER DAY, MOTORIZATION, RELATIVE INCREASE OF PER-CAPITA INCOME IN W. GERMANY FROM 1950 to 1973

Growing motorization is however also a symptom of rising incomes and rising standards of living. This is clearly shown by Figure 2, in which the relative growth of the per capita income, the motorization increase, and the rise in the specific trip frequency per resident per day since 1950 in the Federal Republic of Germany are placed side-by-side.

Increasing incomes and growing standards of living, however, also demonstrably lead to a decline in the number of births and thus in the size of families (Figure 3). The number of trips per household admittedly declines because of this; but not, however, the number of trips per person. The smaller the families, the more mobile each individual becomes. Housewives are no longer bound to the kitchen sink and to the home. Children try to make contact with others who do not belong to the family circle, or go to schools offering a higher level of education which are situated at a distance from the place where they live. This causes new trip demand to come into being (Figure 4).

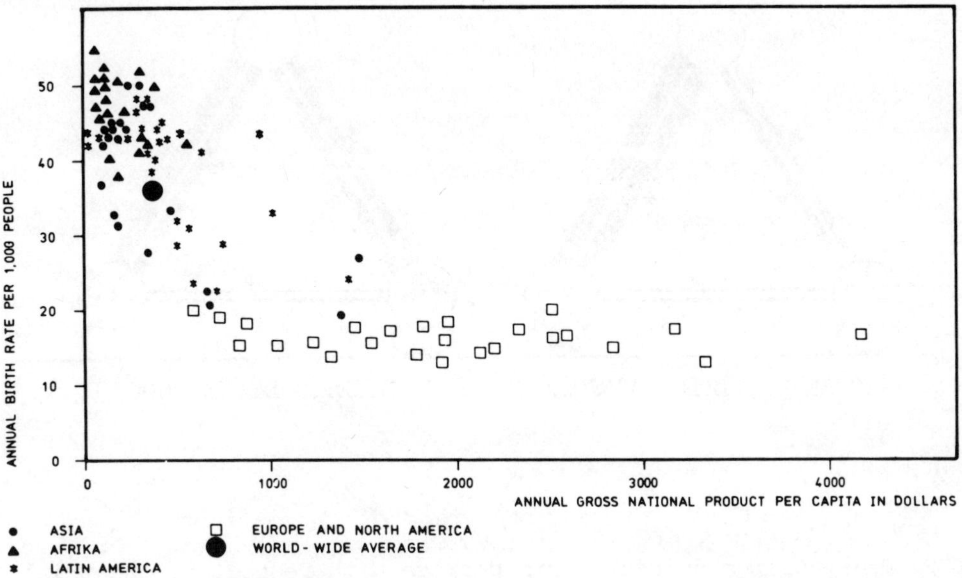

FIGURE 3. ANNUAL BIRTH RATES AND GNP PER CAPITA

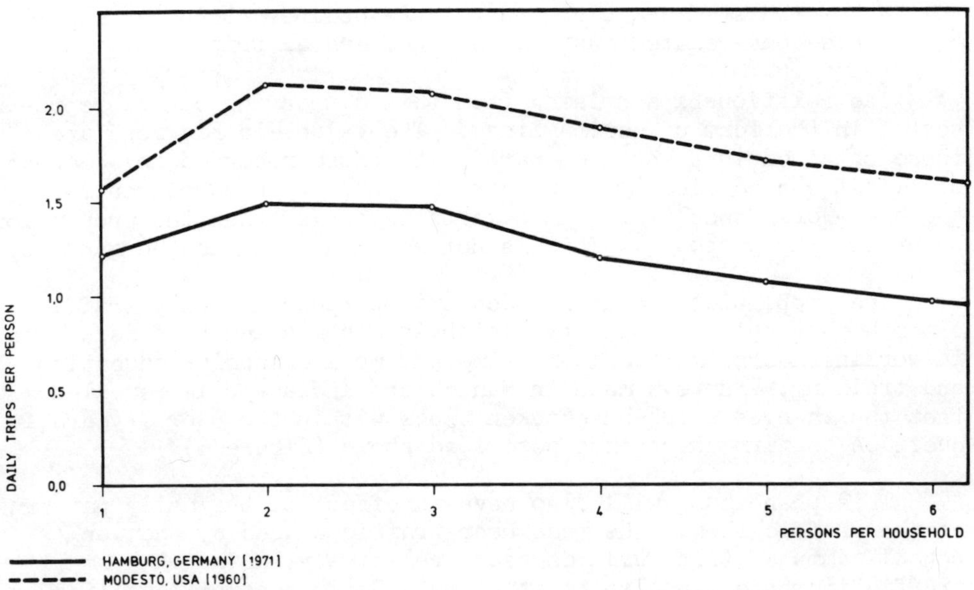

FIGURE 4. AVERAGE NUMBER OF DAILY TRIPS PER PERSON AND HOUSEHOLD SIZE

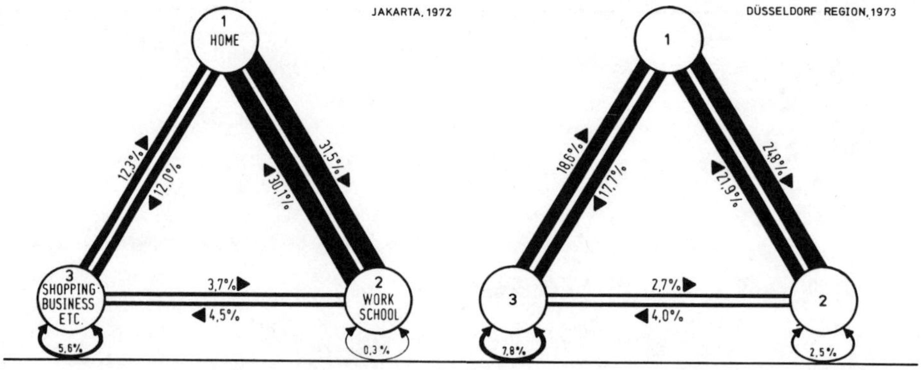

FIGURE 5. TRIP PURPOSE DESIRE LINES, DAILY TRAFFIC

Every trip made has a trip purpose with respect to the trip's origin and destination. Eight basic trip purposes have been analysed in Jakarta, which were clustered in three main groups for the purpose of model calibration:

- home-related work/school commuter trips
- home-related shopping and private business trips
- non-home-related business and commercial trips

The relationships arising from this clustering can be represented in the form of desire lines. The values in Figure 5 are those obtained in Jakarta compared with those recorded in a German city. It can be clearly seen that the proportion accounted for by the home-work/school trips in Jakarta is bigger than the proportion by other trip purposes. This is not so evident in the German city.

The proportions taken by each trip purpose in daily traffic alter with growing prosperity, which is accompanied by a reduction in working hours, more leisure time and more extensive education and training. Surveys made in Munich and Indianapolis revealed that the changes which have taken place within the 5 or 7 years in question confirm the trends described above (Figure 6).

This phenomenon will also have an effect on the daily pattern of the traffic flow. The rush hour traffic caused by the work/school commuter trips will decrease relatively, although a small traffic increase can also be expected. But a general levelling-off in the traffic flow will occur, because of the relatively vigorous growth of private shopping and leisure trips as well as of commercial business traffic.

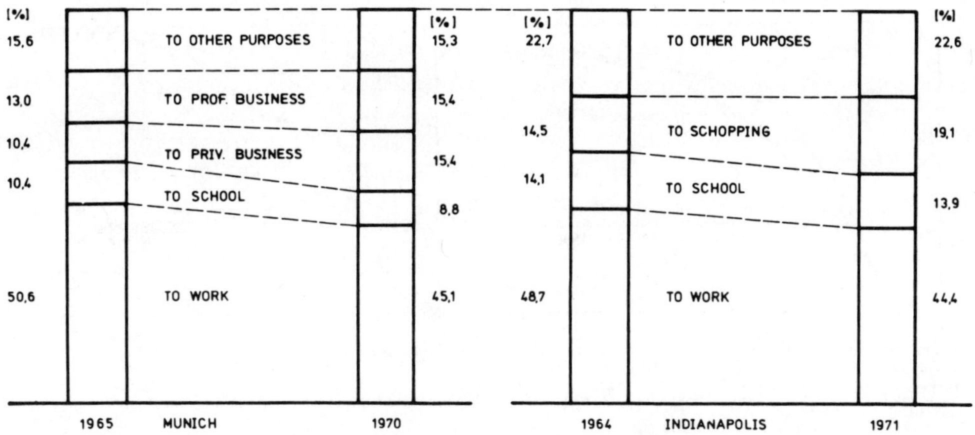

FIGURE 6. CHANGE IN PORTION OF DAILY TRAFFIC TRIP PURPOSES DURING TIME

This development in the population's travel behaviour - which is more or less apparent throughout the world - leads to the conclusion that, in any case, there will be an increase of the specific trip frequency per capita of the population in the developing countries.

When forecasts of traffic volumes are being calculated, the development trends and interrelationships which have been shown above must be taken into account in the model parameters. Any calculation, therefore, which uses the unchanged formula calibrated with the analysis data - where the growth of the traffic volumes is based only on the growth of the major trip-generating factors such as the number of residents, work-places, and motorization - and ignores the behavioural attitudes of these traffic generators, will not be correct.

The same applies to the other steps or models to be implemented and used in the planning process for calculating future traffic demands in a network system. A considerable influence is also exerted on the spatial distribution of the trips by the proportions of the different trip purposes.

Investigations in Jakarta have shown that, as in other countries, the choice of the trip-destination is decisively determined by the time necessary to reach this destination. The importance of the time factor in the choice of destination will alter with growing industrialization and the reduction of unemployment and underemployment. Efforts will be made, particularly in the work and business traffic sectors, to reduce the necessary travel time

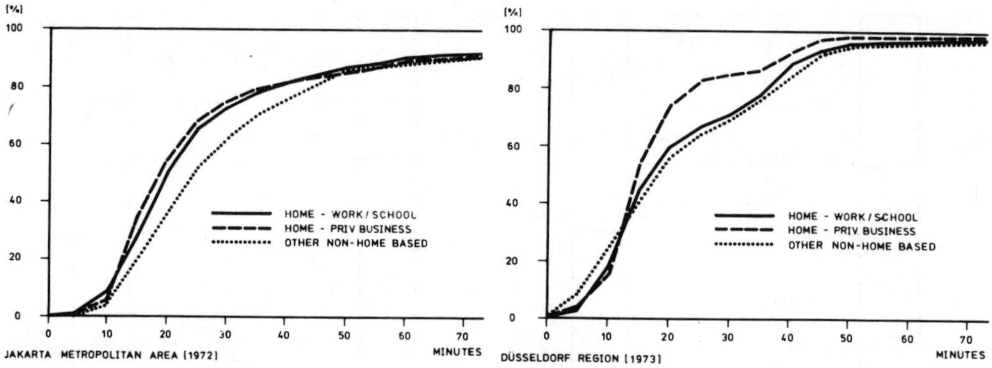

FIGURE 7. TRAVEL TIME DISTRIBUTION BY TRIP PURPOSE (ALL MODES)

so that it does not exceed extreme values. Figure 7 provides information on the travel-time distribution of work-commuter trips contrasted with the other trip purposes analysed in Jakarta in 1972. Conditions in the Dusseldorf area are also shown for comparison. It may be assumed that in the developing nations the accumulated travel-time distribution curve will tend to become steeper.

However, this tendency is being countered by a rapidly growing traffic density and an increasing spatial expansion of the urban centres as well as a change in land-uses. The longer travel-times caused by these facts can only be stopped by an improved transportation network. Otherwise the urban vitality and economic development of these cities will stagnate and decay. Generally, it may be assumed that the average travel-time for all trip-purposes will not increase, or increase only slightly, and will only show insignificant deviations from those of today. However, the average travel length for all trip-purposes will, in certain circumstances, vary considerably from those of today because of the improved supply of transport facilities.

Due to the low motorization in all developed countries the use of mass transport modes (mainly buses) is still very high, despite an uncomfortable bus service not attuned to the actual travel-demand of the population. This will certainly change in the future with growing industrialization, full employment, better education and higher incomes. People will begin to react more sensitively to time advantages existing between private and public transport modes.

The percentage of public mode users will be decisively determined by the proportion of captive riders. That depends, in the first instance, on the availability of a personal motor vehicle and on the possession of a driver's licence: that is, on the population's level of motorization and the average age structure.

The estimated future development of both factors will lead to a relative decline in public transportation in favour of private traffic. Naturally, growing motorization will reduce the proportion of trips made on public transportation in favour of private traffic (Figure 8). In the same way the gradual change of the age structure will have its effect through a future reduction in the size of households, especially the smaller number of children and young people (Figure 9).

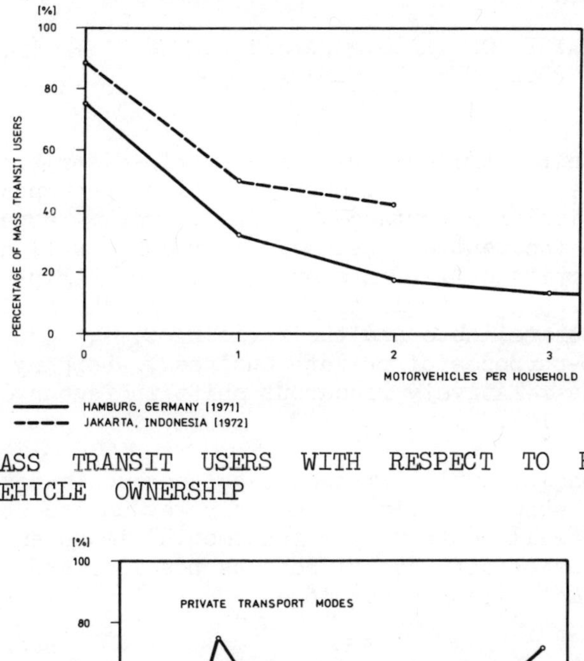

FIGURE 8. MASS TRANSIT USERS WITH RESPECT TO HOUSEHOLD VEHICLE OWNERSHIP

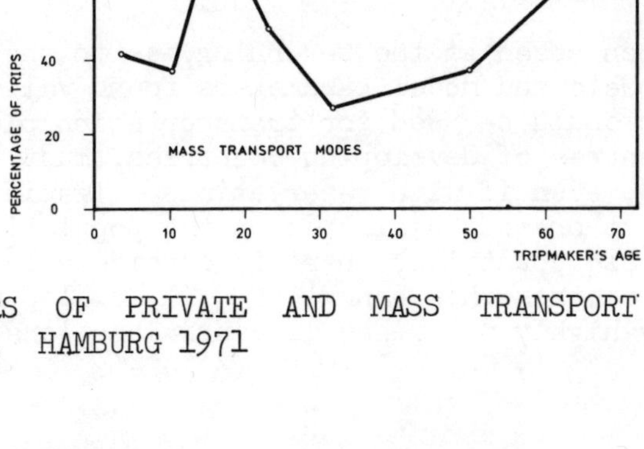

FIGURE 9. USERS OF PRIVATE AND MASS TRANSPORT MODES BY AGE, HAMBURG 1971

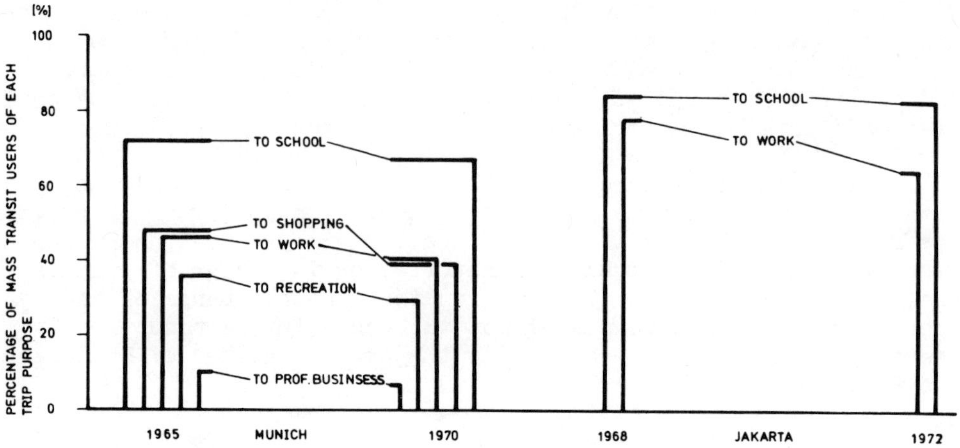

FIGURE 10. CHANGE OF MASS TRANSIT USERS FOR EACH TRIP PURPOSE DURING TIME

As in both the trip-generation and trip-distribution models described above, in the modal-choice model trip-purpose also assumes a particular importance. Trips to work - especially to work-places in the central business districts - will divert relatively few to private traffic compared to other trip-purposes, even with increasing motorization, because of the smaller number of parking lots available and their rising fees. On the other hand, for trip-purposes of private business, shopping and leisure there will be a relatively vigourous shift in favour of private traffic (Figure 10).

Proportionally public transport passengers in developing countries will show a decline. The only way an excessive decline can be avoided will be to make a fundamental improvement in the mass transportation services as soon as possible and before a high level of motorization has come.

CONCLUSION

The question posed at the beginning, as to whether the relationships, models and model parameters found valid in the industrial countries could be used for transportation planning forecasts in the urban centres of developing countries, must be answered in the affirmative, even if with reservations. Despite the lack of basic material at present available on the population's travel behaviour in these countries, there is considerable evidence that in their urban centres this behaviour will develop similarily to that in already highly motorized industrial nations provided that there is an increase in motorization, improvements to the mass transportation system and freedom to develop (i.e. absence of political pressure). There is, after all, no reason to suppose that with an increase in prosperity and a simultaneous improvement in the whole traffic system, the people of the developing countries (provided they are allowed free expression of personality) will not display the same demand for mobility as the people in the Western world have done during the preceding decades. The economic development of each country will determine the speed at which this development takes place.

Nevertheless, forecasts on person-traffic in the built-up areas of these countries are only possible by using the models also employed in Western forecasts. Care must always be taken to ensure that all the traffic models reflect human behavioural traits. The parameters which make the models sensitive to these behaviour traits must be predictable. Trend extrapolations and logistic procedures are to be used here. Developments already ascertained under similar conditions in other countries provide good guidelines and criteria.

TRANSPORTATION AND URBAN LIFE IN DEVELOPING NATIONS

Seyhan Kozanoğlu

Management Department

Boğaziçi University

Today transportation problems can be observed in big cities all over the world. Yet we can note a few differences in the nature of the problem between developed and developing nations.

THE URBAN TRANSPORTATION PROBLEM IN DEVELOPED NATIONS

The traffic problem in developed nations is attributed by some to the growth of the private automobile and neglect of public transit. For example in the United States it has been claimed that the suburban sprawl made possible by the private automobile has made the operation of public transit uneconomical, giving rise to a circle of decreasing service level and decreasing transit patronage. Yet as Owen (1972, p.38) observes the big city transportation problem is also present in European cities where public transit offers a good service and receives more government subsidies than the private automobile.

The basic cause of the problem seems to be related to inefficient urban design; in particular an imbalance between various capacities in the system which can congest wide streets. For example, the traffic generated by the large vertical capacities of high-rise buildings cannot be efficiently handled by the available horizontal capacities of roads leading to and from these high densities. Another capacity imbalance is created by expanding the capacity of the arteries leading into a city without considering how that traffic will be handled by the smaller road capacities within the city (Williams, 1971, p.156).

THE URBAN TRANSPORTATION PROBLEM IN DEVELOPING NATIONS

The causes of the big city transportation problem are not very different in a developing nation. Yet the problem seems more threatening. The reason is that the growth process is far from a steady-state equilibrium. In addition to growing incomes, large migrations from the villages into the big industrial centres result in a much faster growth in the demand for transportation. The forecasting of independent locational choices which is required for planning purposes becomes very difficult.

In contrast to the developed nations the suburban population in developing countries consists of low income groups. Given low densities, economical bus operation is not possible. Without automobiles these people resort to minibus services for transportation to and from the city centre. Within the city jitneys competing with buses can overcrowd the narrow streets. Jitneys provide a much superior service to the bus at rates quite close to bus fares. But no mode can be a substitute for the private car. The car represents a serious threat since, with fast rising city incomes, more and more private cars will appear on the already crowded streets. High rates of migration and the combined demand for transportation and other city services make the traffic problems in a developing country different from the problems of the less structured regions in industrialized nations. In other words, the large urban centres of developing countries are not on the same "ladder" of development as the small centres of developed countries.

STRATEGIES TO DEAL WITH THE URBAN TRANSPORTATION PROBLEM

The situation described above leads us to question the attempts to deal with the big city transportation problem through the expansion of transportation capacity. If we cannot control people's behaviour, if they keep on migrating into the cities, if they keep on making independent locational choices, then whatever investment in transportation facilities is made can only relieve congestion for a short period of time. Traffic will expand to fill available space, and large investments like a subway system, which have especially high opportunity costs in a developing nation, will not produce the expected benefits.

How can we affect people's choices in a democratic world? We can do some planning and provide the right economic incentives to affect their locational decisions. This planning will not include transportation only; but it will involve the planning of whole communities, with the objective of providing better transportation and living standards.

TRANSPORTATION IN CONNECTION WITH OTHER BIG CITY PROBLEMS

Large migrations into the city, besides affecting transportation, also result in problems of residence, water, sewage disposal, and public utilities. The interaction of transportation with these other problems goes beyond their demand on the same resources. Residential building costs will change as the number of floors is increased. Higher densities would require less investment in transportation and public utilities per household. On the other hand, savings can be made in sewage disposal costs when residential density is low enough to use soil absorption instead of a system of sewer pipes. These trade-offs in urban design have been studied by Robert Jones (in Haefele, 1969, pp.49-85). Jones first identifies different types of residential building according to the structural materials used, quality of finish and method of construction (self-help vs. contract). Construction costs per square foot are estimated according to the type of building, number of stories, and size. Total costs, including the costs for houses, streets, utilities, and other improvements, are estimated for different neighbourhoods consisting of one type of building with a specified number of stories and size. As a final step, Jones considers the costs of several city-wide layouts with different degrees of decentralization of work-places, where each layout consists of several neighbourhoods designed according to the previous analysis. Jones makes his cost estimations using data from Venezuela. Similar estimations and design suggestions can be made for other developing nations using prevalent cost figures.

PLANNING WHOLE COMMUNITIES

A consideration of transportation in relation to problems of living standards would involve the planning of whole communities. Communities containing work-places, housing, recreational facilities, and shopping centres would form self-sufficient clusters. All daily traffic would take place within the cluster. The clusters would be separated by free space and connected with the main roads (Owen, 1972, pp.56-61). This system seems to be a significant improvement over a system with dispersed points generating traffic that will enter a common centre; or a system with dispersed points generating traffic that will get into each others' way while entering another set of dispersed attraction points. How it can be achieved is another question.

If we could at least direct new migrants to settle at desired locations we would accomplish a lot. In Turkey the migrants build their houses overnight. These houses can multiply very fast even on land that is publicly or privately owned. Political and social reasons make their later removal very difficult. The main issue is the determination of the factors that guide the locational choices of newcomers. One factor might be the need to be close to relatives and other migrants from the same village or region. Or it might be the need to be close to possible or potential work-places. Urban planners can at least prevent future developments in undesirable places before it is too late. They can identify desirable residential locations, i.e. places at an appropriate distance from industrial establishments that would employ unskilled or semi-skilled labour. The provision of public utilities at these places would give an incentive that could affect locational choices.

INTERVENTION STRATEGIES AT THE MACRO-LEVEL

To avoid suboptimization we should analyze the problem from a systems point of view, identify the intervention points and trace the effects of these interventions throughout the system. At the micro-level we would take migration as given and try to make the best of city transportation and public utility problems. But at the macro-level, including the whole country within the system boundary, we can intervene upon the size of migrations by providing economic incentives away from the big industrial metropolitan centres. This involves a choice between centralized and decentralized growth. Centralized growth requires less investment in infrastructure and enables fast development in a few concentrated centres. According to one observation growth will be profitable if it is concentrated on large cities (Haefele, 1969, pp.53-4):

".... It does not seem feasible to consider every district town or city as a potential focal point for industrial development ..... The reliability of power and communication services, the accessibility of cultural and educational facilities, the resourcefulness of local service technicians, the passability of the roads, and the work outlook of the labor force tend to fall off with movement down the scale of size and outward from the metropolitan center."

This fast growth would provide the means for a later leap back to the previously neglected regions (Hunter, 1968, p.132).

Yet concentrating economic activities and people in one or two places results in congestion and other big city problems, and widens the gap between a limited number of industrial centres and a vast poor hinterland. Since a very large percentage of this poor population is in the agriculture sector, the development of market towns with storage and processing facilities, as suggested by Owen (1968, pp.101-102), would be a means of enhancing agricultural production and local incomes. With added local economic incentives fewer people would migrate into large centres.

Two major flows can occur over a connection between a rural area and an industrial centre. One flow involves the movement of people from the villages into the city. The other flow, which is the desired one, involves the movement of agricultural inputs from the city to the rural area which will come back in the form of increased agricultural output. The resultant increase in agricultural incomes facilitates the flow of consumer goods from the city into the country. Both kinds of flow would be facilitated by good transportation links between the two areas. But what exactly will flow over these roads, people or goods, depends mostly on the relative attractiveness of different regions within the country. Thus, at the macro-level the major intervention in the big city congestion problem involves the development of intercity transportation and the distribution of economic incentives more evenly across the country.

INTERVENTION STRATEGIES AT THE MICRO-LEVEL

The link between the migrations generated by the macro-system and the land-use patterns developing within the micro-system, the big city, involves the distinction between endogeneous and exogeneous work-places (Meyer, 1971, pp 112-113). Exogeneous work-places consist of manufacturing establishments, wholesalers, government offices, hospitals, agricultural and mining operations. In contrast, work-places involving service and market-oriented retail operations are endogeneous to the system since they make their locational choices with the objective of being close to housing. If we can assume that migrations are mainly determined by relative job opportunities in different regions, then the loop in Figure 1 would represent the two way interaction between the macro- and the micro-system.

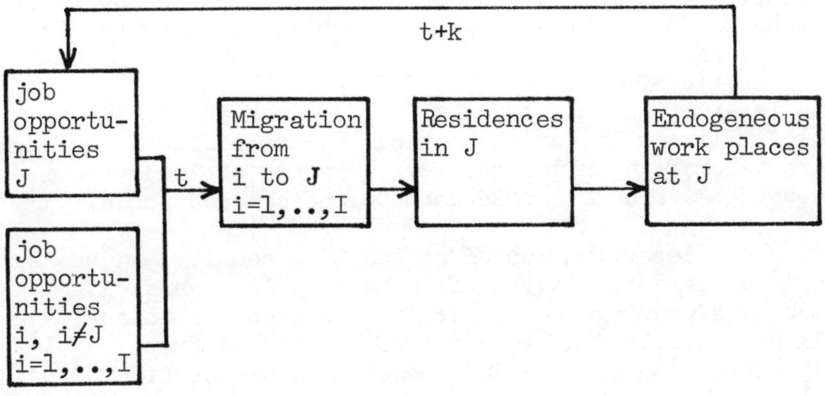

FIGURE 1. FACTORS INVOLVED IN THE DETERMINATION OF MIGRATIONS INTO CITY J.

This distinction between the endogeneous and exogeneous jobs was used by Lowry in a static, cross-sectional model formulated for the city of Pittsburgh (in Meyer, 1971). Direct land-use forecasting was not possible with Lowry's formulation because of the static nature of the model. In this paper the endogeneous-exogeneous job distinction is used in a dynamic context. The subscript 't' in Figure 1 indicates the time period. The lag 'k' should be determined during model calibration. Job opportunities in 'i' are assumed to be mainly determined exogeneously and is part of the intervention at the macro-level. As is apparent the use of the model requires time-series data on such variables as the size of different establishments in urban centres and the number of residences. The estimation of the size of migrations over time can either be made by using parameters obtained from a cross-sectional analysis; or time-series data on regional employment opportunities can be used to estimate migrations that best explain the rate of city growth after allowance is made for natural increases related to births and deaths.

Population concentration in different zones of city J will be proportional to total migration into J. Population growth in zone 'm' will be a function of total migration into city J, amount of empty land in zone 'm', number of floors in buildings, accessibility of zone 'm', availability of public utilities, scenery and prestige factors. The accessibility of zone 'm' is determined in direct proportion to the number of job opportunities in different zones and inversely to the distance of these jobs from zone 'm'. Obviously the above factors will not be equally important for every income group. Therefore, the population growth in each zone should be estimated separately for different income groups and the figure for empty land should be adjusted to correspond to available land for each income group. These relations are depicted in Figure 2. Again the two-way interaction involves the response of endogeneous jobs to the growth in each zone, in a dynamic context.

The intervention points at this stage consist of 1. provision of public utilities in various zones, 2. specification of number of floors allowed in buildings, 3. location of industry, i.e. exogeneous jobs, and 4. road connections between zones.

After the determination of residential density and the number of jobs in different zones, traffic between two zones can be estimated by a gravity model. Traffic between two zones 'ℓ' and 'm' will be a direct function of trips generated from zone 'ℓ', the trip attraction of zone 'm' and an inverse function of trip-time between the two zones. Trip generation is usually estimated by regression methods where the independent variables commonly used are residential density and socio-economic characteristics, like

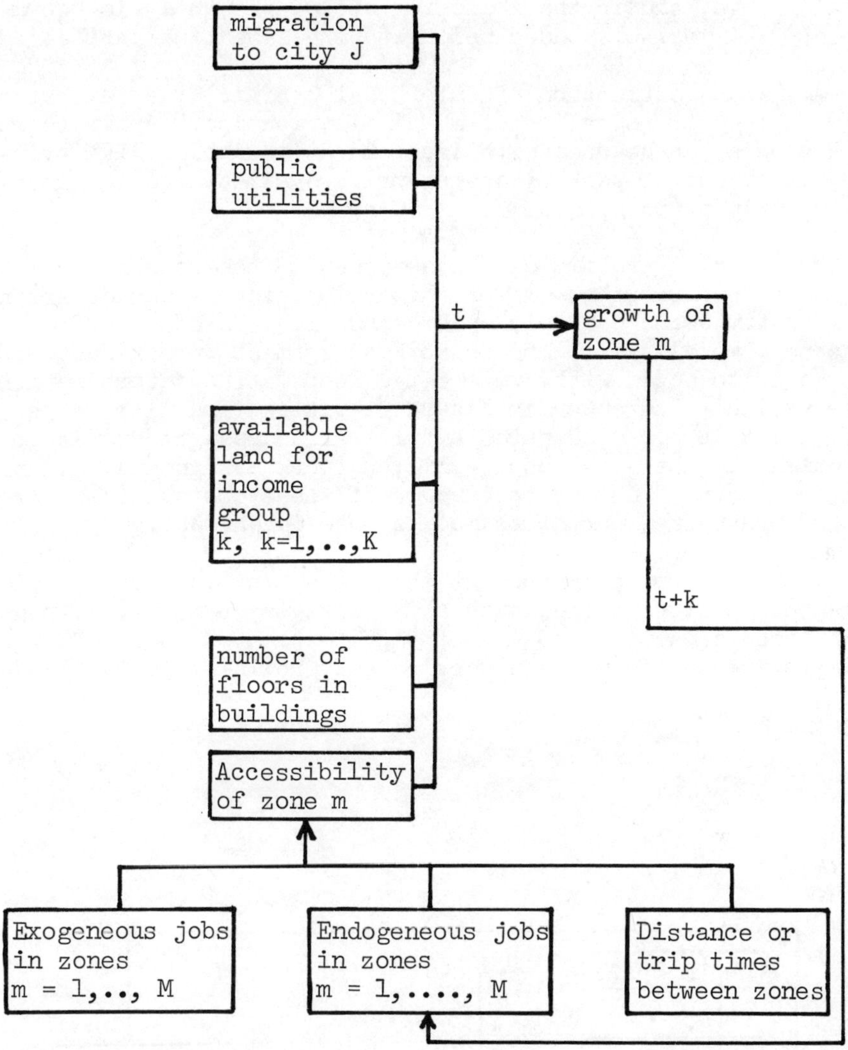

FIGURE 2. MIGRATION AND OTHER FACTORS DETERMINING THE GROWTH OF DIFFERENT ZONES IN CITY J.

income and education. Trip attraction will be a function of variables such as retail sales, school enrolment, and number of jobs. At this stage, and in the following steps, the analysis becomes static: all variables used belong to the same time period.

After the estimation of interzonal traffic, the next step would be the determination of modal distribution. Again regression analysis can be used with independent variables like triptime, modal costs, parking costs, private car-ownership, income levels, trip purpose, etc.

Modal split traffic can be assigned to the road network by a minimum path algorithm. After this allocation we can determine the trip times and costs in each connection. Since traffic between two zones, modal split, and network assignment were a function of trip costs to begin with, we enter a loop in the determination of these variables as shown in Figure 3. Iterations will be performed until the trip costs computed after traffic assignment result in approximately the same modal split, network assignment, and trip distribution used in the estimation of these costs. (For a review of urban transportation models, see Catanese, 1972.)

The intervention points at this level involve 1. provision of new modal alternatives such as a subway system, 2. expansion of the road capacity, and 3. the protection of some modes, as by improving public transit service and providing exclusive bus lanes.

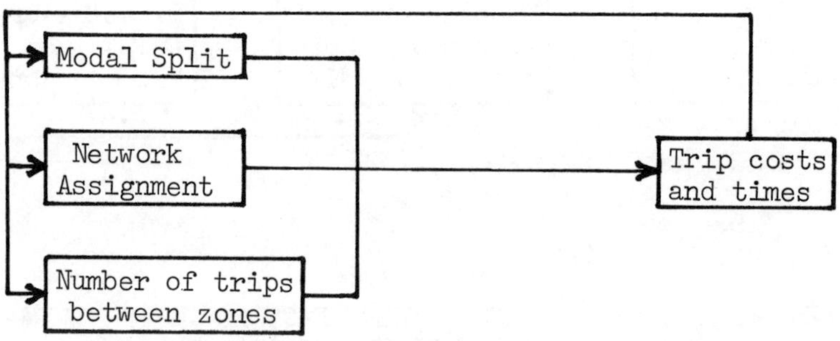

FIGURE 3. DETERMINATION OF TRIP COSTS AND TIMES

CONCLUSION

Intervention policies at the macro-level involve national goals. The basic issue is the choice between centralized and decentralized growth. At the micro-level we can group the intervention strategies under two classes: those affecting land-use patterns and those affecting transportation capacities. When land-use patterns and densities are changing rapidly, then considerably more long-run benefits can be obtained by directing land-use choices than by expanding available capacity. Simulation of future land-use patterns, zonal densities, traffic, and incomes can tell us how long increased transportation capacity would relieve congestion.

If the selected strategies involve different budgets, then the necessary coordination between these budgets should be obtained to achieve the desired objectives. For example, if housing in one community is under the jurisdiction of one municipality, and if its occupants travel into the boundaries of another municipality for work, then the budgets of these two municipalities should not be planned independently.

The determination of migrations into city J and residential growth in the city is formulated in a dynamic context. The model requires time-series data which are difficult to obtain for a sufficient number of years, but the problems are not insurmountable. The analysis presents a potential tool for improving our forecasts of future urban developments by first estimating the size of migrations and then allocating the migrants to available land within the city.

Total cost-minimization in the system would involve the assigment of political and social weights to different costs, depending on how much decentralized growth is desired and how much it is important to offer a decent standard of living to the migrants. The simulation of future migrations, land-use patterns and traffic generation, together with the specification of political and social goals, would be a useful guide in avoiding sub-optimization through lack of planning and coordination. Enforcement of the plans would depend on the success of incentive systems and of prohibitions used to direct growth into desired locations.

REFERENCES

Catanese, A.J. (ed.) (1972) New Perspectives in Urban Transportation Research. Lexington, Mass.: D.C.Heath.

Haefele,E.T. (ed.) (1969) Transport and National Goals. Washington,D.C.: The Brookings Institution.

Hunter,H. (1968) Soviet Transport Experience: Its Lessons for Other Countries. Washington,D.C.: The Brookings Institution.

Meyer,J.R. (ed.) (1971) Techniques of Transport Planning, Pricing and Project Evaluation. Vol.1. Washington,D.C.: The Brookings Institution.

Owen,W. (1968) Distance and Development. Washington,D.C.: The Brookings Institution.

Owen,W. (1972) The Accessible City. Washington,D.C.: The Brookings Institution.

Williams,E.W.,Jr. (ed.) The Future of American Transportation. Englewood Cliffs,N.J.: Prentice-Hall.

THE RELATION BETWEEN SOCIO-ECONOMIC DEVELOPMENT AND TRANSPORTATION IN TURKEY

V.Doğan Sorguç and Erdoğan Şener

Department of Civil Engineering

Middle East Technical University

There are strong indications that transportation and socio-economic development are tightly interrelated. Transportation and communication, as a result of their tangible and intangible effects, have very important socio-economic functions. A given quantity and quality of transportation and communication is thus an indicator of socio-economic development, reflecting its level in a country or region.

This interaction was shown in recent research carried out by the State Planning Organisation (SPO) of Turkey (1973) to determine the priority order of the regions deserving public support for a balanced development of the country. In this work, based on the evaluation of 53 indicators in eight indices (groups), a composite index of socio-economic development is developed and values are calculated for the 67 provinces of Turkey, thus classifying these regions according to their level of development. This has provided an essential criterion of the State's promotional policy for evaluating private and public investments by region. Consequently, the results of the research in use is influencing the overall investment policy of Turkey. In this paper, the influence of transportation and communication is analysed taking highways as the most important element of transportation.

SOCIO-ECONOMIC DEVELOPMENT CRITERIA

The composite index of socio-economic development is calculated using the taxonomy technique, a mathematical method originally used for ranking, classifying and comparing countries. In the

SPO research, 67 cities in Turkey, with their administrative boundaries (provinces), were taken as the "regions" to which the methodology was applied. Eight indices were constructed to represent the component of socio-economic development.

In this study N represents the total number of "regions" and M the total number of "indicators" (variables) to be considered. Each "region" can be considered as a vector (P_i) whose components are the "indicators":

$$P_1\ (1,2,\ \ldots\ M),\ P_2\ (1,2,\ \ldots\ M),\ \ldots\ P_N\ (1,2,\ \ldots\ M)$$

In this way each region (P_i) is represented as a point of a vector in M-dimensional space. All these points (vectors) can be shown in matrix form:

$$Z = \begin{bmatrix} X_{11} & X_{12} \cdots\cdots\cdots & X_{1M} \\ X_{21} & X_{22} \cdots\cdots\cdots & X_{2M} \\ X_{N1} & X_{N2} \cdots\cdots\cdots & X_{NM} \end{bmatrix}$$

where X_{ij} shows the value of the indicator j (j = 1,2 ... M) for the region i (i = 1,2 ... N).

Since each of the indicators is measured on a different scale, X_{ij} values must be standardised in order to eliminate the effects of various units. This is done by a normalisation procedure and the standardised variables D_{ij} are obtained as follows:

$$D_{ij} = \frac{X_{ij} - \bar{X}_j}{S_j} \qquad i = (1,2,\ \ldots\ N),\quad j = (1,2,\ \ldots M)$$

where $\bar{X}_j$ is the arithmetic mean and S_j is the standard deviation.

As a result of the standardisation process a new matrix D is obtained in which each region is represented as a standardised point or vector in M-dimensional space.

The next step is to establish the development level of all regions, and thus the pattern of development or development ranks of the regions. For this purpose the maximum D_{ij} is found in the standardised matrix D for every indicator (for every column j of the D matrix) and this value is subtracted from all other values in that column. If the indicator is negatively influencing the level of development or modernisation, the minimum value of D_{ij} for each column must be subtracted from all other values in that column rather than the maximum value of D_{ij}. A new (NxM) matrix is obtained. By taking the square root of the sum of the squares of every row element for all the N rows one obtains a column vector, C_{io}, which gives the values representing the development pattern. If these values are ordered in an ascending sequence one gets the ordering of N regions from the more developed to the less developed.

In order to find the percentage distribution of the development levels of regions with respect to the hypothetically "ideal" region we need to express the values of all elements of the development column vector $[C_{io}]$ as a percentage of the development value, C_o, of the hypothetically "ideal" region. This value, C_o, is calculated as the sum of the arithmetic mean plus twice the standard deviation of the elements of the column vector $[C_{io}]$:

$$C_o = \bar{C} + 2S$$

One obtains the matrix of the regions' level of development, L_{io}, using the relation,

$$\left[L_{io}\right] = \frac{1}{C_o}\left[C_{io}\right]$$

which can be ordered in ascending order of values, to give the ordering of regions from more developed to less developed.

SELECTED INDICATORS AND INDICES

The indicators used to reflect development indices were, for each province:

1. Industrial and Mining Index (X_1)

   (i) Mining and manufacturing industry investments in the province (planned investments), considered as a percentage of total investments
   (ii) Value of planned investments per person
   (iii) Amount of electrical energy used in industry
   (iv) Input-output ratio of industry
   (v) Value added in the manufacturing sector per employee in this sector

(vi) Number of unskilled workers as a percentage of total employed population
(vii) Value added per firm

2. Commercial and Financial Index (X_2)

(i) Banking and insurance transactions' taxes per bank and agent
(ii) Number of bank accounts per person
(iii) Average value of an account
(iv) Amount of commercial credits per person
(v) Amount of transfers from central administration to local administrative units per person
(vi) Income tax per person employed in industry and services
(vii) Local administration's own income per person

3. Agricultural Modernisation and Development Index (X_3)

(i) Value of agricultural production per person
(ii) Amount of irrigated land as percentage of total cultivated land
(iii) Fertiliser consumption per unit of cultivated land
(iv) Amount of meat and milk production per person
(v) Amount of credit distributed by agricultural credit co-operatives per member
(vi) Amount of agricultural credit per person in agricultural production
(vii) Average farm size in agriculture
(viii) Value of agricultural production per family involved in agriculture
(ix) Value of agricultural production per unit area of cultivated land
(x) Gini concentration ratio (an index related to income distribution)
(xi) Amount of cultivated land per unit tractor capacity (horsepower)

4. Social and Cultural Development Index (X_4)

(i) Number of radio receivers per person
(ii) Amount of electricity used in lighting per person in the municipalities
(iii) Total number of court notifications per person
(iv) Amount of people with social security as a percentage of total population
(v) Number of newspapers read
(vi) Number of parks, hotels and restaurants
(vii) Total construction permits (in square metres) received

5. Health Index (X_5)

(i) Number of cases of infectious diseases per person
(ii) Number of people per physician
(iii) Patient-bed occupancy ratio (treatment period)/(365 x number of beds)
(iv) Number of people who have applied for treatment as a percentage of total population
(v) Amount of health facility investments according to programmes per person
(vi) Number of people per pharmacy
(vii) Number of persons inoculated as a percentage of total population

6. Education Index (X_6)

(i) Amount of literate population as a percentage of total number of people over age 6
(ii) Number of children between ages 7-11 per primary school teacher
(iii) Amount of planned educational investments per person

7. Demographic Index (X_7)

(i) Population per unit area
(ii) Total population of areas having a population of over 10,000 as a percentage of total province population
(iii) Number of people per dwelling
(iv) Number of deaths as a percentage of the total population
(v) Provincial population growth rates

8. Transportation and Communication Index (X_8)

(i) Number of people per vehicle
(ii) Amount of land per kilometre of road
(iii) Amount of postal goods per person
(iv) Total number of people per registered telephone
(v) Freight carried (by kilometre - ton)

SOCIO-ECONOMIC DEVELOPMENT AND TRANSPORTATION

All the 67 provinces included in the taxonomy procedure are ordered from more to less developed, according to each of the sectoral indices calculated as a composite index of their indicators, and according to the composite index of socio-economic development. These orderings of regions are then examined in the reverse direction (from less to more developed ones) and provinces are grouped according to various ranks of priorities, in order to provide for some flexibility in investment decisions and to avoid a rigid implementation of the results, which necessarily cannot be very accurate. Considering the priority rank for each sectoral index as its weight, points are given to groups of provinces. This leads to various promotional measures to be applied by the government to investments, exports and tourism according to the priority level of the region examined. As far as the type of investment for a given region is concerned, the industrial authority in charge of industrialisation policy, namely the Ministry of Industry, considers, in case of application for promotion, the market size, the available capacity, transportation distance, the minimum acceptable economical capacity and the feasibility of the project concerned.

In order to see how far the indicators chosen affect the composite index of socio-economic development and to determine how strong the relation between the variables are, a multiple linear regression analysis was carried out. The relation obtained was:

$$Y = 0.508 + 0.020\,X_1 + 0.043\,X_2 + 0.164\,X_3 + 0.014\,X_4 - 0.090\,X_5 + 0.164\,X_6 - 0.159\,X_7 + 0.172\,X_8$$

where Y = composite index of socio-economic development
X_i are as defined before ($i = 1,2 \ldots 8$)

The multiple correlation coefficient was found to be 0.996 ($F = 769.09$); and the zero-order correlation coefficients were:

$$r_{YX_1} = 0.95 \quad r_{YX_2} = 0.93 \quad r_{YX_3} = 0.95 \quad r_{YX_4} = 0.93$$

$$r_{YX_5} = 0.92 \quad r_{YX_6} = 0.98 \quad r_{YX_7} = 0.98 \quad r_{YX_8} = 0.98$$

Although high correlation coefficients between each of the indices X_i and Y are of limited value, due to the fact that no statistically representative data exists specifically for Y, the values of which are determined by the taxonomy, the relatively higher weight of some indices in the correlation is obvious. These are indices of transportation, education and demography.

For the purposes of this paper a Spearman rank correlation analysis was carried out to shed light on these relationships. The following coefficients were obtained for the Transportation and Communication Index:

$r_{48} = 0.752$ (Social and Cultural)

$r_{28} = 0.719$ (Commercial and Financial)

$r_{68} = 0.527$ (Educational)

$r_{18} = 0.525$ (Industrialisation and Mining)

$r_{28} = 0.345$ (Agricultural Development)

$r_{78} = 0.335$ (Demographic)

$r_{58} = 0.182$ (Health)

From the above it is obvious that a strong relation exists between socio-cultural development and transportation and communication, and also between commercial and financial development and transportation and communication.

CONCLUSION

This paper shows that, according to the Turkish experience, there is a very strong relationship between socio-economic development and transportation, the most important element being highways. This has been indicated by a multiple regression analysis applied to the function of social development criteria obtained through a taxonomy technique; and by a rank correlation analysis.

Similar interrelations have also been revealed by previous research carried out in Turkey. In Sorguç et al. (1972) the Turkish highway network was analysed regionally with respect to a 'road density factor' (kilometres of road per unit area) and a 'road utilisation factor' (kilometres of road per 1000 vehicles). The regions which were found to have a low level of road density and road utilisation and thus identified as socio-economically underdeveloped, coincided with the regions so determined in the research presented above. Another study (Tütengil, 1961) demonstrated the socio-economical changes that took place in Namrun, a remote district in southern Turkey, as a result of thirty years of development in highway transportation. The period covered the years 1930-1960, at the beginning of which horses and camels were the only means of transportation. In the second half of this period a highway was built enabling buses and trucks to run between Namrun and its sub-provincial centre, Tarsus. Qualitative evaluation of the socio-economic development that occurred puts forward the same important interrelation between transportation and socio-economic development. One must point out that the indicators incorporated in the present study have been severely limited due to the lack of statistical data or data which can be quantified. Therefore, from the point of view of transportation one ought to add further criteria, as in Adelman & Morris (1967):

a. per capita GNP

b. rate of growth of per capita GNP

c. abundance of natural resources

d. overall investment rate

e. change in industrialisation since a given year

f. degree of improvement in agricultural productivity since a given year

g. structure of inter-regional trade.

Similarly some qualitative indicators might be added:

a. extent of the indigeneous middle class

b. degree of administrative efficiency

c. extent of leadership commitment to economic development.

For a more exhaustive and meaningful explanation of the nature of the relation between transportation and socio-economic development, further research work dealing with the development and use of the above indicators is needed, as they may play a prominent role.

REFERENCES

Adelman,I., & Morris,C.I. (1967). Society, Politics and Economic Development. Baltimore: The John Hopkins Press.

Sorguç,D., Toner,I., & Şener,E. (1972). Evaluation of Highway Networks. In Application of Operational Research to Transport Problems. North Hollywood: Western Periodicals.

State Planning Organisation (1973). Determination of Regions with Priority in Development and Investment Promotion Measures in these Regions. (Turkish). Ankara.

Tutengil,C.O. (1961). Turkish Highways from the Social and Economical Point of View. (Turkish). Istanbul Matbaasi.

SECTION 3. MODELLING AND QUANTITATIVE TECHNIQUES

The three papers in this section are each concerned with aspects of advanced quantitative techniques as applied to transportation planning. In the first and the last papers, however, transportation is properly considered as only a part of the system which needs to be modelled. The papers range in their treatment from a mathematical model at the initial development stage, to the application of numerical techniques in a recent design project for a new town.

The computerised version of Ensminger and DiCesare's model of suburban land development is intended to facilitate the assessment of demographic, fiscal and land-use impacts which result from particular capital investment and regulatory policies. The actions, motivations and constraints of land-owners, developers, and building customers are modelled to show how their choices may affect the characteristics of a small, growing community. With its on-line implementation for interactive use by elected officials and interested citizens such a model could prove to be a significant technological aid to the participatory processes described in Section 8 of this volume.

One of the advantages of computerised models is that they make possible the rapid evaluation of alternative courses of action by a standard procedure. This facility is as useful educationally to planners as to the lay person. Truelove's paper reports on the use of a simplified transportation planning package, as part of a student project in a planning school. Model runs were repeated for identical road and public transport networks, while varying assumptions were made about traveller behaviour. Apart from the value of the work as a teaching aid, Truelove draws attention to the conclusion reached from the model runs that relatively simple transport models may be adequate for many purposes, and that greater refinement and precision might be unproductive.

A degree of precision was required, however, in the criteria developed by Navin and his associates in specifying the street system for the new town of Flower Mound, Texas. The intention was to plan and develop a street system which would be efficient for all classes of use and allow for future changes in community density, land-use and transportation technology. In this latter respect the plan responds to one of Richrath's main pleas for urban planning in Section 1. And as in Wicht's development plan for Kaufbeuren, the provision of safe and enjoyable walking and bicycle pathways was a major element in the project.

ANALYSIS TOOLS FOR EVALUATING THE EFFECTS OF LAND-USE AND FISCAL CONTROL POLICIES IN DEVELOPING COMMUNITIES

Daniel A. Ensminger and Frank DiCesare

Electrical and Systems Engineering Department

Rensselaer Polytechnic Institute

This paper discusses research in developing new mathematical models to aid planners and elected officials at the county and township level in planning for rational suburban growth over the next ten to fifteen years. These models will enable local decision-makers to assess the demographic, fiscal, and land-use impacts of development that result from the application of certain land-use and fiscal policies by the community. The final computerized version of these tools will be implemented on-line to enable the model to be used in an interactive mode by interested citizens and officials. In this way alternative controlling actions, input as growth scenarios, may be evaluated instantaneously. The theoretical underpinnings for these models have been obtained through analysis of the literature and by consultation with local officials. The data to calibrate and validate the mathematical relationships are obtained using as a case-study the suburban township of Brunswick in the Albany-Schenectady-Troy metropolitan area of New York State.

Motivation for the Research

In recent years, the need for more sophisticated planning tools at the local level in the United States has become increasingly obvious for a number of reasons. Communities at the fringe of metropolitan areas have been inundated with "sprawl" growth that is frequently induced by external actions such as the construction of interstate highways and the relocation of employment centres. This has frequently resulted in great diseconomies of scale in providing water, sewer, school and municipal services to a dispersed residential area. A means for controlling or coordinating such development must be sought.

Federal revenue sharing in the U.S. has transferred the decision of how to allocate large block grants of federal monies to local officials. The long- and short-term effects of their budgetary allocation decisions must be understood in advance. These tools will provide a means for evaluating such decisions.

Existing planning tools, both computerized and manual, have not adequately treated the specific socioeconomic and fiscal problems faced by relatively small developing communities.

Summary of Previous Work

This research seeks to build on other efforts that have been documented in professional journals, reports, and theses. Muller and Dawson (1972), in considering a specific large subdivision with residential and commercial development, formulated a useful methodology for fiscal impact analysis. An extension of this was used by the Montgomery County (Maryland) Planning Board 1974 to evaluate the fiscal impact of an amended master plan for Germantown, Maryland. MUNIES (3), developed by Marcou, O'Leary, and Associates (Gale, 1973) is a systemized and computerized version of a methodology similar to that of the Urban Institute. Santa Clara County, California, was the site of two research efforts directed toward the development of comprehensive computerized land-use-related models in conjunction with a regional planning board. Cristiani et. al (1973) developed a system, utilizing interactive graphics for input and output, to aid in evaluating planning policies. This work using econometric as well as other models, concentrated on the prediction of urban development and its distribution within the county. The other Santa Clara County effort, reported in a doctoral dissertation by Pellegrini (1973), provided a tool for testing the transportation and fiscal impacts of proposed land-use plans. This work provided short-term and long-term equilibrium impacts. McLaughlin (1972) reported a model which dealt with the dynamics of land conservation vs. land development for a small New England town. This model was of the Forrester Urban Dynamics (1969) type.

The authors' previous work in this area (DiCesare et al., 1975) presents an interactive model that evaluates the demographic and fiscal impacts of development for a small city. The present work is extended to include the generation and impact of development as a function of time.

THE MODEL

Overview

The model employs the "operational" point of view in treating the actions of the suburban land development process; that is, it replicates the actions and motivations of the participants in equation form. The actors shaping the conversion of land include farmers and other land-holders, speculators, developers, and new home purchasers. They are guided in their actions by both external and township-influenced forces, as well as variables generated by the system (i.e. the price of land). Their choices of action in turn affect the town's demographic, economic and fiscal characteristics.

Model Inputs and Outputs

Controls outside the township's jurisdiction. Among these are federal/state highway construction, regional housing demand, location of employment centres, availability of capital, inflation, federal/state income tax on capital gains, federal/state categorical grants and formula aid, and federal/state-mandated services and constitutional tax limits.

Controls within the township's jurisdiction. These include property tax assessment and exemption policy, zoning, subdivision regulations, sewer/septic tank regulations, and service pricing.

Township's impact variables. The model will display outputs over time such as revenues and tax rates for the township, school districts, and special districts; township, school, and special district expenditures; population by age and income group; total township full and assessed value; number of units of single-family and apartment housing; units of commercial development; number of public, school, and tax-exempt structures present; and number of acres of vacant land.

The Two Submodels

The model traces both the land development process and the resulting fiscal and demographic impacts in two separate submodels.

The land and housing submodel. This submodel simulates the actions of the farmer, speculator, developer, and homebuyer in response to the system's control and system variables. (A single asterisk (*) denotes a factor controllable by the township, while a double asterisk (**) denotes a factor outside local jurisdiction.)

The land balance equation requires that the total acreage of residentially-zoned land in the township held by farmers, speculators, developers, homeowners and the government in a given year add up to the township's total acres of residentially-zoned land in that year.

$$R(t)* = F(t) + S(t) + D(t) + \sum_i \frac{H_i(t)}{\delta_i*} + G(t)* \qquad (1)$$

where:

$R(t)*$ = Total land in study area zoned residentially (acres)
$F(t)$ = Total residentially-zoned land held by farmers (acres)
$S(t)$ = Total residentially-zoned land held by speculators (acres)
$D(t)$ = Total residentially-zoned land held by developer (acres)
$H_i(t)$ = Total of housing units of type i (units)
δ_i* = Density of housing units per acre by type (units per acre)
$G(t)*$ = Total residentially-zoned land held publicly (acres)

The farmer decides whether or not to sell land to a speculator or developer based on a weighted consideration of his age and length of tenure on the farm, the price per acre offered for his land, his income per acre from farming, and his holding costs (property taxes) (Kaiser & Weiss, 1970; Clawson, 1971; Kaiser et al., 1968).

$$F(t)-F(t-1)= a_0+a_1 \cdot PL(t)+a_2 \cdot AHH(t)+a_3 \cdot TEN(t)+a_4 \cdot HC(t)+a_5 \cdot INCPA(t) \qquad (2)$$

$$HC(t)=PL(t) \cdot EQRL(t) \cdot \sum_j (TAX_j(t)*)/1000 \qquad (3)$$

where:

$PL(t)$ = Price per acre of undeveloped residentially-zoned farm land in township (Twp) (dollars per acre)
$AHH(t)$ = Mean age of head of household among farmers in Twp (years)
$TEN(t)$ = Mean tenure at residence among farmers in Twp (years)
$HC(t)$ = Annual holding costs per acre of vacant land in Twp (dollars per acre)
$EQRL(t)$ = Equalization rate for farm land (assessed value divided by full value)
$TAX_j(t)*$ = Tax rate per thousand dollars of assessed valuation in tax jurisdiction j (dimensionless)
$INCPA(t)$ = Mean farm income per acre in Twp (dollars per acre)

The speculator buys land from farmers or sells it to developers based on the land's present price versus its perceived potential price and the annual taxes he bears per acre of land held.

$$S(t)-S(t-1)=b_0+b_1\cdot(PPL-PL(t))+b_2\cdot HC(t) \quad (4)$$

where:

PPL = Potential price per acre of vacant land in Twp (dollars per acre)

The developer buys land from farmers and speculators based on the land's accessibility to areas of activity, his availability of capital, the perceived housing demand in the area, and the availability of a sufficient number of contiguous parcels of land with suitable topographic features (Case,1967; Neutze,1968).

$$D(t)-D(t-1)=c_0+c_1\cdot T(t)^{**}+c_2\cdot PLR(t)^{**}+c_3\cdot PL(t)+c_4\cdot DEM(t-1)+c_5\cdot CONTIG(t) \quad (5)$$

where:

T(t)** = Mean travel time from available vacant land parcels in Twp to nearest major highway (minutes)
PLR(t)** = Prime lending rate (i.e., rate at which developer must borrow) (dimensionless)
DEM(t-1) = Township housing demand during previous year (units sold per year)
CONTIG(t) = Max. number of contiguous acres of vacant land in Twp (acres)

Two equations are presented to compute land prices. Equation (6) determines the price of undeveloped residentially-zoned land at a given point in time t, while equation (7) derives the potential value of such land in terms of its environmental characteristics (Samuelson,1964; Adams et al.,1968; Clawson,1962; Shoup, 1970).

$$PL(t)=d_0+d_1\cdot DEM(t-1)+d_2\cdot\left(R(t)^{*}-\sum_i \frac{H_i(t)}{\delta i^{*}}\right) \quad (6)$$

$$PPL=e_0+e_1\cdot T(t)^{**}+e_2\cdot IC^{*}+e_3\cdot INC+e_4\cdot INFL(t)^{**} \quad (7)$$

where:

IC* = Avg. cost per acre of installing streets, water, sewage and utilities in undeveloped land (Schmid, 1968) (dollars per acre)
INC = Mean income of population (people aged 18-35) in the region (dollars per year)
INFL(t)** = Projected annual percentage rate of inflation (dimensionless)

The developer develops the land he owns at a rate that will maximize his profits subject to the constraints of availability of labour supply in the construction trades, availability of capital, and the taxation of capital gains accrued through his sale of houses (Kaiser & Weiss, 1970; Case, 1967).

$$H_i(t)-H_i(t-1)=f_0 + f_1 \cdot HPROP_i(t) \cdot D(t) \cdot \delta_i * \quad (8)$$

$$HPROP_i(t)=g_0+g_1 \cdot LAB(t)**+g_2 \cdot PLR(t)**+g_3 \cdot CAPG(t)** \quad (9)$$

where:

$HPROP_i(t)$ = Fraction of developer's land held at time t that is developed into i-type housing (propensity to develop) (dimensionless)
$LAB(t)**$ = Available labour supply in building trades for constructing housing in Twp (man-hours)
$CAPG(t)**$ = Capital gains tax as a percentage of subdivided lot value (dimensionless)

The person seeking to buy a home in a region chooses to settle in a particular community in response to the amenity he perceives in the community (which can be measured by the number of acres of publicly-owned open space available or by per-pupil expenditure) weighed against the price of housing and local taxes (Harris et al., 1968; Berry & Horton, 1970).

$$HB_i(t)-HB_i(t-1)=h_0+h_1 \cdot HBPROP_i(t) \cdot (H_i(t-1)) \quad (10)$$

$$HBPROP_i(t)=K_0+K_1 \cdot PPEXP(t)*+K_2 \cdot \sum_j TAX_j(t)*+K_3 \cdot G(t)*+K_4 \cdot PH_i(t) \quad (11)$$

where:

$HB_i(t)$ = Number of i-type housing units purchased in Twp during year t (housing units)
$HBPROP_i(t)$ = Fraction of new i-type houses (built and unoccupied) sold in Twp during year t (dimensionless)
$PPEXP(t)*$ = Average per-pupil expenditure in Twp (dollars per pupil)
$PH_i(t)$ = Price of i-type housing in Twp (dollars per unit)

The fiscal impact submodel (Muller & Dawson, 1972; Pellegrini, 1973). Once development is generated over time in the township, the revenues returned to the community and the expenses necessitated by it are computed by the fiscal submodel. Revenues returned to the community by development include property tax revenues, state and federal formula aid, and sewer and water rents. Expenses incurred due to new development may be divided into population-sensitive expenses that vary predictably with the number, age and income of residents (e.g. recreational facilities, schools, health facilities, social services); expenses sensitive to the physical characteristics and acreage of new development (e.g. water and sewer expenses, street maintenance, school busing); and expenses that respond only slightly to development (e.g. salaries of municipal employees). The outputs from this model include measures of income, expenditures (capital and operating), and tax rates for the community at discrete points in time over the ten to fifteen year model run.

The equation describing the annual expenses incurred by the township in providing sewer and water service to its citizens is now presented as an example of a population-sensitive expenditure.

$$EXP_j(t) = COSTPG_j(t) \times GALPERS \times POP_j(t) \times 365 \qquad (12)$$

where:

$EXP_j(t)$ = Sewer (or water) district j's operating budget for year t (dollars)
$COSTPG_j(t)$ = Cost per gallon of effluent (or water) treated in district j (dollars per gallon)
GALPERS = Gallons per person per day of effluent produced (or water consumed) (gallons per person)
$POP_j(t)$ = Population in district j (persons)

To illustrate an acreage-sensitive expenditure, the equation for annual town highway expenses is used.

$$HEXP(t) = COST \times \sum_i (ACRES_i(t) \times MPA_i) \qquad (13)$$

where:

HEXP(t) = Annual highway operating expense (dollars)
COST = Cost per mile per year of resurfacing roads (dollars per mile)
$ACRES_i(t)$ = Number of acres of land zoned for i-type usage (acres)
MPA_i = Number of miles of roads per acre of i-type land (miles per acre)

The revenues accruing to the township are also sensitive to population (e.g. per capita state and federal aid) and acreage developed (e.g. property taxes). A general equation that sums up the revenues accruing to the town and its special (fire, sewer and water) districts is as follows:

$$TREV_j(t)=OBL_j(t)+\sum_i (TAX_{ij}(t) \times ACRES_{ij}(t))+AID_j(t)+RESERV_j(t) \quad (14)$$

where:

$TREV_j(t)$ = Total revenues received by district j (school, fire, police, township) (dollars)

$OBL_j(t)$ = Obligation revenues accruing to district j (dollars)

$TAX_{ij}(t)$ = Tax rate per i-type acre in district j (dollars per acre)

$AID_j(t)$ = Formula and categorical aid accruing to district j (dollars)

$RESERV_j(t)$ = Reserves in district j budget (dollars)

Techniques Used in the Model-Building Process

Formulation. The variables shown in the operational relationships just presented are derived from the literature and through consultation with local planners and elected officials. The form in which they are presented, namely with the dependent variable expressed as a linear combination of its independent variables, is preliminary. The first part of the data analysis and calibration process will be to test these relationships for non-linearities and term-by-term significance by use of hypothesis testing, goodness-of-fit criteria, and examination of residuals. Multi-collinearity among independent variables will be eliminated, as necessary, by combining independent variables or by eliminating them from the equation. These tests may suggest new (and possibly non-linear) relationships which will again be evaluated by the above procedure.

Model Quantification. Calibration and validation of the model requires reducing time-series (longitudinal) data from the case-study community, as well as cross-sectional measures from similar communities, into a cogent data set. This set is being assembled from a diverse number of sources including computer tapes from all levels of government, maps, aerial photographs, published reports and local records. For depicting spatial longitudinal relationships over time for the case-study area, large base maps of the township are being prepared, one for each year in the study period. These will allow existing and potential areas of development to be laid out to permit the modeller to view the relationship over time of these plots to special and school districts, census blocks, geological features (e.g. elevations and soil percolation conditions), and accessibility to areas of economic and social activity via highways and transit systems. These visual relationships must then be reduced to appropriate numerical measures and coded.

In organizing this data set, care will be taken to suitably partition the data into two sets, one of which is to be used for the model's calibration and the other for its validation. This can be done in one of two ways: either by partitioning the township into two physical portions and dividing each year's data accordingly; or by using all the township's data each year from year t to year t+k to calibrate, and from year t+k to year t+n to validate (where $0 < k < n$). It is expected that the former method will be the one employed.

Calibration and Validation. The data being used in the calibration process will be carefully separated into two categories: nominal-scale, denoting those measures having no true numerical value scale; and interval-scale, denoting those measures having ordered values along some scale. The first category of data will be assigned an appropriate dummy variable, while the second group's variables will be normalized with respect to their means and standard deviations to facilitate subsequent computations. Multiple stepwise regression analysis is to be the chief tool employed in the model-building and calibration process.

The final relationships obtained will be tested for accuracy using both statistical criteria (confidence intervals of the estimated parameters, estimation of the equations' predictive error, analysis of variance) and parametric methods (sensitivity analysis via perturbation).

Model validation will be accomplished by running the model on that portion of the partitioned data not used in the calibration phase. Deviations between model projections and known data will thus establish a measure of the model's error through the proceedure of retrospective prediction.

Implementation. The model will be programmed for the computer in such a manner as to allow interactive use. To this end, the model will be designed to permit public officials and interested citizens to input control actions, review the impacts generated and resubmit controls in an iterative fashion.

Status of the Research

Presently, work is proceeding on the data assembly, coding and calibration phases of model formulation. Once calibrated, the model validation, implementation and on-line testing steps will be carried out.

EXAMPLES OF APPLICATION OF THE MODEL

Transportation Policy

The community will be able to evaluate the development impacts caused by the location of major state and local arterials and to test various options of providing local access to arterials.

Zoning and Subdivision Regulations

Local citizens and officials will be able to evaluate the impacts of proposed changes in the town's zoning and subdivision ordinances to guide future growth in the face of external factors such as highway construction. Policies they may wish to test include re-zoning for higher density of land use, granting special use permits for open space sub-divisions, and zoning for timed development.

Open Space Preservation

Various open space policies may be tested such as outright purchase by the local government of vacant land with or without lease-backs, large-lot zoning, creation of agricultural or conservation districts, seeking easements or land dedication from land owners, and granting tax abatements to landowners in return for keeping land vacant.

Taxation and Service Pricing

The model will allow evaluation of the impacts of alternate forms of financing township, school, and special district expenditures. Some of these are controllable by the local government, others by the state and federal government. The town may require the developer to completely or partially finance the new schools, fire, sewer, and water facilities needed by his development, as opposed to taxing the entire community. The state may change the present relationships that determine federal and state formula aid to localities. The town may wish to investigate staging of large-scale developments in the township over time so as to minimize the local tax burden. The local government may consider adoption of the "land value" tax or a local income tax to replace the present property tax.

FUTURE RESEARCH DIRECTIONS

There appear to be at least two promising directions in which this work might be extended. The first is the use of optimization in generating strategies to control future growth. This would involve determining optimal staging of capital investments and regulations such that the goals of the community are achieved over time. The second involves the introduction of uncertainty into the modelling. The relationships presented in this paper are basically deterministic and deal with aggregated phenomena. Uncertainty enters this process of development in many ways, among which are uncontrollable events, bias in the data, and errors in the assumptions. The ability to analyse development strategies in terms of a range of probable outcomes and their expected impacts would be a useful extension of this work.

REFERENCES

Adams,F.G., Green,E.W., Mansfield,C., & Milgram,G. (1968) Undeveloped land prices during urbanisation: a micro-empirical study over time. Review of Economics & Statistics, 50, 248-258.

Berry,B.J.L., & Horton,F.E. (1970) Geographic Perspectives on Urban Systems. Englewood Cliffs,N.J.: Prentice-Hall

Case,F.E. (1967) Construction and real estate market interrelationships. Annals of Regional Science, 1, 145-146.

Clawson,M. (1962) Urban sprawl and speculation in urban land. Land Economics, 38, 104-111.

Clawson,M. (1971) Suburban Land Conversion in the United States: an Environmental and Governmental Process. Baltimore: The John Hopkins Press.

Cristiani,E.J., Evey,R.J., Godman,R.E., & Mantey,P.E. (1973) An interactive system for aiding evaluation of local government policies. IEEE Transactions on Systems, Man, and Cybernetics, SMC-3, 141-146.

DiCesare,F., Ensminger,D., Van Buskirk,P.G., & Carlisle,K. (1975) Policy analysis modelling for community development. (Submitted for publication).

Forrester,J. (1969) Urban Dynamics. Cambridge: MIT Press.

Gale,D.E., (1973) The Municipal Impact Evaluation System: computer-assisted cost/revenue analysis of urban development. ASPO Planning Advisory Service. Report No.294.

Harris,R.N.S., Harrell,G., & Tolley, G.S. (1968) The residence site choice. Review of Economics & Statistics, 50, 241-247.

Kaiser,E.J., & Weiss,S.F. (1970) Public policy and the residential development process. Journal of the American Institute of Planners, 36, 30-37.

Kaiser,E.J., Massie,R.W., Weiss,S.F., & Smith,J.E. (1968) Predicting the behaviour of predevelopment landowners on the urban fringe. Journal of the American Institute of Planners, 34, 328-333.

McLaughlin,D. (1972) The use of simulation to study the implications of public policy - an example. Proceedings of the 1972 International Conference on Cybernetics and Society. IEEE Society of Systems, Man and Cybernetics.

Montgomery County Planning Board (1974) Fiscal impact analysis: Stage I - Germantown Master Plan. Research Division Final Report.

Muller,T., & Dawson,G. (1972) The fiscal impact of residential and commercial development: a case study. The Urban Institute.

Neutze,M. (1968) The Suburban Apartment Boom. Baltimore: The John Hopkins Press.

Pellegrini,A.J. (1973) A model for multiple-criteria analysis of land-use plans. PhD thesis. Stanford University.

Samuelson,P. (1964) Economics. New York: McGraw-Hill.

Schmid,A.A. (1968) Converting Land from Rural to Urban Uses. Baltimore: The John Hopkins Press.

Shoup,D.C. (1970) Optimal timing of urban land development. Regional Science Association Papers, 25, 37.

THE SENSITIVITY OF TRANSPORT MODEL OUTPUT TO VARIATIONS IN INPUT ASSUMPTIONS ABOUT TRAVELLERS' BEHAVIOUR

Paul Truelove

Department of Architectural, Planning & Urban Studies

University of Aston in Birmingham

The conventional transport modelling procedure requires as input information about transport networks and transport costs. It also requires assumptions to be made about patterns of social behaviour:

e.g. - to what extent people share their cars on journeys to work
- how far people are prepared to walk to reach public transport
- what value people place upon time spent waiting for public transport.

Normally, a transport model uses a single set of assumptions about such behaviour patterns, and applies these to a very complex analysis of the use of road and public transport routes. So sophisticated is this process of traffic generation, distribution, modal split and assignment that consideration of alternative assumptions about behaviour patterns is usually prohibited by cost and time limitations. However, the development of simplified transportation planning packages, such as COMPACT 2, has enabled complete model runs to be undertaken comparatively quickly and cheaply.

This paper reports on the use of such a package to repeat model runs for identical road and public transport networks, while varying one of the assumptions made about the behaviour of a fixed number of travellers. The scope for variation in the assumptions is considerable. Certain changes, such as the level of car occupancy, would affect the overall network loadings throughout the area studied; whereas other changes, such as those relating to the acceptance of waiting for public transport, might have effects which would be highly dependent upon the nature of the area studied - in terms of housing density, car ownership levels among the residents, etc. As an example of possible effects of variations in behaviour patterns, it was decided to repeat model runs in which the value placed upon time spent waiting for public transport was varied, whilst all other assumptions were kept constant. Examination of the output suggested that even this simplified model was fairly sensitive to wide variations in one of the many assumptions that have to be made in the transport modelling procedure. Whether variations in other assumptions would have produced similar effects of significance is a matter for further research. However, it is the context within which the transport model is to be used that should determine whether greater refinement and precision should be aimed for. The student project of which this transport modelling work formed a part, suggested that there may be circumstances, e.g. in formulating local planning proposals, when traffic forecasts may be needed, but when great precision is not required in providing the guidance necessary, e.g. on environmental planning policy.

DESCRIPTION OF THE MODEL, THE STUDY-AREA, AND THE PARTICIPANTS

The COMPACT 2 transport modelling package takes as input trip productions and attractions for a zone system containing up to 98 zones. These trips are distributed between pairs of zones using a gravity model, are assigned to a mode of transport on the basis of the 'generalised cost' of making the journey by private or public transport, and are finally assigned to particular routes through the network on a minimum cost-route basis.

The area chosen for study was a sector of the Birmingham conurbation. Trip-end forecasts were obtained from the West Midlands Transportation Study. Within the sector study-area, zones for detailed study corresponded with W.M.T.S. zones or with small groups of these zones. Immediately beyond the sector boundary, zones were formed from groups of 8 to 10 W.M.T.S. zones, while in sectors of the conurbation located at some distance from the study sector large areas were treated as single zones, compiled by aggregating W.M.T.S. zones. Within the study sector, transport networks corresponded with the major road and public transport network, while beyond the sector the representation of roads and public transport became more notional, indicating desire lines for movement between zones (a 'spider network') rather than a realistic representation of particular roads or services. Assumptions about the links to be included and the link times were made by students.

The students undertaking the project, as part of the course-work for an M.Sc. in Environmental Planning and Design, came from a variety of disciplinary backgrounds. The course aims to equip students for work in metropolitan planning authorities and is not aimed at training experts in transport modelling. It is an objective of the course that students should be able to communicate successfully with other specialists involved in urban planning and transport policy making. The particular advantage of the use of a package of transport modelling programmes such as COMPACT 2 is that it enables students to gain experience in the use of transport models and in consideration of the output, without requiring a specialist knowledge of computing procedures.

MODEL RUNS UNDERTAKEN

The initial objective of the model runs was to provide evidence on the likely future use made of transport facilities in the Aston/Witton sector of Birmingham, to be used in the formulation of local planning proposals for the area in a subsequent student project. This guidance on planning policy could take three forms:

- Predictions about those routes where traffic flows were likely to exceed the traffic capacity of the roads
- Predictions about those routes where traffic flows were likely to exceed the environmental capacity of the roads
- Predictions about the likely local consequences of conurbation transport strategies, e.g. for the upgrading of rail services or the restriction of parking in the city centre by a pricing policy.

It was necessary to identify those residential and shopping streets likely to experience traffic flows in excess of their capacity and to undertake model runs reflecting different strategies.

It was immediately apparent that the output from the model would be used qualitatively, and that the answer to questions of local planning policy did not depend upon precision in the model run output. A decision on the future proposals for an overloaded street through a long-established shopping centre did not hinge upon whether the route was overloaded by a factor of 70% or 71%, but rather on the general statement that the route was seriously overloaded, or more/less overloaded than comparable routes through other centres. This evidence could only be used in combination with other evidence, e.g. on the number of pedestrians attempting to cross the road, the number of people experiencing the noise and other detrimental effects of traffic. This other evidence might well be expressed in much less precise terms than traffic model output. Even if forecasts are made with precision, the aggregation of different kinds of evidence must be a matter of judgement.

Model runs were undertaken to examine the consequences of different conurbation transport strategies upon traffic in a particular suburban area. For example, in one model run a doubled city centre parking charge was included in the elements of the generalised cost of travel to the centre. The effects on traffic flows in the sector were measurable, but extremely small. Certainly, the changes did not have an effect that was so marked as to influence local policy plan-making.

A further model run was undertaken to examine the consequences of an improved suburban railway service. There was a considerable effect on highway loadings near newly-modelled suburban railway stations. The link loading on street "A" (see over) illustrates this. Interestingly, this extreme case illustrates just how crude transport forecasts need be for making forecasts of certain environmental conditions. Whether the traffic flow is 1000 P.C.U. per hour or 1500 P.C.U. per hour, kerbside noise levels would be relatively stable. While traffic noise increases with traffic volume at small traffic flows (less than 1000 P.C.U. per hour), above this level noise increases are far less marked.

Subsequent model runs were undertaken varying the value ascribed to the time spent by public transport travellers waiting at a bus stop or a railway station. It is usual to assume that this time should be ascribed twice the value of travel-time, and this assumption is backed up by some evidence (Quarmby). It was decided to test the effect of extreme variations in the valuation of travel-time, and to undertake model runs in which waiting time was valued the same as travel-time, and then three times the value of travel-time. The results are shown in Table 1.

Waiting time valuation	Average journey cost by P.T.	Total public transport trips	Total highway trips
= travel time valuation	64.15	29,207	37,910
2x " "	68.52	27,515	39,589
3x " "	72.42	26,342	40,750

TABLE 1 VARIATION OF WAITING TIME VALUATION: EFFECTS ON NETWORK AS A WHOLE

			Link load Street "A" (Node 100-101)	Link Load Street "B" (Node 103-100)
	1.	Basic model run	3070	375
	2.	Network refined to make more rational "trees"	3077	373
Transport Strategy Changes	(3.	10p city centre parking charges imposed	3018	400
	(4.	City centre parking charges doubled	3014	407
	(5.	Suburban rail service improved	2028	397
Time Valuation Changes	(6.	Waiting time valued the same as in-vehicle time	1932	376
	(7.	Waiting time at 2 x in vehicle time	2028	397
	(8.	Waiting time valued at 3 x in-vehicle time	2079	416

TABLE 2. TYPICAL MODEL OUTPUT USING DIFFERENT ASSUMPTIONS

The effects of these variations may be interpreted in different ways. In terms of the balance between total private and public transport usage the changes are significant, indicating the importance that is rightly attached to the frequency of public transport services, and indicating also the need to make good assumptions about the value of waiting time in evaluating transport strategies with transport models.

However, if the transport model is to be used to give guidance to local plan-makers, the important output is the loading on specific links. Used for these purposes, the variations in loading assume a lesser significance (See Table 2).

Street "A" is a heavily trafficked secondary radial road, a two-lane single carriageway, through an inner suburban shopping centre some two miles from the city centre. Street "B" is a non-radial secondary road in the same area and passes through obsolete housing. The flows represent morning peak-period (two hour) journeys, derived from 1981 trip-end forecasts. As can be seen, traffic flows in Street "A" remain at about 2000 P.C.U., and in Street "B" at about 400 P.C.U., for quite extreme variations in the valuation assumption for time spent waiting for public transport.

Every model run undertaken indicated unsatisfactory environmental conditions in shopping street "A" and modest traffic volumes in residential street "B". The area, containing much obsolete property, will clearly be the subject of planned change in the course of the next few years.

CONCLUSIONS

Like many attempts at research, even those on such a modest scale as this student project, this work raises more questions than it answers. In particular it provokes two questions:

(i) How can the output from a transport model best be used?

(ii) Can a simplified modelling package such as COMPACT 2 help planners who wish to take into account possible changes in social behaviour?

To the first of these questions, a number of answers are possible (e.g. Starkie, 1974). But if attention is limited to simplified packages such as COMPACT 2, the problem can be expressed more specifically: is it worth attempting to use such transport models to give policy guidance on questions of planning local areas or small sectors of cities? The need for simplified models to evaluate broad strategies for whole cities is well established (O.E.C.D., 1974). However, where the guidance required is on a more local scale, the variations in traffic flow forecasts obtained using different assumptions proved insufficient to alter any of the inferences that could be derived. Perhaps a better approach would have been to examine a small number of links rather than whole networks, and to first come to value-judgments as to hypothetical flow levels at which different policies should be considered. Standards of performance (e.g. delay, noise levels, ease of crossing roads, etc.) would be set before frequent model runs were undertaken to forecast traffic flows under various assumptions. In parallel with this, development work should be undertaken in the use of strategic transport models for whole cities, using a greater variety of behaviour assumptions, and devoting particular attention to that combination of assumptions that led to extremes in the output - maximum and minimum forecasts. Inferences from such modelling procedures might then be used to help in deciding whether particular local transport proposals remained valid for the full range of behaviour assumptions used in studying the travel patterns of the inhabitants of the area covered by the strategic model.

To the second question, the limitations in scope and in detail of the work undertaken clearly preclude a precise scientific answer. However, it is possible to say that:

(i) There may well be models which are better suited than COMPACT 2 to take into account changes in behavioural assumptions; but the output produced is indicative of the importance of one, namely waiting time.

(ii) There are uses for transport models such as COMPACT, in which further refinement of assumptions may not be productive or necessary. In particular, for local plan-making purposes a high degree of precision in traffic flow forecasts is not important.

(iii) Perhaps the real value of the project was as a teaching aid. It left the students, who might well be recipients of output from transport models in the course of their professional careers in town planning, with a much clearer view of the value and the limitations of mathematical transport models in the planning process.

REFERENCES

O.E.C.D. (1974) Urban traffic models: possibilities for simplification. Road Research Group. Paris.

Quarmby,D.A. (1967) The choice of travel mode for the journey to work. Journal of Transport Economics and Policy, 1, 273-314.

Starkie,D.N.M. (1974) Transport planning and the policy-modelling interface. Transportation, 3, 323-334.

ROADS FOR NEW COMMUNITIES

Francis P.D. Navin

Department of Civil Engineering

University of British Columbia

In the modern community the major transportation problem is that pedestrians and cars are expected to take the same routes. But they are very different creatures and do not mix. Rather they clash, and this problem increases as communities grow along their traditional lines.

The most permanent feature of the community is its road network.

Historically, road patterns have evolved in response to transportation technologies, land tenure, governmental requirements, and examples of other cities. Early North American communities had a fine grid system of streets which gave a good approximation to the radial pattern of medieval towns. However, as this system began to respond to the pressure of the automobile, it was mutated, until it came to serve the purposes of the automobile and expressways came to dominate cities, such as Los Angeles. Despite the great changes in density and type of traffic, houses for the most part have remained close to and fronting on major roads. The result of this pattern has increased the problem surrounding neighbourhood quality, pedestrians and automobiles and their interactions.

New towns offer the opportunity to make the local roads safe and enjoyable for pedestrians and the regional roads efficient for vehicles. This paper describes the analytical techniques used by the author and his associates to achieve such a development in Flower Mound (new town) Texas.

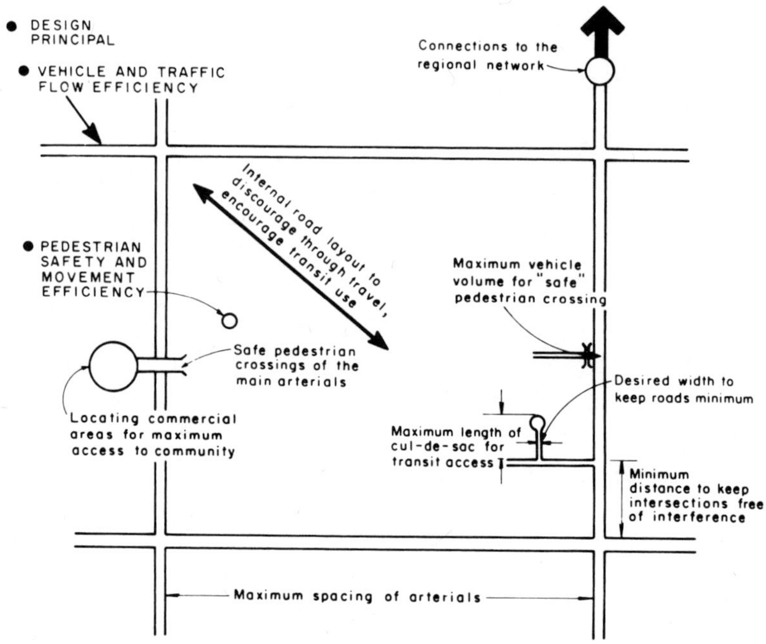

FIGURE 1. PEDESTRIAN AND VEHICLE DESIGN ELEMENTS FOR ENVIRONMENTAL AREAS.

The techniques used were in part derived from the designs of others. Figure 1 illustrates many of the community objectives articulated by Buchanan (1963) and developed further by Llewelyn Davies (1973). The concept has a grid of arterial streets for the efficient movement of vehicles. Another set of roads provides access to dwellings and is designed primarily for the safety of pedestrians.

THE COMMUNITY

The area contained by a set of arterial streets is the domain of the pedestrian and residents of the area. All the engineering and design must consider his or her perception of noise, safety and convenience.

Through Traffic

A first requirement for a livable and safe residential environment is to reduce "through" vehicle travel. To do this, through travel must be made inconvenient. Highway diversion curves indicate that 80 percent of the traffic will take a route that is 20 percent faster. Vehicle travel time, "t", is made up of three components; free flowing travel, stop time and acceleration or deceleration which may be represented algebraically as:

$$t = \frac{D}{v} + \frac{n\,v}{2}\left(\frac{1}{a} + \frac{1}{b}\right) + nt \qquad (1)$$

where

D = distance travelled, feet
v = expected speed, feet/second
n = number of stops
t = time at each stop, seconds
a = acceleration, feet/second^2
b = deceleration, feet/second^2.

The only factors under the control of the designer are the speed limit and the number of stops within the area.

Noise

The level of noise inside and outside a dwelling is an important indicator of the quality of life. The exact influence of traffic noise on human behaviour is not well understood. It is known that continued exposure to noise levels beyond 80 dBA causes loss of hearing (Berland, 1970; Bragden, 1971).

Figure 2 represents a first approximation of a noise surface. The surface is a function of vehicle density (K, vehicles/mile)

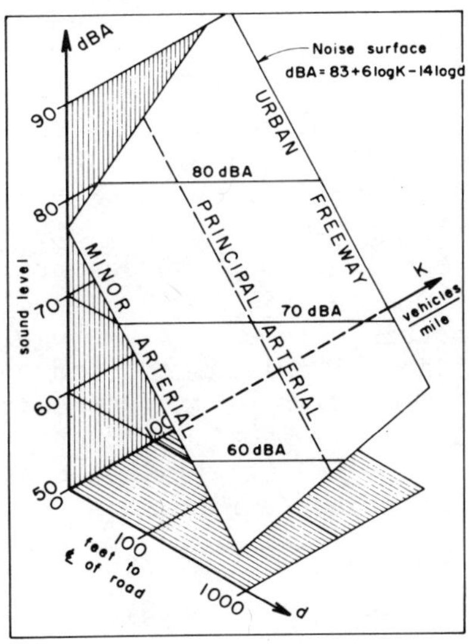

FIGURE 2. NOISE, VEHICLE DENSITY AND DISTANCE

and distance from the road (d, feet) and has the equation:

$$y = 83 + 6 \log K - 14 \log d. \qquad (2)$$

If, for example, an acceptable noise criteria is 65 dBA, and from the regional computerized traffic assignment we obtain a traffic density of 70 vehicles/mile, then from Equation 2 a setback of 125 feet from the road centre is required.

The type of house construction needed may be determined by drawing "iso-noise" bands by using Equation 2 rearranged as follows:

$$\log d = 5.93 + .43 \log K - .07y. \qquad (3)$$

Automobiles were the majority of vehicles in the traffic for Equation 2. If large numbers of trucks and buses are expected, the surface should be raised by 5 dBA and by an additional 10 to 20 percent if steep grades exist on the proposed road alignments.

The house setback recommendations for most new communities is 100 feet along major arterials when windows are double glazed and earthern noise barriers are employed. Traffic volumes greatly in excess of 10,000 vehicles/day should be reduced by redesigning the regional transportation links and land use.

Transit

Extensive literature exists on the influence of walking on transit use. Petersen (1968) found that transit patronage beyond 500 feet was confined to persons with little mode choice. Studies by age groups in Clearwater (1971) Forida and to feeder bus service in Bay Ridges, Ontario (Bonsall, 1971; Navin, 1974b) also support a reasonable competitive limit of 500 to 600 feet walking and an acceptable upper limit of 1000 feet. Many authors (Lisco, 1967; Pratt, 1970; Navin, 1974a) have shown walking, when considered by the transit patrons, to be two or three times more important than the time spent riding. Thus walking distance to the bus should be limited to 600 feet to encourage transit.

The street and housing system must recognize this pull of access and push of noise. Dwellings located on cul-de-sacs or side roads, and apartments with parking areas between the building and the road, are suitable compromises for the transit noise-access problem. The cul-de-sacs give the transit operator two or three times the number of potential patrons of a conventional road-route arrangement. Cul-de-sacs also increase pedestrian safety by reducing the tendency for drivers to speed up as they tend to do on long straight or gently curved roads.

The bus, if a conventionally routed vehicle, should operate such that potential passengers can hail the bus at any street corner. This increases the number of stops and thus the bus travel time through the environmental area. The through transit travel time is given by:

$$T = \frac{D}{v} + \left(\frac{N+n}{2}\right) v\left(\frac{1}{a} + \frac{1}{b}\right) + nt_s + Nt_p \tag{4}$$

where

T = total travel time, seconds
t_p = time to load or unload passengers, seconds
N = number of bus stops.

Using the previous example data and setting, $a = b = 4$ feet/second^2, and a loading time of 10 seconds at 5 stops through 4 intersections gives T = 396 seconds = 6.2 minutes.

This is the travel time along one route in the area. All the potential routes may be investigated and those selected to minimize the transit travel time past the greatest number of potential patrons selected for possible use.

Roads

The amount of concrete (or asphalt) should be minimized in a residential area. Thus the width of roads becomes an important consideration. To develop the mathematics, assume a road which has vehicles arriving from opposite directions as two independent Poisson distributions, with mean arrival rates of V_a and V_b. The length of the road is L and the speed is S. The length of time any vehicle is in the section is L/S and the average time the road section is occupied is $T = V(L/S)$. The probability that no cars meet is

$$P\left(\frac{\text{cars not}}{\text{meeting}}\right) = P(V_a=0)\,P(V_b\neq 0)+P(V_a=0)P(V_b=0)+P(V_a\neq 0)P(V_b=0) \tag{5}$$

$$P\left(\frac{\text{cars}}{\text{meeting}}\right) = 1 - e^{-V_a\frac{L}{S}} - e^{-V_b\frac{L}{S}} + e^{-\frac{L}{S}(V_a+V_b)} \tag{6}$$

where

V = vehicles per minute, in direction a or b
L = section length in feet
S = speed in feet/minute.

By way of example, a typical, 600 foot, cul-de-sac may have 30 homes which generates 15 outbound and 5 inbound vehicle-trips in the peak 15 minutes with an average speed of 1320 feet/minute.

The probability of meeting is 0.05 or a 5 percent chance. Because the probability is so low the need to provide a full two travel lanes is not necessary and a narrower roadway may be used. The narrower cul-de-sac will also slow the traffic somewhat. The roadway widths recommended for Flower Mound, based upon the above method of analysis, are shown in Table 1.

Pedestrians

Good access to services such as schools, shops and recreation is essential. Travel time to social-recreation and shopping is approximately 15 minutes from the home or an effective walking radius of 2400 feet. Bicycles have a comfortable operating radius of 8000 feet. The design and layout of pedestrian and bicycles routes must be such as to attract usage. The pathway design parameters found useful in specifying routes are:

- Major destinations should be within a 10 to 15 minute walk.
- The pathway characteristics should be:

1. most direct alignment,
2. avoid grades greater than 8 percent,
3. safe, and
4. interesting.

The most direct pedestrian route will produce a pathway system that is similar to a medieval town's road network and not a grid pattern. Studies of pedestrian use of underpasses and overpasses found that virtually no one used the route which required 50 percent more walking time (Moore & Older, 1965). To assure use of the overpass the time advantage had to be approximately 10 to 20 percent less. Steep grades demand considerable energy and those greater than 12 percent (ramped) are a safety hazard (ITE, undated). Ideally the grade should be less than 8 percent. A trade-off between steps and the directness of a route and rate of grade change must be made in the detailed route location.

Safety in a park like setting is an increasing community concern. A well lighted "alternate" path should be available, this path should pass within easy observation of dwelling units and, if necessary, follow the edge of the local access roads within the environmental area. Interest is provided by retaining trees, passing playing fields, and any natural amenities. The minimum width of the pathway should be five feet.

Bicycles may also use the pedestrian pathway systems if precautions are taken to minimize hazards. Hazards are reduced if the following pathway standards, based on recent research by the author, are employed:

Roadway Classification	Average Daily Traffic	Design Speed mph	Moving Lanes ft.	Parking Lanes ft.	Median ft.	Sidewalk ft.	Separator ft.	Subtotal Right-of-way ft.	Transit ft.	Total Right-of-way ft.	Remarks
Connecting Driveways	-	-	2@10	0	0	0	0	20	0	20	Up to approx. 25 dwellings
Local Access	to 250	25	1@10	2@8	0	5	8-1/2	53	0	53	Up to approx. 25 dwellings
Secondary	250- 2,500	30	2@10	2@8	0	5	8-1/2	63	0	63	Up to approx. 250 dwellings
Secondary	2,500- 5,000	35	2@12	2@8	0	5	8-1/2	67	0	67	Up to approx. 500 dwellings
Minor Arterial	5,000-20,000[a]	40	4@12	0	0	0[d]	16[f]	80	24[g]	104	100' setback to bldg.line
Minor Arterial Divided	20,000-30,000[b]	45	4@12	0	20	0[d]	16[f]	100	24[g]	124	due to vehicle noise. If volume is
Major Thoroughfare	25,000-40,000	55	4@13[c] 2@12	0	20	0[d]	16[f]	128	24[g]	152	greater than 10,000 ADT

(a) Requires special channelization at locations having frequent turns (b) Over 25,000 ADT (Average Daily Traffic) reserve principal arterial right-of-way (c) 13' lanes adjacent to the curb and median (d) Sidewalks discouraged along arterials but if necessary then 5' to be included within separator area, 11' from curb (e) Includes 1/2' for curb on each side of road (f) Fixed guideway transit reservations are along selected routes that should be firmly established with further investigations at the appropriate time (g) Variable up from the minimum for aesthetics

TABLE 1. ROADWAY DESIGN STANDARDS

- minimum width of 8 feet,
- maximum grade of 10 percent for short distances and ideally 5 percent,
- minimum radius of curve 26 feet,
- minimum sight distance of 120 feet (10 mph) and 220 feet (15 mph).

BOUNDARIES

Conflicts are often experienced when two different systems must be joined together. Road systems are no exception, and in this case collector roads must be safe for pedestrians and cars.

Critical Exit/Entry Volumes

The criteria selected for a "safe" local street is the existence of gaps in the traffic on the local road, acceptable to pedestrians, at the point it joins the arterial road system. In the first instance an acceptable pedestrian gap is assumed to be 10 seconds, a value established by Di-Pietro (1968) and the Road Research Laboratory. Assume it is desired that a person has a 75 percent chance of finding an acceptable pedestrian gap. Let "m" be the average number of cars per time interval. Then

$$m = \frac{t}{3600}(V_{ob} + V_{ib}) \tag{7}$$

where t = gap between cars, seconds
V = hourly traffic volume, "ob" is out, "ib" is inbound.

Assume that the vehicles arrive in a Poisson distribution and since we are interested in no cars we have $P(0) = e^{-m}$ and

$$V_{ob} + V_{in} = \frac{3600}{t}(\ell n\ P(0)) = 104 \text{ vehicles/hour.} \tag{8}$$

The gap criteria is: 10 seconds with a 75 percent chance, then the total volume is 104 vehicles per hour. If the criteria is set to a fifty-fifty chance of a 10 second gap then the "safe" traffic volume is 250. The number of dwelling units, which produce 0.5 vehicle trips per hour, attached to this "safe" road, ranges from 200 to 500 depending on our criteria.

The results may be arrived at by viewing the situation from the pedestrian perspective. The 10 second gap between automobiles must provide sufficient time for the pedestrian to cross the street and a buffer or "safety" factor. Thus the gap time as viewed by the pedestrian may be rewritten as:

$$t = t_w + t_s = \frac{D}{v} + t_s. \qquad (9)$$

where

t_w = walking time for a distance D (feet) at a velocity of v (feet/second)

t_s = the "safety" time (seconds).

We may replace "t" in the previous equations to get:

$$P(0) = e^{-\frac{1}{3600}\left(\frac{D}{v} + t_s\right)(V_{ob} + V_{ib})}. \qquad (10)$$

The critical number of gaps, G_c is calculated by:

$$G_c = \frac{3600}{t} P(0)$$

$$= \frac{3600}{\frac{D}{v} + t_s} e^{-\frac{1}{3600}\left(\frac{D}{v} + t_s\right)(V_{ob} + V_{ib})}. \qquad (11)$$

Solving for the critical traffic volume yields:

$$V_{ob} + V_{ib} = \frac{3600}{\left(\frac{D}{v} + t_s\right)} \left(\ell n\, 3600 - \ell n\left(\frac{D}{v} + t_s\right) - \ell n\, G_c\right). \qquad (12)$$

A maximum value of the traffic volume may be obtained by assuming $t_s = 0$ and allowing V = 3 feet/second. The equation becomes:

$$V_{ob} + V_{ib} = \frac{10800}{D} (9.29 - \ell n D - \ell n\, G_c). \qquad (13)$$

This equation is plotted in Figure 3.

We may also calculate the length of time that a person must wait to cross a road if "G" acceptable gaps appear each hour, or the fraction of unit time in which a gap of "t" seconds appears if a wait of "W" seconds was required. The formulation for this is:

$$\frac{3600}{t} \left(\frac{t}{W + t}\right) = G_c. \qquad (14)$$

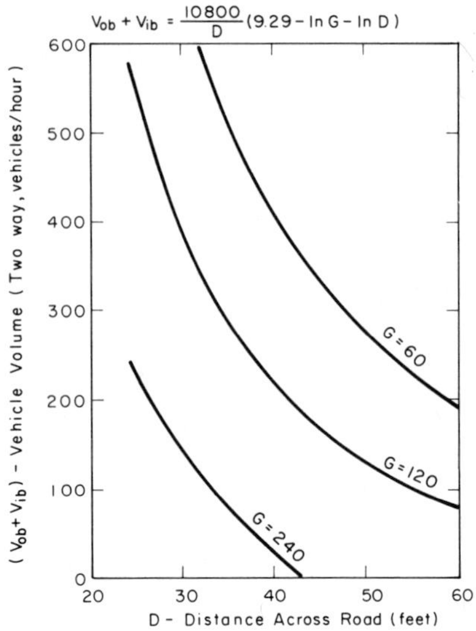

FIGURE 3. CRITICAL TRAFFIC VOLUMES

Let t = D/v as before and then the average wait time is:

$$w = \frac{1}{v}\left(\frac{3600\ v}{G} - D\right). \tag{15}$$

In the example street we have used the average wait for the slowest person would be 20 seconds.

The designer may wish to limit the waiting time to some other value. To do this he would set the design waiting time then given "D" and "v", calculate "G". With G estimate ($V_{ob} + V_{ib}$), then knowing the trip generation rate estimate the number of dwellings.

The recommended maximum volumes through any exit is 150 vehicles giving approximately 300 dwelling units to be associated with each exit. Another consideration is to keep entry volumes low enough to remove the need for signalized intersections.

FUTURE ORIENTATION

The housing within any area may eventually be replaced by developments generating more vehicle trips. The community must insist on certain roadway reserves to help ease future roadway costs during redevelopment. Earlier it was pointed out that the arterial roadway is constrained by intersections. Small wedges of land reserved during the first development provides an inexpensive hedge against errors in traffic estimation, changing land use, or both.

Personal rapid transit is controversial, but may in the future come into being. Provision may be made for such systems by superimposing a road in the sky that has the geometric characteristics of a 20 to 30 mph road with limited superelevation. This idea has been used in Columbia, Maryland, U.S.A., where transit reservations follow the green-belt. Flower Mound, Texas, already mentioned, used, in addition to the road system, small reservations for one-way transit routes and utilities.

DESIGN MODEL

The movement system objective for a new town calls for both efficient vehicular operations and provision of safety and convenience for pedestrians. The major design principles of the movement system's sub-elements are shown graphically in Figure 4 (Pratt & Voorhees, 1972; Barnett, 1973).

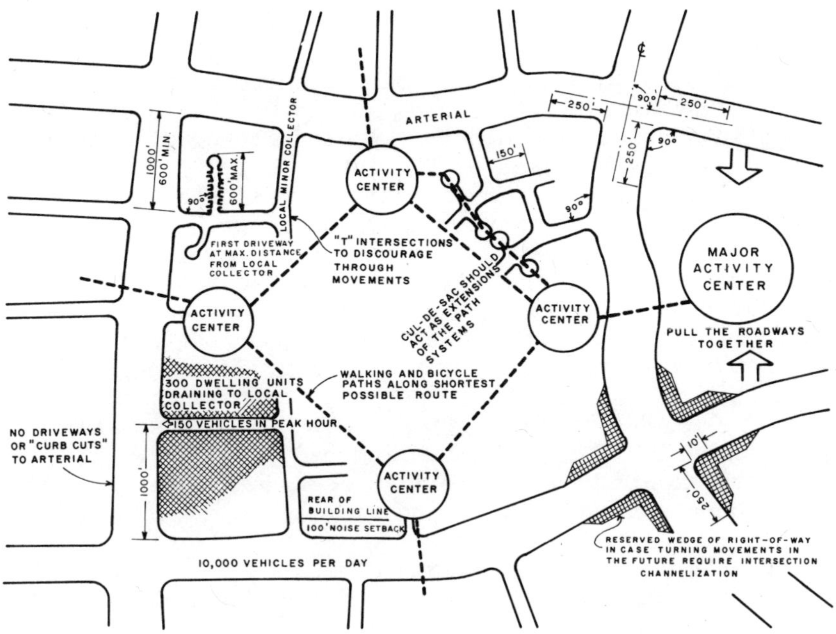

FIGURE 4. TRAFFIC APPLICATIONS OF THE ENVIRONMENTAL AREA CONCEPT

To encourage safe and efficient operation of intersections the following design principles should be followed. All roadways should intersect at or close to right angles, and should not jog or change direction in the intersection. For efficient traffic engineering operation, arterial streets should desirably have a straight section of road for a minimum of 250 feet from the intersection. The roads may follow a natural course between intersections consistent with the recommended geometric design standards. If high turning volumes are anticipated, strips of land may be reserved from potential intersection expansion. These strips should be at least 250 feet long and should flare out to a minimum width of 10 feet.

No major exit from the area enclosed by the arterials should be closer than 1,000 feet to the intersection of two arterials, to allow for efficient intersection operation. This suggests locating activity centres somewhere towards mid-block to provide equal accessibility. Along arterials with traffic volumes anticipated to exceed 10,000 vehicles per day, a 100 foot noise setback to the building line is recommended.

Within the area contained by the arterial system the roadways and pathways must be designed first for safe pedestrian and bicycle operations and then for vehicular movement. On any internal street, it is desirable that the volume of traffic entering the arterial system not exceed 150 vehicles per hour. The first residential street entrance should ideally be 250 feet minimum from the arterial-minor collector intersection. Subsequent minor collector intersections with residential roads should be spaced at a minimum of 150 foot intervals. The minor collector should discourage through movements by a series of "T" intersections. The driveway into the first dwelling on a residential street should be placed at the maximum distance from the minor collector.

In the design of the walkway system, convenience of a pathway may be measured by its directness. Safety in pathway systems is more difficult to define but at least one alternate route should be provided along well lighted streets with many dwellings. This may be accomplished by connecting cul-de-sacs end to end as shown in Figure 4. The minimum sight stopping distance on the pathway itself should be 150 feet, with minimum radius of curve at 26 feet. The path width should be at least five feet, and ideally eight feet, to accommodate pedestrians and bicycles.

Cul-de-sacs should be a maximum of 600 feet in length, and all other streets should have dual entry for fire and police protection. The basic street system can provide for services such as garbage collection, but special studies need to be done on the exact means for each location. General standards for street lighting design should be drawn upon but should be reviewed for amendments appropriate to New Town design. Street lights should be located prior to final land plotting so that electrical conduit easements may be properly located.

REFERENCES

Barnett, J. (1973) Designing new communities. Architectural Record.

Berland, T. (1970) The Fight for Quiet. Englewood Cliffs N.J.: Prentice-Hall.

Bonsall, J.A. (1971) The Bay Ridges dial-a-bus experiment. Department of Highways. Ontario.

Bragdon, C.R. (1971) Noise Pollution. Philadelphia: University of Pennsylvania Press.

Buchanan, C. (1963) Traffic in Towns. London: HMSO.

Clearwater (1971) Demonstration Bus Project. State of Florida.

Di Pietro, C. (1968) Pedestrian gap acceptance at midblock crossing. Unpublished thesis. West Virginia University.

ITE (undated) Traffic Engineering Handbook.

Licso, T.E. (1967) The value of commuters' travel time: a study in urban transportation. PhD thesis. University of Chicago.

Llewelyn-Davies, R. (1973) Architectural Record, 134.

Moore, R.L., & Older, S.J. (1965) Pedestrians and motors are compatible in today's world. Traffic Engineering.

Navin, F.P.D. (1974a) Demand activated transit. Center for Urban & Regional Affairs, University of Minnesota.

Navin, F.P.D. (1974b) Urban transit ridership in an energy supply shortage. Transportation Research, 8.

Petersen, S.G. (1968) Walking distances to bus stops in Washington, D.C. residential area. Traffic Engineering.

Pratt, R.H. (1970) A utilitarian theory of travel mode choice. Highway Research Record, No. 322.

Pratt, R.H. Associates & Voorhees, A.M. Associates (1972) Flower Mound New Town movement systems report. For Raymond D. Nasher Co., Dallas.

SECTION 4. THE DEVELOPMENT AND OPERATION OF PUBLIC TRANSPORT SYSTEMS

The design and operation of urban public transportation systems is discussed at various scales and levels of specificity in the following five papers. Spencer gives a synoptic picture of recent improvements made to public transport by the Australian Government. Apart from a long-term decline in the significance of public transport, they have had to cope with a population 70% of which lives in the major cities and with peculiar urban densities. Projects to provide a new urban passenger train and a standard urban passenger bus are briefly described.

The introduction of new systems such as these must rest on previous investigations of technical, economic and environmental facets, and of operational performance and capability. A very early step in ascertaining the benefit to be derived from introducing new urban transportation systems is to determine to which towns they are likely to be most suitable. Their introduction should depend on the system characteristics and spatial and social structure of candidate towns. A standard typology of West German towns has been developed by Krampe as a basis for selecting representative examples in which new systems might be tried out.

A different type of prior investigation is represented by Ashford and Kirk's report. They describe a human factors examination of the effect of passenger behaviour on the design of vehicles, stations and system parameters for a Minitram system. Differing compositions of the travelling public, including the very young, elderly and physically handicapped, were incorporated in the research design, particularly in relation to station access, security and safety. Possibilities for crime and vandalism were also taken into account.

The final two papers turn our attention from the development of future public transport systems to their present-day operation. Rice examines the quality of services offered to travellers on the London underground railway system in terms of expected waiting-time and the degree of overcrowding experienced. A model is derived to evaluate the extra social costs of travel as a result of random variations in train headways. Rice believes that headway control by computer might smooth headways, reduce waiting-times and eliminate some of the peaks in loading which are caused by irregularities in the service.

By contrast, Engelbrecht offers a perspective on public transport systems from the viewpoint of the manager. He briefly summarises the tasks, processes and problems which are involved in managing a large system.

AUSTRALIAN GOVERNMENT INVOLVEMENT IN URBAN TRANSPORT

John W. Spencer

Urban Transport Branch

Australian Department of Transport

Australia has experienced a change of attitude towards urban public transport in recent years. This has been influenced by two recent social changes. Firstly, setting the political scene was a change of Federal Government in December 1972, with an integral part of the new Government's election platform being a commitment to accept a share in the responsibility for public transport in the cities. Also, about this time there was an awakening of the environmental conscience in Australia which has caused a revision in thinking about the motor vehicle as the prime means of urban transport.

This paper briefly traces some details of the evolution of urban transport systems in Australia and then discusses the Australian Government's recent involvement in improving urban public transport.

BACKGROUND

As a background, let us consider the particular problems facing transport in Australia and the circumstances leading to the present status of urban public transport.

Many difficulties arise through the size of the country. The immense area of Australia can be illustrated by comparison with Europe (excluding USSR) which is only 64% of Australia's land mass. Apart from communication implications, this means that our transport systems have to traverse great distances and are subject to wide climatic variation. Our country also has inherent peculiarities of population distribution. Our startling urban concentration is displayed by the fact that approximately 70% of Australia's population lives in ten cities of over 100,000 people.

Australia is a federation of six States, each with its own parliament. Each State has responsibility for its own transport system. Most forms of transport within Australia have suffered staggered development because of the different rates of growth of the States and the need for the transport systems to provide not only for city development but to serve the inland areas and to provide links over what often are vast distances between cities. Growth of our cities has been influenced by such events as the discovery of gold. The control of transport originated in the colonial days and was preserved in our Constitution when the separate colonies were federated in 1901.

The legacy of all this has been fragmented national transport with a motley of railway gauges and road networks, a kaleidoscope of legislation and a variety of administrative and operational techniques, all adversely affecting the economies of transport operation.

Associated with this general transport situation has been a long-term decline in the significance of urban public transport in Australia. Prior to the present initiatives, the last major investments in urban transport were in the 1930's when Sydney and Melbourne electrified their rail systems. The decades of neglect since then are evidenced by the antiquated nature of much of the rolling stock. Half the rail carriages in Sydney and Melbourne are over 45 years old and more than half the Melbourne tram fleet is over 40 years old. This neglect has resulted in a steady decline in patronage, illustrated by the drop of 16% on all government-operated urban public transport services in the decade 1961/62 to 1970/71, despite an increase in urban population of over two million. Table 1 is a specific example of the falling patronage on urban public transport services for the period 1963/64 to 1973/74.

YEAR	TRAINS	TRAM	GOVERNMENT BUSES	PRIVATE BUSES	TOTAL
1963/64	148	160	32	86	426
1968/69	141	119	24	80	364
1973/74	110	109	22	70	311

TABLE 1. MELBOURNE URBAN PUBLIC TRANSPORT SERVICES PASSENGER JOURNEYS ('000,000)

YEAR	REVENUE	WORKING EXPENSES	CAPITAL CHARGES	SURPLUS/ DEFICIT
1963/64	59	63	4	- 8
1968/69	70	76	5	-11
1973/74	96	131	12	-47

TABLE 2. OPERATING RESULTS OF URBAN PUBLIC TRANSPORT BUS AND TRAM SERVICES OPERATED BY PUBLIC AUTHORITIES ($M)

The repercussion of fare increases is illustrated by events in the State of New South Wales a few years ago. In 1970/71 there had been an increase of 2.5 million in the number of passenger journeys by rail over those in 1969/70; but in 1971/72, when increased fares were introduced, the number fell by over eight million.

Proportionally small increases in revenue and rapidly escalating losses, as illustrated by Table 2 for the period 1963/64 to 1973/74, has prompted compensatory fare increases and reduced services; thus disenchanting even more commuters who turn to their private automobiles for their urban transport. The effect, of course, has been to further aggravate traffic congestion, delays, air pollution and parking demand. Adding to the situation has been the facilitation of the demands of private transport through massive expenditure on road and freeway construction, which has only hastened the demise of public transport. For example, during the years 1961/62 to 1970/71, while no Australian Government finance was made available for urban transport $1,323 million was made available for roads. Table 3 further displays the emphasis on urban roads expenditure.

YEAR	TOTAL GROSS CAPITAL EXPENDITURE ON ROADS IN AUSTRALIA	ESTIMATED PROPORTION SPENT ON ROADS IN URBAN AREAS OF SIX STATE CAPITALS
	($M)	%
1968/69	578	30
1969/70	634	38
1970/71	695	43
1971/72	705	53
1972/73	772	47
1973/74	864	45

TABLE 3. CAPITAL EXPENDITURE ON ROADS IN AUSTRALIA

THE SITUATION

The rundown condition of urban public transport has led to a situation in which in many instances it is now considered a last resort by commuters. While freeway development has enabled mobility to be retained, their predominantly radial patterns have resulted in the channelling of more traffic through central business districts.

Congested roads, polluted air, excessive noise and road accidents have come to be accepted as norms rather than exceptions. Our larger cities have become less attractive places to live and work with levels of pollution, for the major offenders carbon monoxide and oxidants, often four to five times the World Health Organisation's long-term goals.

All of this has been added to the dilemma of transport authorities which have the responsibility of reconciling the conflicting demands of growth and preservation of the environment.

THE AUSTRALIAN GOVERNMENT'S APPROACH TO A SOLUTION

The Australian Government saw the solution to the present situation lying in a transition from the traditionally fragmented transport systems of the individual State Governments to a nationally integrated system. After the present Federal Government came into office in 1972, it drew together its own previously diversified responsibilities for transport into one new Department of Transport, under one Minister, to co-ordinate Federal involvement in transport.

AN URBAN PUBLIC TRANSPORT ASSISTANCE SCHEME

In the urban transport field the Government negotiated an agreement with the States to provide for sharing the cost of approved urban transport projects. This Agreement, signed by the six State Governments and the Australian Government in June 1974, is designed to assist the States to improve their urban public transport systems, by providing for financial assistance of two-thirds Australian Government contribution to the capital cost of approved projects, in the form of non-repayable grants over an initial five-year period.

The main objective of the scheme is to achieve some immediate improvement in urban public transport by revitalising the systems in the capital cities. Consequently, the initial emphasis has been on passenger rolling stock such as new rail cars, trams and buses. The upgrading of existing systems will involve such projects as the quadruplication of rail-lines, increasing ferry services, electrification of the entire rail system in Adelaide and Brisbane and construction of separate busways. Other projects include the provision of new installations and services in the form of terminals, passenger interchanges, workshops and complete new railway systems.

Constitutionally, the Australian Government has no direct power over transport in the States. To ensure that its objectives are met in the most effective manner, the UPT Agreement imposes certain conditions upon the grants. This affords control over the programmes, a critical part of which is the provision in the Agreement for Australian Government representation in each State on a committee concerned with the forward planning and development of policies relating to urban transport.

Implementation of the Agreement was by the States Grants (Urban Public Transport) Act 1974, which included funds of $31 million for the first year of the Agreement, 1973/74, marking the first allocation of finance for urban public transport by an Australian Government. In 1974/75 an additional $66 million was allocated. Because of the present economic situation, for the current financial year the Government has limited its funding to projects already approved in earlier years; the budget allocation for 1975/76 being some $42 million.

Administration of the Agreement is carried out by the Australian Department of Transport in co-operation with the State authorities, through the representation on the State committees previously mentioned and by instruments such as the Notes on Administration for UPT Financial Assistance which cover operational aspects such as the awarding of contracts, making of advances and auditing procedures.

Under the UPT scheme, assistance can be provided for provincial cities of over 100,000 population and for the corridors linking them to their State capitals. Assistance has been provided in three such cities.

Projects considered for assistance are, as appropriate, subjected to cost/benefit analysis prior to admission to a programme. Consideration is also given to social, environmental and urban and regional planning factors. Evaluations are generally carried out by the States themselves or within the Department of Transport, in accordance with a methodology developed by our Bureau of Transport Economics. All programmes put forward are reviewed by a Federal Inter-Departmental Committee to ensure compatibility with overall Government planning.

THE AUSTRALIAN URBAN PASSENGER TRAIN

Of the other urban transport initiatives and projects of the Australian Government, the development of a new urban passenger train incorporating the most modern technical advances available, is foremost. The aim of the Australian Urban Passenger Train (AUPT) project is to develop a standard, modern, comfortable passenger train which, with some variations, can be used in all the capital cities which have electrified systems.

The problems emanating from the evolution of separate rail systems in each of the six States set the basis of the AUPT development. Being convinced that the present urban rail systems do not offer optimal services, the Australian Government embarked upon a co-operative effort to advance technical knowledge in this field. A study team, which included seconded State railway officers and private consultants, was organised within the Department of Transport to undertake the project in four stages: preliminary studies, the development of specifications for a prototype, the construction and testing of the prototype and, finally, the procurement and commissioning of a production series train.

The second stage of the project has now been completed. A nation-wide research programme, involving the use of a special rail-car display in the capital cities, was undertaken to gauge the expectations of commuters, assess public interest in better trains and determine what features users prefer. This market-oriented emphasis on comfort has led to new methods of defining comfort standards for comparing train options. One user requirement which showed through firmly was for air-conditioning. This has technical and economic ramifications, and points to potential conflict between user-requirements and operator/maintainer interests. Nevertheless, the finished product is expected to incorporate the best available features in proven rail technology. We expect to be running prototypes of the new train on State urban systems in 1977.

Trade union co-operation in the AUPT was sought at the outset by three-cornered liaison machinery between the project team, the unions and railway management. A full scale mock-up of a driver's cab was used to validate the engineering and ergonomic judgments bearing upon the crew's workplace.

Detailed inputs and expert advice from railway authorities concerned with engineering, purchase, operation, maintenance and repair of urban trains have been vital. Each city's transport needs are thus being related to existing rail cars, including the Australian version of the two-decked electric railcar AC/DC traction power alternatives, the different gauge of tracks in the various State systems, limitations imposed by features such as existing tunnels and bridges and the optimisation of performance.

The detailed design of the AUPT is still being developed and many lessons are being learnt. It is hoped that this venture will provide a basis for future joint ventures between federal and State Authorities.

OTHER PROJECTS

Apart from the UPT and the AUPT, the Australian Government has undertaken other complementary projects and initiatives having direct impact upon urban public transport in Australia.

The Australian Standard Bus

Consequent upon the favourable response to the AUPT, the State and Federal Transport Ministers agreed that the need for a standard urban passenger bus should be examined. The principal obstacle to this project is the differing legislation in each State governing the dimensions, layout and operation of buses. The Department of Transport in co-operation with the States and Union officials have produced a user specification. Based upon this, discussions are shortly to take place with manufacturers on the development of a design specification and an examination of its economic worth.

Research and Planning

Financial assistance is being given by the Federal Government to the States, again on a two-thirds contribution basis, for essential research and planning in the field of urban public transport and roads. This enables proper programme formulation by providing more uniform and rational bases for decisions on projects, preventing domination by any one mode through research of alternatives, and facilitating evaluation of the latest advances in transport technology.

At the Federal level the Bureau of Transport Economics and the Bureau of Roads have been established by the Australian Government to undertake research into transport matters. However, much valuable research is still being undertaken by the individual States with the assistance of federal finance.

After a period during which urban transport planning and research and roads planning and research were separately funded, legislation was passed which provided for rationalisation of future planning and research by the integration of the previously separate assistance measures. Under the Australian Government's Planning and Research Act 1974, the Australian Government will provide two-thirds of the cost of approved studies by the States to the extent of $26 million over a three-year period. Greater flexibility in re-allocation of funds and priorities between roads and public transport is being realised than in the previously separate programmes. Efforts are being made to encourage a greater exchange of information between interested research bodies and to help the States develop more comprehensive planning strategies.

Roads

Extensive assistance for construction and maintenance of urban arterial and local roads is also being given to the States. This action is supported by the Australian Bureau of Road's statistics which show that although urban arterial and sub-arterial roads constitute only 2% of all roads in Australia, they actually carry 50% of total traffic. The $358 million to be spent in the period 1974/75 to 1976/77 on these categories of road, as distinct from national highways and rural arterial roads which also receive assistance, is based on detailed investigations into the environmental, social and economic aspects of road proposals and their integration with public transport needs.

Growth Areas

Another urban public transport initiative of the Australian Government concerns the inauguration of complete new transport systems to promote new growth areas. The Australian Government has offered to finance entirely, including operating losses, the new railway network in the western suburbs of Sydney. It is also investigating a similar proposal for Melbourne.

Rail Transfers

The Australian Government is also raising with State Governments the question of the transfer of their railway systems which habitually run at heavy losses. Two States, South Australia and Tasmania, have already signed Agreements to amalgamate their systems with the Commonwealth Railways and to form the new Australian National Railways Commission.

Other Investigations

Other investigations such as low density transport alternatives, electric cars, use of liquefied petroleum gas as a motor vehicle fuel, transport for the disabled and the role of private operators in urban public transport have been initiated.

THE FUTURE

Looking to the future, the current roads and urban public transport financial arrangements expire in 1976/77 and 1977/78 respectively. The Government's intention is to introduce legislation at the end of 1975 to unify its present programmes. This integration of all land transport expenditure within one piece of legislation will replace the present separate programmes of assistance. This legislation is being framed to ensure that as far as is possible all transport planning is undertaken on a multi-modal basis and that finance is channelled to that mode which performs the transport task in the most efficient and economical manner.

CONCLUSION

The projects and initiatives outlined are having a physical impact upon, and are regenerating interest in, urban public transport. But the tangible gains will come with increased public transport commuting. There are hopeful signs that the trend away from public transport is slowing down and that given adequate finance and sound policy development the trend can be reversed.

The Australian Government's firm conviction is that urban public transport must be made an attractive alternative to the private automobile by offering improved quality, capacity, efficiency, frequency and expanded services. It is therefore the Government's aim that it co-operate with the States in the Federation in the planning, financing and implementation of programmes designed to meet this objective.

CLUSTER ANALYSIS OF WEST GERMAN TOWNS WITH SPECIAL REFERENCE TO PERSONAL RAPID TRANSIT SYSTEMS

Gerd Krampe

Battelle - Institut e.v.

Frankfurt/Main

In view of the present increasingly difficult traffic situation in West German towns and cities, it is very likely that existing urban transportation systems may be unable to cope with future requirements and will become less and less attractive to users.

To solve urban traffic problems, numerous new urban transportation systems have been developed in Germany and abroad. In West Germany the research and development in this field is focussed on the personal rapid transit (PRT) system, which represents a quantitive and qualitative improvement of public passenger transport facilities. The development of PRT systems has now entered into a crucial phase: the construction of test and pilot lines will probably be decided on in the very near future. On the one hand, great hopes are pinned on the new systems being introduced; on the other, equally great uncertainty exists about their actual usefulness. The West German government is faced with the question of whether in the long run the benefits to society of new urban transportation systems will justify the high expenditure involved.

Previous studies have largely been confined to investigating the technical and economic facets of the problem, or to assessing the operational performance and capability of new systems. Still open questions relate to the actual demand for urban transportation and the effect of new systems on the various sectors of society and on future urban development. In this context it must be assumed that it will become increasingly difficult for new systems to gain acceptance, especially if they have social effects. Certain technologies may even fail because of such difficulties. In the case of urban transportation systems it is necessary to consider not only their acceptance by potential users but also their toleration by possibly affected non-users.

To answer these questions it appears expedient to assess the advantages of new transportation systems within a 'technology assessment' study and including a cost-benefit analysis. The Federal Minister of Transport commissioned Battelle-Frankfurt, Prognos AG Basel, and the Studiengesellschaft für Nahverkehr Hamburg to investigate these problems in a joint research effort. The general objective of the project is to ascertain the social benefit involved in the introduction of new urban transportation systems with special reference to the potential users' attitudes, the most important cost factors, and the effects of new systems on the social and physical environment of our towns.

To achieve this overall study objective, the following individual research steps have to be carried out:

Determination of typical applications of new urban transportation systems as a function of system characteristics, and of the spatial and social structure of West German towns.

Development of modal split decision models which consider the special conditions resulting from new urban transportation systems. These investigations will be based particularly on sociological and socio-psychological approaches to determine specific factors influencing the users' modal split decision. This research step is to clarify to what extent new urban transportation systems will be accepted, and whether they will help to achieve the ultimate aim, i.e. to increase the share of public transportation in the total volume of urban traffic.

Cost-benefit analyses of new types of urban transportation systems as part of integrated transportation networks, using selected planning models.

Elaboration of fundamental planning concepts for new urban transportation systems, with special reference to their effects on urban development.

Investigation of tolerance towards new urban transportation systems and identification of compensatory measures (social planning), which may be necessary especially in the case of elevated systems. Induction of learning processes to shorten the initial period of acceptance.

These research steps illustrate the wide scope of the research project. This paper relates to the first phase of the investigations: the determination of typical applications of the urban transportation systems with special reference to the spatial and social structure of towns and cities. The cluster analysis, which is described in some detail in the following, has been conducted at the Battelle-Institut by my colleagues, Dr. Krieger and Dr. Tonn, and myself.

PROBLEM

When investigating the possible application of newly developed urban transportation systems in West German towns and cities it is found that these systems could be used in at least 250 different locations. There exist no criteria of selection which would enable Town A to be considered suitable and Town B unsuitable for the introduction of a new kind of urban transportation system. To be able to make accurate judgements in this respect, local traffic conditions in the individual towns would have to be investigated in detail. According to rough estimates, this would take about five years with the available manpower. Since this method would obviously be too expensive and time-consuming, a more practicable method had to be found which, as with random sampling, investigates only selected potential applications (towns) and permits fairly accurate inferences to be made about applications overall. A method of this type requires that the most important indicators of social and city structure are compiled and their effects on urban traffic analysed. Hence, it is necessary:

- to pinpoint the interactions between the indicators of urban spatial and social structure and those of town traffic;
- to standardise the towns and municipalities in accordance with the problem at hand, so that, as in random sampling, detailed case studies can be carried out in selected towns and the results obtained be considered as applicable to other towns and municipalities of comparable structure.

APPROACH BY CLUSTER ANALYSIS

To begin with, the relevant data on urban spatial and social structure and on the traffic situation in more than 200 towns and municipalities in West Germany were compiled and converted into useful variables, as shown in the list below (Table 1).

Although town traffic is a serious problem in West Germany, insufficient information is available in this field. The traffic situation in our towns can only be improved if the data base is improved.

TABLE 1

Variable Number	Designation	Statistical value
1	resident population potential (linear)	residents 1972/1000
2*	resident population (logarithmic)	log. residents 1972
3	total manpower potential in services sector (linear)	manpower in services sector 1970 (census figures)
4*	total manpower potential services sector (log.)	log. manpower in services sector
5	gross population density	residents 31.12.1971/municipal land area 31.12.1971
6*	net population density	residents 31.12.1969/built-up area 31.12.1969
7*	relative relief	standard deviation of altitude values
8	building density	residential buildings 1968/ built-up area 1969
9*	old buildings	residential buildings after 1949/ total residential buildings
10	new buildings	residential buildings after 1949/ total residential buildings
11	furnishing of residential units	units with bath, WC and central heating/total residential units
12*	occupation density of residential units	persons in residential units/ total residential units
13*	average rent	average rent in DM
14	gainfully employed, general	total number of employees/residents age 15 to below 65

TABLE 1 (cont.)

Variable number	Designation	Statistical value
15*	female employees	female employees 1970
16*	employees in services sector (percentage)	employees in services sector 1970/ total employees 1970
17	medical specialists	specialists 31.12.1970/total residents 1970
18	dentists	dentists 31.12.1970/total residents 1970
19	commuter balance	ingoing-outgoing commuters 1970/ total employees 1970
20	financial potential	revenue from trade tax per inhabitant 1970
21	percentage of workers	percentage workers of total employed persons 1970
22	elementary education	residents with elementary school education only 1970
23	university education	residents with university education 1970
24	foreign residents	foreigners 31.12.1971/total residents 31.12.1971
25	percentage of children	residents age 0 to below 15, 1970/total residents 1970
26	percentage of old people	residents age 65 and above 1970/ total residents
27*	captive riders	residents age 0 to below 15 and 65 and above/total residents
28	migration balance	migration balance per 1000 inhabitants (1970/71/72)
29*	natural population trends	excess of births/deaths per 1000 inhabitants (1970/71/72)

TABLE 1 (cont.)

Variable number	Designation	Statistical value
30*	ingoing "long distance" commuters	ingoing commuters travelling 30 min. or more, 1970/total ingoing commuters
31*	commuters using private car	car commuters travelling 10 km. and above/total commuters
32*	commuters using railway	railway commuters/total commuters
33*	commuters using tram or underground	tram and underground commuters/ total commuters
34*	commuters using bus	bus commuters/total commuters
35*	private car ownership	cars 1.7.1971/inhabitants 31.12.1971
36*	percentage traffic area	traffic area/built-up area 31.12.1969
37*	net road network density	road length 31.12.1969/built-up area
38*	gross road network density	road length 1.1.1971/municipal land area 31.12.1971
39*	parking space indicator	cars 1.7.1971/parking spaces 1.1.1971
40*	demand for public transportation	trips using public transport facilities/inhabitants 1971
41*	travel distance for public transportation	mean travel distance for public transportation 1971
42*	utilisation of public transport facilities	user-km/seat-km 1971
43*	share of transport by bus (public transportation)	seat-km on bus/total seat-km
44*	manpower intensity of public transportation	employees in public transportation per 1 million seat-km

* These variables were included in the cluster analysis

The complete set of data for the 200 towns and municipalities in West Germany was subsequently subjected to a factor analysis. This included the analysis of the interactions between urban, spatial and social structure and town traffic. We then carried out a cluster analysis by which the towns and municipalities were standardised to form about twenty groups of comparable structure. We used for this a hierarchical grouping method (complete linkage), which ensures a largely homogeneous clustering in individual groups.

RESULTS

Figures 1 and 2 illustrate the relationships between the various indicators considered in the study. Figure 1 shows the internal relationships between the indicators of urban traffic, the strength of the correlation being indicated by the width of the connecting bars. Dots (: :) denote positive and dashes (-) negative correlations. Figure 2 shows the internal relationships between the indicators of urban, spatial and social structure. Because of their complex nature the manifold negative correlations

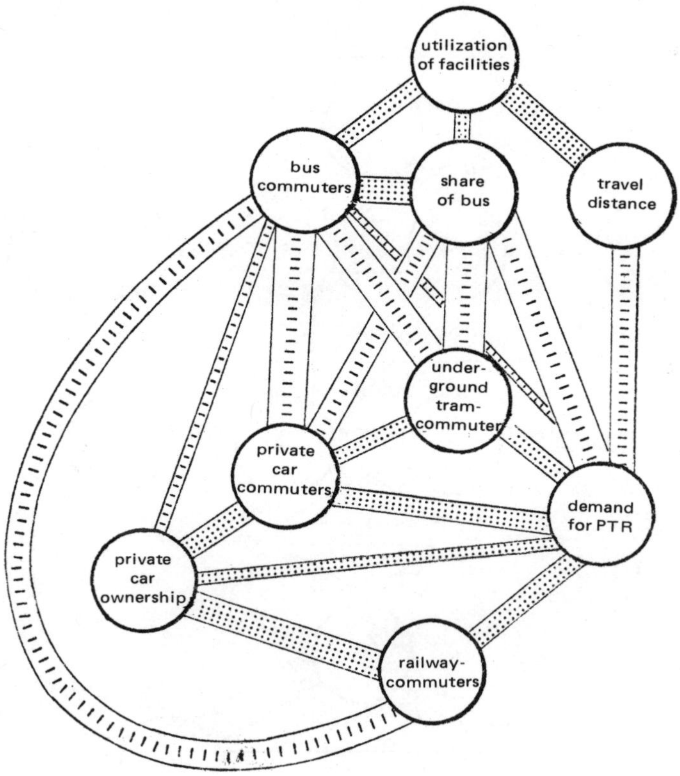

FIGURE 1. INTERRELATIONS OF URBAN TRAFFIC VARIABLES

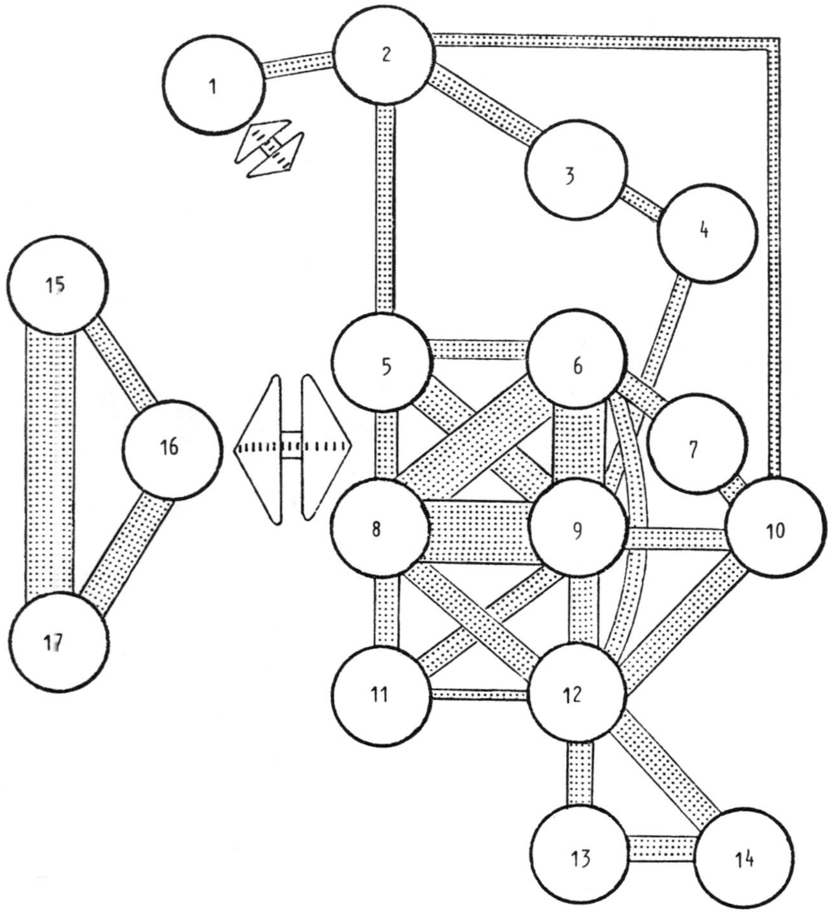

1 = captive rider
2 = old buildings
3 = relative relief
4 = commuter balance
5 = employees in services sector (percentage)
6 = ingoing "long distance" commuters
7 = average rent
8 = resident population potential (log.)
9 = total manpower potential in services sector (linear)
10 = female employees
11 = gross road network density
12 = net population density
13 = percentage traffic area
14 = net road network density
15 = natural population trends
16 = percentage of workers
17 = occupation density of residential units

FIGURE 2. INTERRELATIONS OF URBAN SPATIAL AND SOCIAL STRUCTURAL VARIABLES

between the three variables, percentage of workers, occupation density in residential units, and natural population trends, and the other variables of urban, spatial and social structure are indicated only schematically.

Table 2 shows the interactions between the indicators of urban, spatial and social structure and those of urban traffic, subdivided by categories. The following categories of indicators have been shown to be central for urban traffic: preferred commuting means; number of private cars; demand for public transportation and share of transport by bus.

The urban, spatial and social structure has been classified by the following indicator categories: potential of persons employed in the services sector and resident population potential; social stratification; structure of built-up areas; and importance as a central town. In addition, the proportion of captive riders has been investigated with respect to its effect on public transportation.

When subdividing the complex interactions between the urban, social and spatial structure and urban traffic, very interesting interactions have been identified. The strength of the correlation between two indicators is indicated by the size of the circle, and the type of correlation by the sign in it. The classification of West German towns and municipalities was based on this network of interactions between various indicators.

The largest cities and towns in West Germany were combined in transportation-specific groups formed with the aid of a cluster analysis, based on the 28 selected variables and the interactions characterised by them. In the cluster analysis the cities and towns are classified according to their different structures and gradually grouped in relatively homogenous clusters. At first each individual town forms a "cluster", and at the end of the process the 199 largest of the selected cities and towns are united in one single group.

Between these two extremes, the single-town cluster and the largest cluster, there exists the greatest structural difference, the so-called "maximum distance". This maximum distance is a non-dimensional parameter describing the structural distance over the entire spectrum of selected variables. The maximum distance is assumed set at 100. It serves as a measure or standard for all the other structural differences between towns and groups of towns, with the aid of which the towns can be combined step-wise to form clusters.

Some towns could not be classified or assigned to a specific

TABLE 2. INTERACTIONS BETWEEN URBAN SPATIAL/SOCIAL STRUCTURAL AND URBAN TRAFFIC VARIABLES

Variables of urban spatial and social structure \ Variables of urban traffic	commuters using private car 31	commuters using railway 32	commuters tram or underground 33	commuter using bus 34	private car ownership 35	demand for public transportation 40	travel distance for public transport 41	share of transport by bus (public transportation) 43
2 resident population potential (logarithmic)	+	+	+	−		+		−
4 total manpower potential services sector (log.)	+	+	+	−	+	+		−
12 occupation density of residential units	−	−			−	−	+	+
21 percentage of workers	−	−			−	−		
29 natural population trends						−	+	
6 net population density		+				+		−
36 percentage traffic area		+						
38 gross road network density						+		−
7 relative relief					+		−	
13 average rent	+	+	+	−	+	+		−
15 female employees		+			+	+		
30 ingoing "long distance" commuters	+	+	+	−	+	+		−
19 commuter balance		+	−	+	+			
16 employees in services sector (percentage)	+	+			+	+		
27 captive riders		−	−	+	−	−		+

TABLE 3. TYPES OF TOWN IN EACH CLUSTER

Cluster no.	Town types	Population percentage	Degree of suitability for new systems (point scale)
1	metropolis-type cities	19.4	62
2	large administrative cities	13.0	71
3	administrative and university towns	13.2	61
4	large industrial cities	16.2	58
5	industrial towns of the Ruhr area	4.1	42
6	average-type towns (medium size)	2.9	55
7	small administrative towns	3.0	49
8	small average-type towns	2.0	43
9	monostructured industrial towns (medium-sized)	1.4	45
10	small industrial towns in Southern Germany	1.5	46
11	small towns in Northern and Western Germany	4.7	36
12	small industrial towns at the fringe of the metropolitan areas	1.0	37
13	small industrial towns in rural areas	2.5	47
14	working-class towns	3.0	44
15	medium-sized administrative towns	4.3	44
16	medium to small administrative towns	1.7	46
17	health resorts	0.8	46
18	average-type towns on hilly area	1.9	54
19	average-type towns with little public transportation	1.2	51
20	suburbs close to the centre of metropolitan areas	0.8	47

group. The following towns represent such exceptions, owing to their urban traffic situation and spatial and social structure: Wolfenbuttel, Neu-Isenburg, Bensberg (near Cologne), Garbsen (near Hanover), Porz (near Cologne), Rodenkirchen (near Cologne), Kornwestheim. If case-studies for new urban transportation systems were carried out in these towns, translating their results to other towns would be extremely problematic.

For practical reasons the number of clusters to be used in the study was limited to twenty. It has been found that with a structural distance of 17 percent of the maximum distance, a reasonable number of relatively homogeneous clusters is obtained.

In Table 3 the different clusters are characterised and their percentage in the total sample of towns is shown.

Finally, the clusters are subjected to a multi-attributive selection process, in which they are examined for their suitability for new rail-bound urban transportation systems in order to select case-study candidates for the main study. Suitability was determined on the basis of 23 weighted criteria, using a standard of 100 points.

If a cluster were actually rated at 100 in this process, this would mean that it comprised towns, which, for all the selection criteria, should be ideal for new rail-bound urban transportation systems. Table 3 shows that in none of the clusters do ideal conditions exist. However, the large administrative cities in cluster 2, which has been rated at 71 points, may offer the most promising potential for applications of new systems.

This method of selection can by no means be substituted for the complex assessment which is the overall objective of the study. If it is used for selecting the case-studies for this main study, however, it ensures that the results can be generalised to the largest number of towns possible and that a fairly large number of criteria is taken into account when determining the suitability of towns for new transportation systems.

In conclusion, it can be said that this method of classification and selection can serve as a prototype method for other municipal technologies where limited research funds call for pilot studies from which limited generalisations are required.

HUMAN FACTORS AND ERGONOMIC CONSIDERATIONS IN THE DESIGN OF AUTOMATIC TRANSIT SYSTEMS

Norman Ashford and N.Stuart Kirk

Department of Transport Technology and Institute for Consumer Ergonomics
University of Technology, Loughborough

Over the last few years, the Transport and Road Research Laboratory of the Department of the Environment, in Britain, has been engaged in the in-depth assessment of innovative forms of public transport. Initial interest centred on small capacity Personal Rapid Transit systems. After an intensive study of the proposed "Cabtrack" PRT, which was rejected, mainly from the viewpoint of environmental damage, later work has been chiefly concerned with a development programme for an intermediate capacity automatic light transit system known as "Minitram". Similar in concept to a number of other urban light rail transit systems, Minitram is designed to run on its own segregated guideway with automatic operation, guidance and control. In concept, the systems would have medium penetration networks, with overall system capacities suitable for the British provincial towns with populations typically in the 250,000-700,000 range. With automatic operation, it is conceived that the system would provide reliable urban transport similar in service level to that of the bus, but requiring less operating staff. Service levels would necessarily be lower than for the car or PRT systems.

In its initial concept, Minitram would have a theoretical line capacity in the area of 5,000-10,000 passengers per hour depending on the final selection of headways and vehicle capacity. While below theoretical bus service capacities, the possible Minitram capacities would be a significant improvement on bus line capacities as currently observed in large provincial towns which are in the region of 3,000 passengers per hour. Service levels would be high by reason of the provision of small comfortable vehicles with a nominal capacity of between 12 and 30 persons. Under automatic control on a dedicated right-of-way there could be, furthermore, a guaranteed flexible service with a high frequency of arrivals. If provided on a city-wide basis, such a service is capable of providing a suitable basic public transport alternative to the use of the private car.

PURPOSE OF PASSENGER BEHAVIOUR STUDY

Any automatic system with no driver in the vehicle and no attendants on the stations, will, if it is to run efficiently, be required to consider the human factors requirements as part of total system requirements. In comparing travel on an automatic system with that on its manned counterparts, it can be observed that automation does not allow the same tolerances. Employees on a manual system can, within certain limits, adapt the operation of a system to the idiosyncracies of passenger behaviour. An automatic system, on the other hand, must be designed in such a manner that it is acceptable and attractive to the public at large. It is possible that in time Minitram will both compete with and replace to some degree, other forms of transport, both public and private. This will require public acceptance, entailing tailoring the system in a manner which reflects observed behavioural preferences. Additionally, new public transport systems, where possible, must cater for handicapped persons, under the provision of the Chronically Sick and Disabled Persons Act, 1970.

By studying passenger performance on a simulated Minitram system, the authors have been able to examine observed performance in relation to proposed vehicle and system requirements in the following areas:

a. Space requirements of passengers while travelling and the implications on vehicle dimensions for vehicle designers.

b. Passenger behaviour at stations and the behavioural implications on station dimensions.

c. Behaviour of passengers entering and leaving vehicles and the implications on door design and operation.

The findings of the human factors study have been incorporated in the prototype vehicle and infrastructure designs for a planned public demonstration programme. This programme, currently scheduled for the early 1980's will provide an ideal situation for further human factors evaluation, since the system, while running as a regular but limited transit service will still be in the development stages, where modifications can be incorporated and monitoring of modification effects can be carried out.

Currently, there appears to be little passenger behaviour data relevant to urban transport systems. Those data which are available are not easily adapted to specialized systems such as Minitram. On the other hand, human factors data are available on environmental characteristics such as noise and vibration (Hanes, 1970 ; Kryter, 1970), lighting (IES, 1968) and the thermal environment (Applied Ergonomics Handbook, 1970); all of these are important in the design of vehicle interiors.

EQUIPMENT

The experimental site at Loughborough University of Technology consisted of a purpose-built structure consisting of a central track, flanked by broad platforms; these in turn were flanked by portable modular buildings serving as waiting rooms and offices. The test vehicle was able to move into and out of the platform area along a track with maximum velocities in the region of 15 km/hr. Vehicular movement was controlled by a pre-programming device which controlled the velocity profile, giving controlled accelerations and decelerations of 1.25 m/sec^2. Longitudinal jerks were also controlled to ensure that no values in excess of 1.25 m/sec^3 (or $\frac{1}{8}$ g/sec) were obtained. The use of preprogrammed traction ensured that experiments could be carried out under conditions of controlled velocity, acceleration and jerk, using limiting values at values found tolerable in previous research.

VEHICLE DIMENSIONS

The design of the vehicle evolved from considerations that for small capacity vehicles the "seats-facing" arrangement shown in Figure 1 allows the maximum manoeuvrability in the central standing space. Vehicle width must permit three seats side by side with adequate individual seat width. Ergonomics literature has set minimum seat widths of 0.48 m (McCormick, 1964). McFarland (1969) has suggested that urban transport should provide elbow-to-elbow width of 0.56 m to allow for street clothing. The authors were satisfied that a 0.5 m minimum seat width would suffice. This was later slightly increased to allow better manoeuvrability in the central open space, giving an overall inside width of 1.7 m.

Seat depth was set at 0.85 m to give seat and knee room with adequate protection of feet. Standing space has been recommended by McCormick as 0.255 m^2 (0.46 m x 0.56 m); the experimental vehicle was designed to provide minimum space of 0.25 m^2 for standees. Determination of the final test vehicle size was made from the 11 experimental vehicles shown in Table 1.

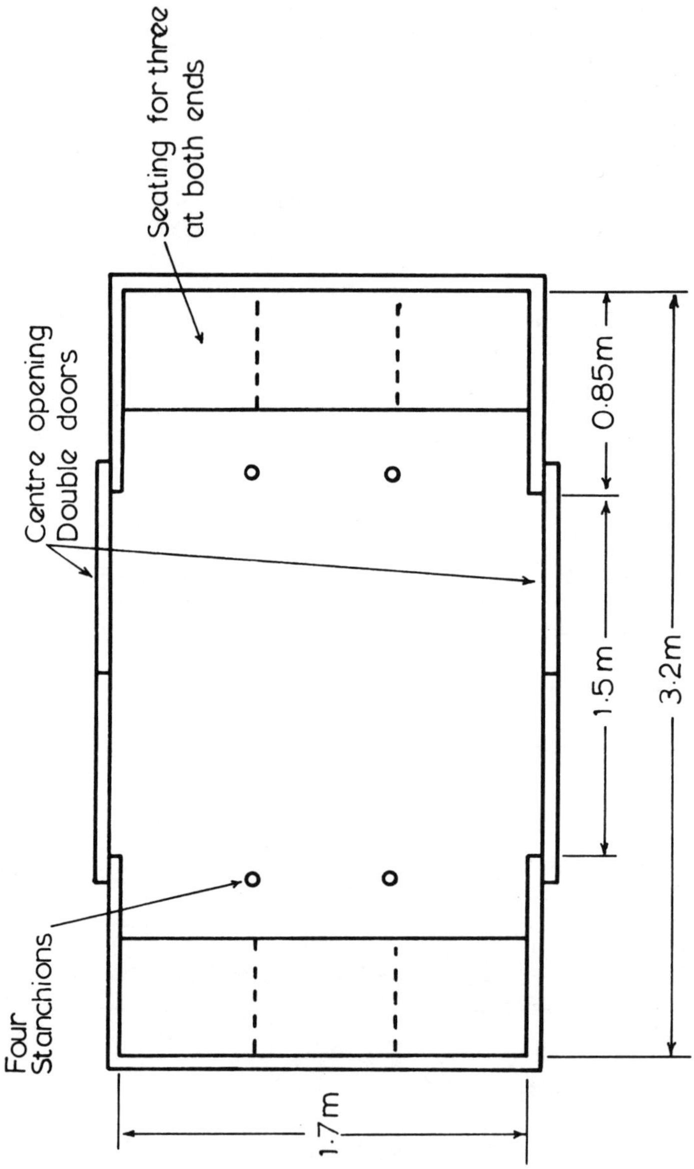

FIGURE 1. FINAL TEST VEHICLE DIMENSIONS

	Length (m.)	Width (m.)	Nominal Capacity (persons)	Door Widths (m.)	
A	2.8	1.6	12	1.2(A1)	1.4(A2)
B	3.2	1.8	12	1.2(B1)	1.6(B3)
C	3.2	2.2	16	1.2(C1)	1.6(C3)
D	3.6	2.2	20	1.2(D1)	1.6(D3)
E	3.0	1.7	12	-	1.4(E2)
F	3.2	1.7	12	1.4(F2)	1.6(F3)
Selected Vehicle	3.2	1.7	12	1.5	-

TABLE 1. INITIAL EXPERIMENTAL AND SELECTED VEHICLE DIMENSIONS

TRAVEL SUBJECTS

It was the intention throughout the research programme to reproduce typical travel situations on an urban transport system designed to cater for a broad spectrum of the community using the system for a broad range of travel purposes. It was, therefore, considered that individuals travelling on the system could be broadly designated into two categories:

1. Commuters travelling to and from work or shopping during peak travel hours. It appeared reasonable to suppose that commuters are largely aged between 18-60 years, carry little luggage and for the purposes of this study were assumed to be active with no major physical handicap.

2. Non-commuters travelling in non-peak hours, principally for non-peak purposes. This population was typically assumed to be constituted of mothers with children, shoppers and elderly persons. It was assumed that the non-commuter group was constituted of a large range of types of person, travelling for a variety of purposes, often encumbered with luggage, or infant children in push-chairs. A major sub-set of this population is the handicapped or disabled who are more likely to travel in off-peak periods.

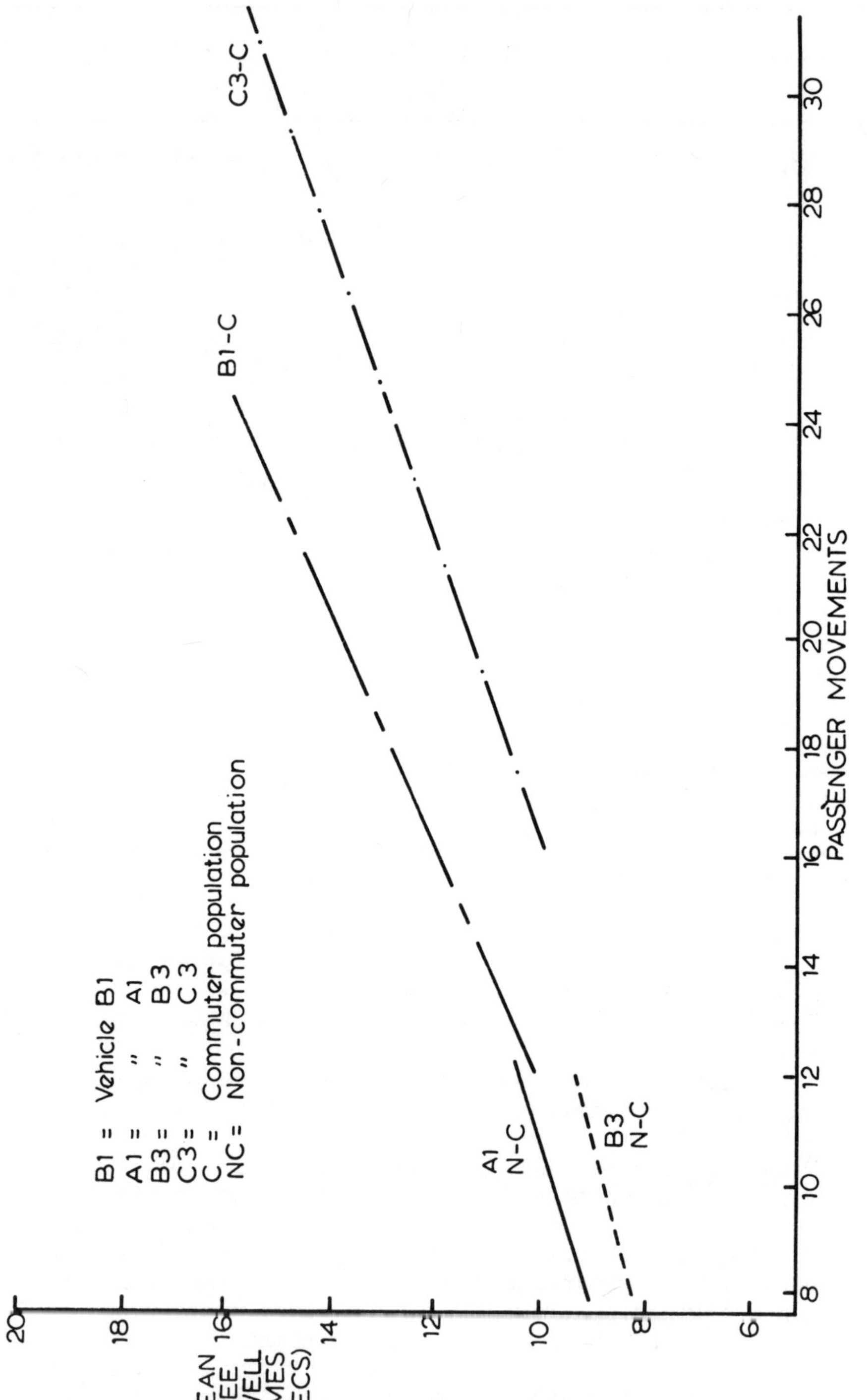

FIGURE 2. VARIATION OF DWELL TIME WITH PASSENGER MOVEMENTS AND VEHICLE SIZE

RESULTS OF PASSENGER BEHAVIOUR STUDIES

The overall findings of the passenger behaviour studies have been reported in depth to the Transport and Road Research Laboratory (Richardson & Stroud, 1974); only the principal results can be summarized here.

The relationship between dwell-time and the number of passenger movements through the doors was found to be reasonably described by linear relationships over the range of passenger movements investigated. In all eleven different configurations of vehicle and door size were investigated as shown in Table 1. The worst and best regression lines are shown in Figure 2. When passengers moved under conditions of time constraint times improved by approximately 2 seconds over the range of movements examined.

Tests using handicapped individuals under free dwell-time conditions showed substantial increases of required time for individuals on crutches or in wheelchairs to negotiate automatic doors. Remarkably, however, under the constraint of fixed dwell times, it was found that, provided the vehicle dimensions were sufficiently large to allow room for handicapped persons to manoeuvre, their progress in and out of the vehicle was no slower than that of other passengers. The final vehicle dimensions were chosen to be 3.2 m x 1.7 m with a 1.5 m door with this manoeuvrability criterion in mind.

It was found that static test times could be reliably converted to dynamic times by an added time constant of two seconds for the commuter population and a variable time of three to five seconds for non-commuter populations.

Figure 3 indicates the effect of through-flow operation with either simultaneous or staggered door opening time. It can be seen that significant time savings can be achieved by the adoption of through-flow, with one door operating for exit only, while the opposite door is for entering passengers. The superior performance from the view point of the vehicle must be balanced against the added cost and environmental intrusion of the larger platform requirements for this sort of operation. It is interesting to note that staggered door opening did not improve overall dwell-times. The difference in time between the two sets of doors opening was 1.7 seconds. It is apparent however that when doors are opened simultaneously in a vehicle of this size, the space vacated by the alighting passengers is large enough to allow free access for the boarding passengers. The extra space provided by staggered door opening is unecessary and dwell-times increase.

The implications of platform design were examined from the

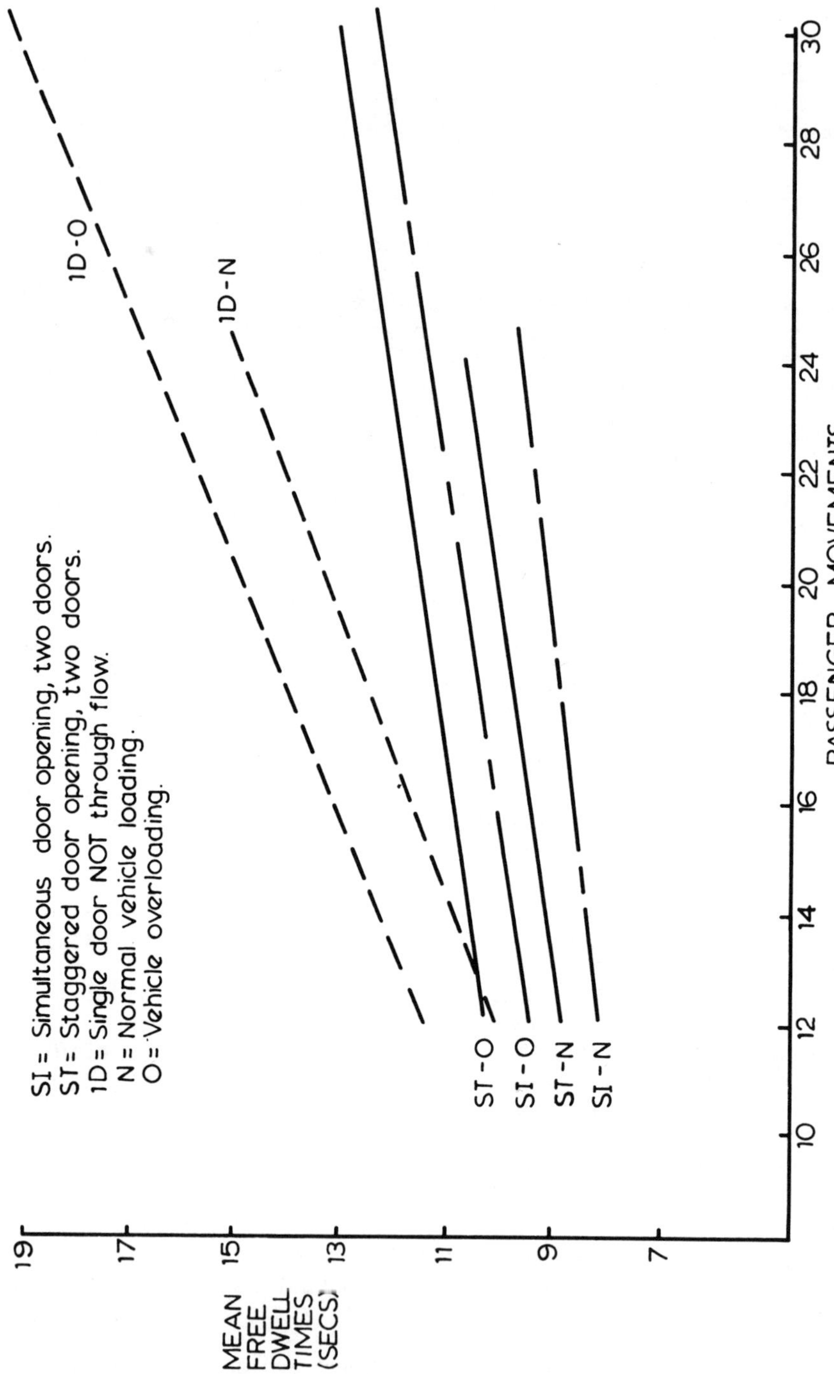

FIGURE 3. DWELL TIMES UNDER CONDITIONS OF THROUGH FLOW

viewpoints of size, location of egress and entry and the use of platform barriers to channel flow. Fruin's (1970) work gives standards for personal space for walkways and queues but these are not entirely applicable. For testing purposes space standards of 1.5 and 1.25 m^2/person were tested. In the nominal twelve person vehicle, these standards gave 6 m x 3 m and 5 m x 3 m platforms, with exit and entry either at the same end of the platform or at different ends.

The clearance times for the four different arrangements were very similar, with a difference of only $2\frac{1}{2}$ seconds between the worst and best times for 24 passenger movements. This was felt to be marginal.

Minitram stations are expected to be constructed in built-up areas, with a consequent restriction on platform size. The cost of land in such areas is high and if street space is to be used, visual intrusion will be high. If there is a single platform it will be used by passengers boarding and alighting from both directions of travel. Cross flows and collisions occur as various groups of passengers attempt to walk in different directions in confined space. The experimental sessions investigated the possibility of passenger separation by a variety of barriers shown in Figure 4. It was found, however, that there was only one second of difference between vehicle dwell-times for the various configurations of barrier tested, both for 24 commuter passenger movements or 12 non-commuter movements. No barrier appeared to have any distinct advantages in terms of dwell-time or ease of passenger flow and there was no clear evidence that a system of barriers would improve platform efficiency. In dealing with handicapped passengers there appeared to be some disadvantage incorporated with installing fixed barriers.

SOME PROBLEMS RELATED TO AUTOMATIC SYSTEMS

During the course of the testing programme, some of the problems of designing automatic systems became apparent and were considered even though design solutions were beyond the remit of this study. The following would appear to rank among the most important.

1. Vehicle overloading. Where vehicles have been designed on an ergonomic basis allowing comfortable standing room and adequate manoeuvring space for wheelchairs, it is possible that urban travellers in the peak hour would attempt to board the vehicle even though it already carried its design capacity. Where overloading is permitted to occur, not only are vehicle propulsion units seriously overloaded (requiring unnecessarily spare motor capacity),

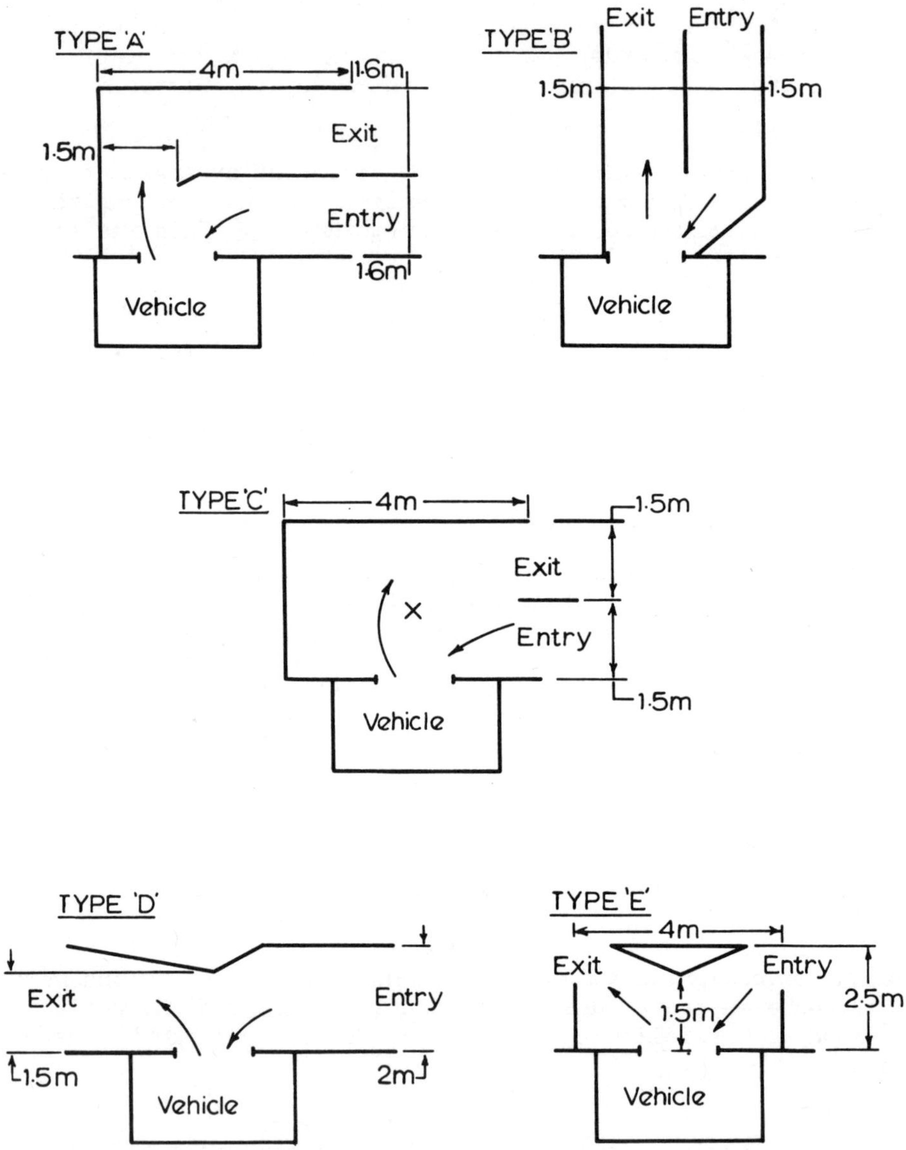

FIGURE 4. BARRIER CONFIGURATIONS

but station dwell-times are increased due to extra passenger movements through the doors. Tests were carried out on the overload case. High station dwell-times were recorded, with strong probabilities of unacceptable delays to system headways.

2. Access. Where a system is to be designed for maximum access to the handicapped and disabled, special care must be taken in the design of fare collection systems and access barriers. Whereas vehicles and platforms are relatively easy to design to accommodate the disabled, the access to these areas presents systems designers with challenges. These difficulties are likely to be most pronounced in older urban areas where the infrastructure design must be integrated without severe visual intrusion into the town centres. Special ramps and liftshafts present difficulties in visual design. Equally difficult is the design of ticket monitoring devices on automatic systems which permit access of wheelchairs, while excluding non-paying riders.

3. Vandalism and Crime. It is now some years since Jane Jacobs (1961) pointed out the dangers of creating unmonitored space in urban areas. Such space rapidly becomes unsafe. Public transport systems are becoming increasingly the site of urban crime and violence. Transport designers are increasingly aware that vandalism and crime must be considered in the design and operation of public transport systems (TRB, 1974). It is essential, where systems are unmanned, both on the vehicle and in the stations, that there is adequate design of remote monitoring systems ensuring passenger security. Recent studies have shown that, not only must passengers be safe, but they must also feel safe before they ride the systems. Perhaps the greatest difficulty facing the introduction of automatic systems is the provision of safe, monitored vehicles and waiting areas which will attract rather than repel users.

REFERENCES

Applied Ergonomics Handbook. (1970) Chapter 8. Thermal comfort in industry. In Applied Ergonomics, 4, 210-216.

Fruin, J.J. (1970) Pedestrian Planning and Design. Metropolitan Association of Urban Designers and Environmental Planners. New York.

Hanes, R.M. (1970) Human sensitivity to whole body vibration in urban transportation systems. APL/JHU - TPR004. John Hopkins University, Boston.

Illuminating Engineering Society (1968). Recommendations for Lighting Building Interiors.

Jacobs, J. (1961) The Death and Life of the Great American Cities. New York: Vintage Books.

Kryter, R.D. (1970) The Effects of Noise on Man. New York: Academic Press.

McCormick, E.J. (1964) Human Factors Engineering. London: McGraw-Hill.

McFarland, R.A. (1969) Human factors in ground transportation with special reference to passenger comfort and safety. High Speed Ground Transportation Conference. Carnegie-Mellon University, Chicago.

Richardson, J., & Stroud, P.G. (1974) Passenger behaviour study. Institute for Consumer Ergonomics. University of Technology, Loughborough.

Transportation Research Board (1974) Crime and Vandalism in Public Transportation. Record No. 487. Washington D.C.

PEAK PERIOD UNDERGROUND RAIL TRAVEL - THE LONDON EXPERIENCE

Phillip Rice

Traffic Studies Group

University College London

The spectacle of a prosperous London whose very prosperity has brought in its wake major problems of employment, housing and mobility has been the subject of numerous enquiries and studies; for instance, the Royal Commission on London Traffic of 1905 and presently the Greater London Transportation Study (GLTS), a full-scale strategic model of transport in London.

Something of the complex interaction in city life of employment, housing and transport can be seen in the historical development of London from 1900 onwards. In Figure 1 the trend in population in the London Transport area is set out showing the migration of population from the inner area in the 1920's and the sharp rise in the outermost area in the 1950's in the era of the motor car. Figures 2 & 3 show the car miles run in service and passengers carried annually on public transport, with a sharp increase in the 1920's reflecting the expansion in commuting from the suburbs to the Central Area. Figure 4 demonstrates the historical decline in all modes of public transport against the rise in motor car ownership after the Second World War.

A long time-series has the advantage of showing the transportation mode life-cycle. Clearly the influence of existing inherited transport infra-structure will modify the range of techniques possible for urban movement and the timing of new investment. An intriguing speculation may be made as to whether London would have had its extensive underground railway network had the petrol bus come a few years earlier.

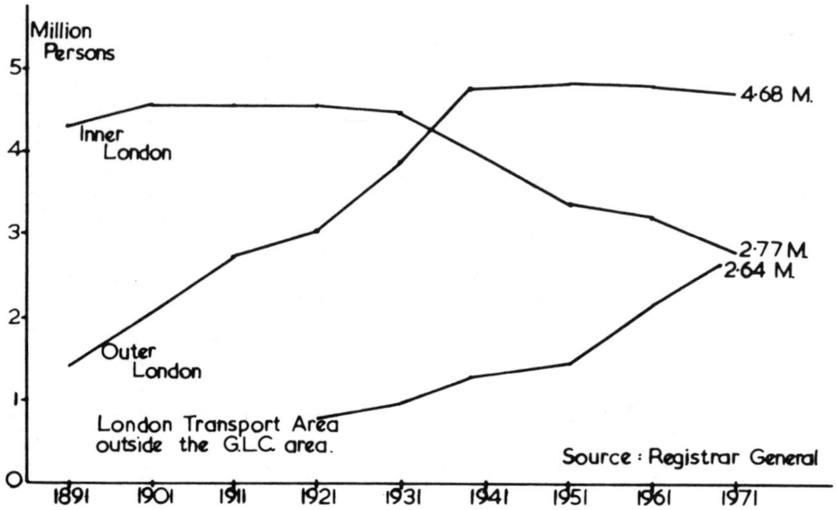

FIGURE 1. POPULATION TRENDS, 1891-1971, LONDON

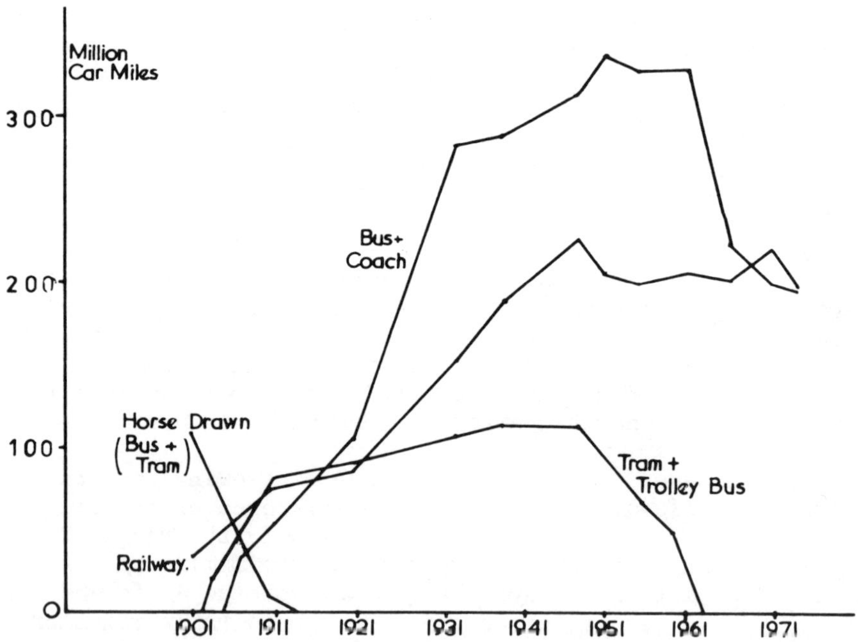

FIGURE 2. CAR MILES RUN IN SERVICE 1901-1973

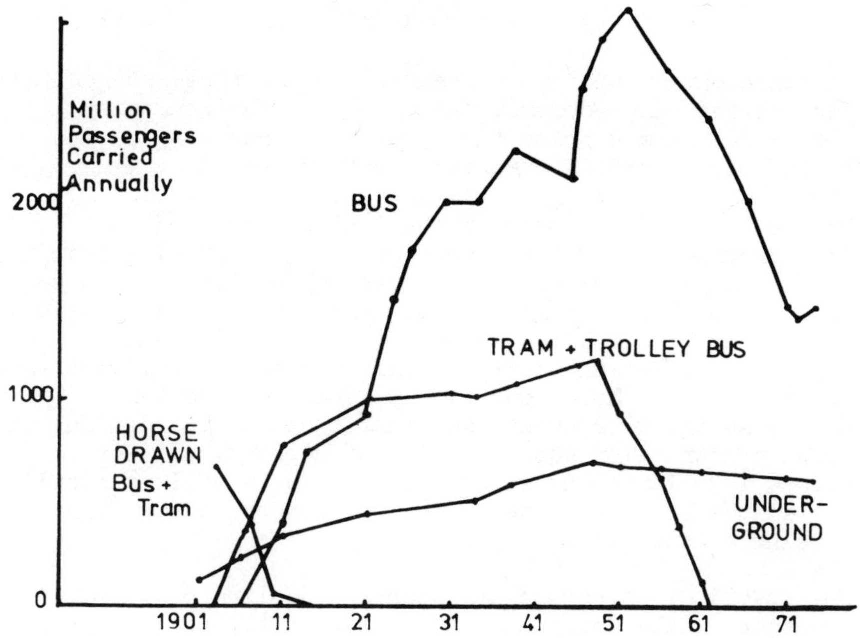

FIGURE 3. PASSENGERS CARRIED BY EACH MODE ANNUALLY

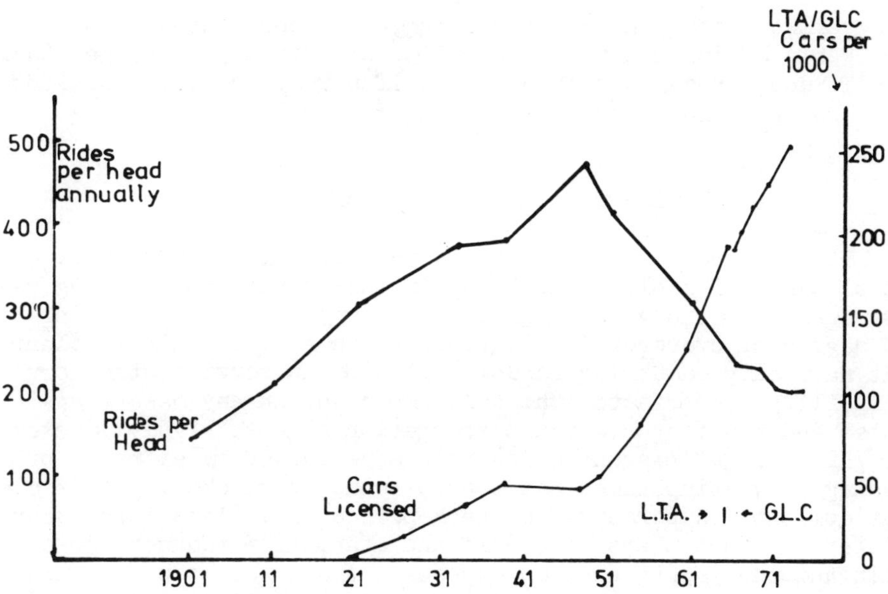

FIGURE 4. COMPARISON RIDES PER HEAD (PUBLIC TRANSPORT) WITH CARS LICENSED LTA/GLC

Changing Economic Criteria

The development of London Transport shows three stages, reflecting the changing economic philosophy of the community. Following Barker and Robbins (1974) and concentrating on the development of the underground railway they are:

(1) Pre-1933. From the turn of the century independent underground railway lines were promoted as profit-making commercial enterprises. Heavy capital outlay and fierce competition led to collaboration and finally amalgamation.

(2) 1933 - 1962. At the start of this period the railways were municipalised. More importantly advantage was taken of Keynesian thinking about the benefit of public spending on large construction works in time of depression. Consequently before 1939 large investments were made in railway facilities in the suburbs to serve the new London County Council housing estates.

(3) Post-1962. This period has seen the decline of public transport while motor car congestion in the city centres has worsened, degrading still further road-based public transport services. However, since 1962 a new era in the appraisal of investment decisions has begun with the acceptance of social costs and benefits across the community as being the basis for justifying new investments. A new railway line, the Victoria line, was constructed on the basis of its net social benefit, principally passenger savings. It would never have been built if private benefits to London Transport had been the deciding factor.

Peak-Period Travel in an Era of Social Costing

Given a social cost framework the problems of peak-period travel can be freshly evaluated. In the morning and evening peaks passengers experience sometimes extreme variations in waiting time and degree of overcrowding in their train service. It is clear that naturally occurring random variation in service, the service reliability, is a factor which is important in any mass transit system and the factor which passengers rank most highly. However the valuation of passenger inconvenience caused by extremes of loading or waiting times has tended to be overlooked. It is clear that comfort, as perceived by peak-period travellers is closely related to the numbers of seated and standing passengers in a train and the length of time exposed. A stochastic programming model is here proposed which incorporates a full social cost treatment of overcrowding and waiting time costs for varying levels of stochastic variation in service. This model is applicable to the usual peak-period situation in and around the central area of London and shows how to optimise an underground rail service where the social costs associated with overcrowded trains predominate.

CONSIDERATIONS IN MODELLING THE PEAK-PERIOD

The adequate treatment of peak-period travel in a transportation model such as the GLTS model requires the 24-hour average day approach to be refined so as to incorporate the following factors:

(1) Differing journey purpose - working-time, commuting, leisure (see Table 1).
(2) Variations in public transport frequency throughout the day.
(3) The impact of restraint measures on shifting journeys from peak to off-peak and within the peak.
(4) The problem of modelling capacity at any time of day.
(5) The growth in non-work journeys emphasising off-peak travel.
(6) The importance of modal choice, the possibility of passengers changing modes due to overcrowding as a comfort consideration. (The underground railway as typical of mass transit public transportation is very sensitive to this).

Purpose	0700-1000	1000-1600	1600-1900	1900-2200
	%	%	%	%
Non Home-Based	3.1	37.2	10.4	11.5
Home-Based Work	87.1	32.9	74.5	43.1
School-College	7.8	12.0	7.7	12.1
Shopping	0.4	6.6	1.5	1.1
Sport-Entertainment	0.2	2.5	2.5	17.0
Other	1.4	8.8	3.4	15.2
Total Travellers (1000's)	642.4	480.0	666.2	120.0

Notes This table does not give entrants to Central London. However, of the above 495,000 in the morning peak enter or use the underground in Central London. Those who use the underground to cross the cordon of Central London are 387,000 out of 1,111,000 entrants in the morning peak.

(Source London Transport Planning Office 1971-2 O-D Survey)

TABLE 1. PRESENT UNDERGROUND RAILWAY USE (ALL JOURNEYS IN LONDON) BY TIME OF DAY

Havers (1973) notes that in peak-period modelling there is usually an absence of information on generalised cost matrices for public transport. If peak-period trip-ends are generated from 24-hour data for use in peak-period distribution and assignment models, it is difficult to know how to model the relief of an overload for a specific time-period.

Specification of Social Costs in the Underground Railway Context

The concept of generalised cost in time or money units is the basis of the present philosophy of transport planning. The major advance in planning in London over the past ten years has been the marrying of the methodology of traffic prediction models with the principles of economic assessment.

Let us consider a link in the underground network and specify the components of cost (behaviour or resource) incurred in a journey undertaken at a particular time of day. They are:

- y_1 in-vehicle time,
- y_2 walking and booking time consequent on use of underground,
- y_3 waiting and transfer time in the system,
- y_4 travel cost,
- y_5 ratio of passengers to seats - the overcrowding measure,
- y_6 unit train operating cost based on the engineering assessment of marginal cost.

Let us adopt a framework similar to that of McIntosh and Quarmby (1970), but with provisions for variations in waiting times and loadings of trains. We initially adopt linear functions of cost for behavioural b, modified behavioural u, and resource r specifications.

$$b = B_1y_1 + B_2y_2 + B_3y_3 + B_4y_4 + B_5y_5 \quad (1)$$

$$u = U_1y_1 + U_2y_2 + U_3y_3 + U_4y_4 + U_5y_5 \quad (2)$$

$$r = R_1y_1 + R_2y_2 + R_3y_3 + R_5y_5 + R_6y_6 \quad (3)$$

The behavioural parameters B_i, the modified behavioural parameters U_i and the resource parameters R_i, convert the equations into a consistent base of either time units or money units. For predicting passenger behaviour the present practice is to use generalised time units; for economic assessment it is money units.

Observations on the Calculation of Overcrowding Cost

It is quite clear that overcrowding is an important consideration for travellers in peak-periods. Passengers will often be faced with the decision of whether to board an already overcrowded train or to wait for the next train. This is a behavioural problem. If we wish to set a money cost on this travel, representing society's valuation of it, it is a modified behavioural assessment. The area where more study is needed is in the resource cost specification of travel in an overcrowded train. It remains to be seen how important effects of thermal strain or tiredness can be directly attributed to travel in overcrowded conditions. In this paper the modified behavioural cost of overcrowding is hypothesised to be a monotonically increasing function of y_5 denoted by $h(y_5)$ (Figure 5). Its shape is such that the cost is zero when the train is empty and rises sharply in the region of four passengers per seat, the crush-load as specified by L.T. standards. The present procedure put forward in L.T. O.R. memo 244 (1972) allows the overcrowding cost to rise linearly up to a maximum at four passengers per seat. The difficulty here is that in modelling the relief of an overloaded link at a peak-time it is necessary to have a model structure which will accurately predict when passengers cease to travel in crush conditions, but wait for the next train at an acceptable loading. In practice, the random variations in time-intervals between trains and in passenger flows give rise to fluctuations in loadings which can only be adequately represented in stochastic terms.

Justification for a Scheme of Headway Control

An operator (London Transport) may adopt the strategy of maximising passenger miles travelled subject to a budget constraint. But the strategy is weak in that no account of comfort or overcrowding is included explicitly. We may imagine a peak-period service with equal headways and compare it with the same number of trains operating with a long gap in the service in the middle of the peak when the maximum demand for transport takes place. In the second case it is argued that the delay and overcrowding experienced by passengers after the gap will in social cost terms be greater than any small time-saving to other passengers who travel in uncrowded conditions. Both cases have the same passenger miles run for the same budget constraint.

Although the above argument is rather crude, it does make the point that where random variations in service are tolerated because of non-linearities in the way the overcrowding cost builds up and because the expected waiting time also increases sharply with the variance of the service, the social cost will be greater than if a more regular service had been run. A scheme of headway control

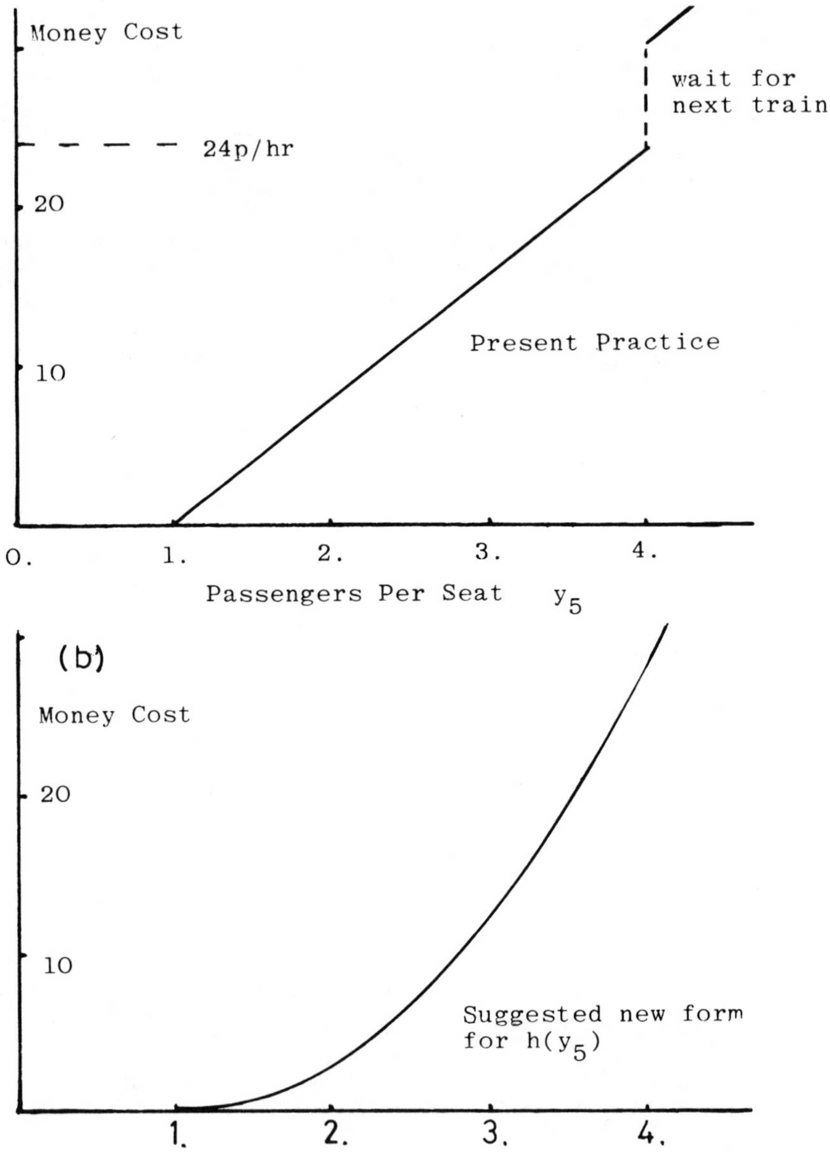

Source Present Practice L.T. O.R. memo M244 (1972)

FIGURE 5. MODIFIED BEHAVIOURAL COST $h(y_5)$ IN MONEY UNITS

for a train service by computer could have a large monetary or social benefit. The question is how to evaluate the time and overcrowding saving which can result when a strategy for controlling a string of trains during the peak-service is conceived. At present London Transport are actively pursuing this possibility which represents a major new innovation in the principles of railway signalling. The final model proposed in this paper develops an approach to minimising the generalised cost of travel in a peak-period for inputs of mean and variance of the train service at, say, quarter-hour periods.

PRESENTATION OF SOME OPERATING DATA FROM L.T. RAILWAYS

The magnitude and importance of the variations in time-interval between trains, headway variation, is best seen in Figures 6 & 7. The Northern line is a predominantly feeder line into Central London in the morning peak, carrying a peak-hour flow of 13,500 passengers between 8 and 9 a.m. at Kennington, the place for which the data in Figure 5 is given. It is also the line which is thought to be operationally the most unreliable. The Victoria line in Figure 6 by comparison has automatic driving of trains, against the manual operation of the Northern. In the direction considered it is also a feeder line into Central London in the morning peak, when slightly over 20,000 passengers are carried between 8 and 9 a.m. at Euston, for which the headways in Figure 6 are given. The Victoria line has considerably less variance in the quarter-hour periods of the peaks than the Northern line. However, it is evident that considerable variations can occur within a quarter-hour period. The midday off-peak and evening off-peak data are given for completeness.

Modelling Stochastic Variations in Train Service

The cause of the stochastic variation in service headways is due to two major causes. Cancellations of trains for staff shortages or failures tend to open large gaps in an otherwise fixed, timetabled service. Natural randomness in the service depending on place and time within the peak is also apparent. In practice it is difficult to isolate the two effects. The major problem is the fixed timetable; until computer control of the service can be undertaken there is little prospect for improving the reliability by 'headway smoothing'.

Let us divide the peak into periods T_i, $i = 1 \ldots 4$

Let x_i be the planned mean headway in period T_i

This requires $n_i = T_i/x_i$ trains (We neglect the integer problem)

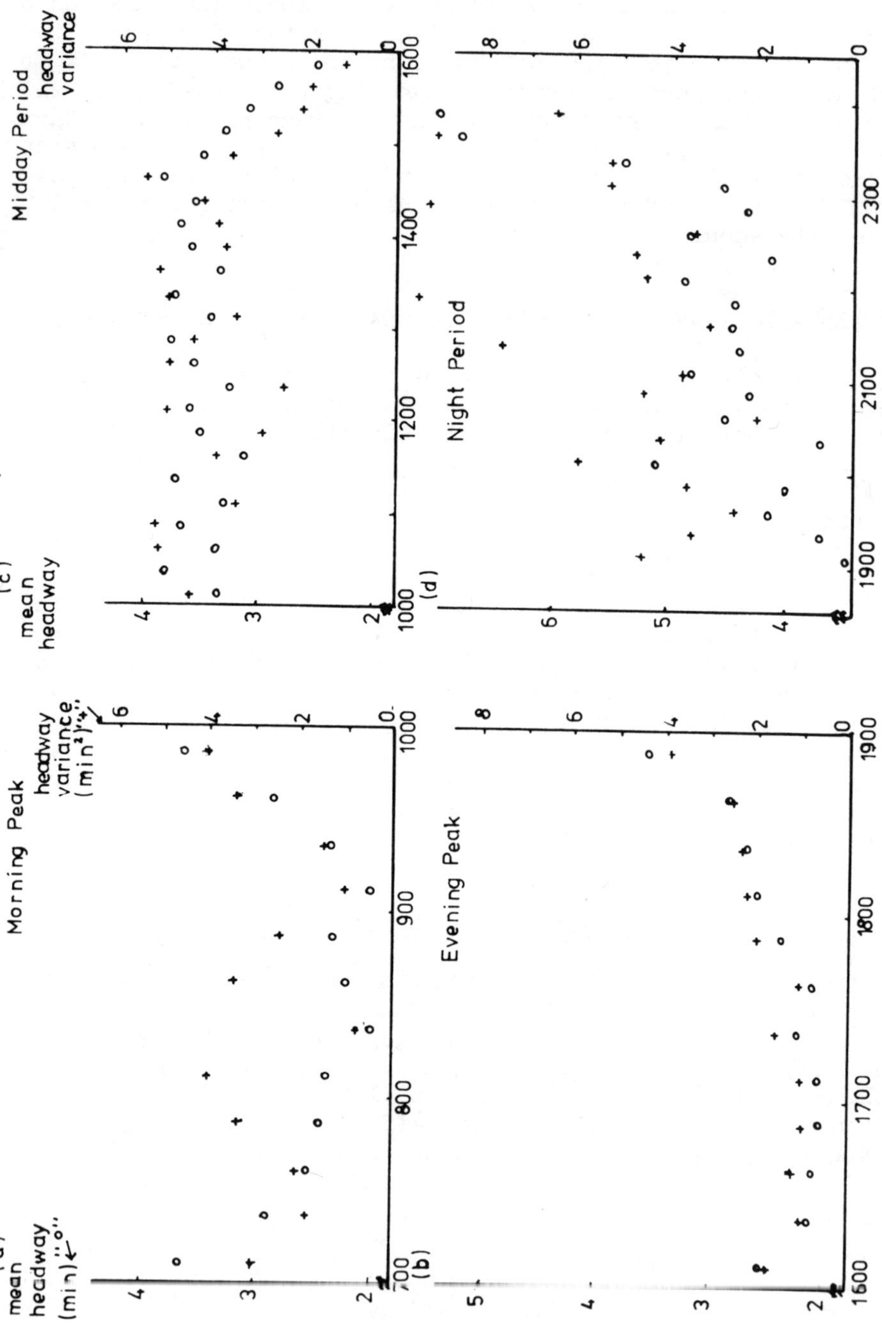

FIGURE 6. NORTHERN LINE, NORTHBOUND, KENNINGTON, OCT-NOV 1972

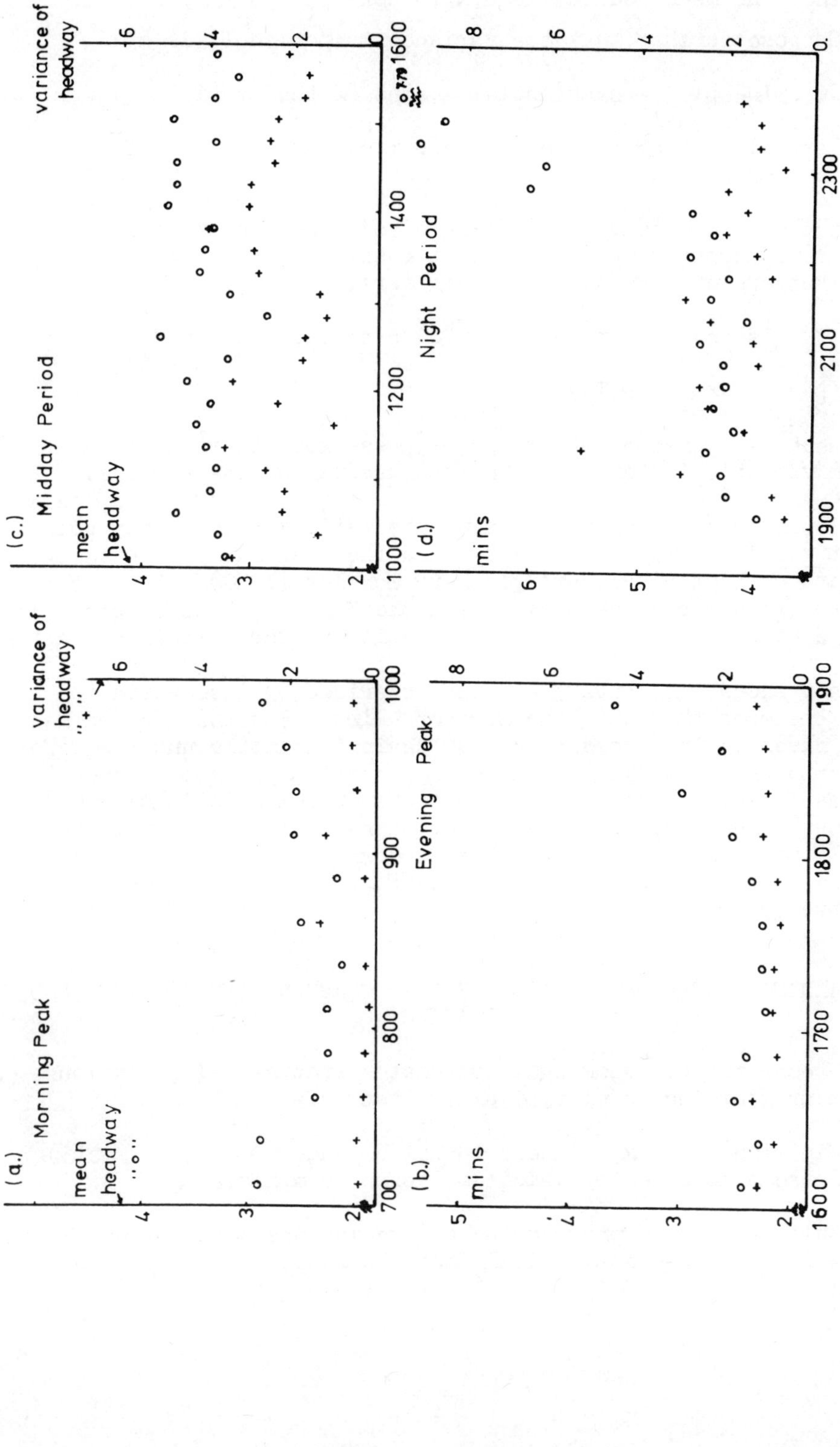

FIGURE 7. VICTORIA LINE, SOUTHBOUND, WARREN STREET, JAN-FEB 1975

Suppose that only π_i of the trains run in period T_i

Then the mean headway is $p_i x_i$, where $p_i = n_i/\pi_i$

Suppose further that the variance in period T_i is $k_i^2 x_i^2$.

In principle, a substantial amount of the statistical variation in service quality can now be incorporated into a programming model by calibrating the headway distribution in terms of p_i and k_i.

Let us now consider modelling of overcrowding in the peak that is represented by the ratio of passengers to seats y_5. Suppose F_i passengers wish to use the train service in T_i.

If C is the seating capacity of a train, then in T_i we obtain:

$$y_5 = F_i p_i x_i \; / \; (C.T_i) \qquad (4)$$

We have expressed the ratio of passengers to seats as a random variable in terms of the mean headway of the service.

It is interesting to see how this ratio varies at key points and at various times. Figure 8 shows the mean ratio of passengers to seats on historical data given by Menzler (1950). This is compared with more recent data from the Northern line at one point, but the bar lines indicate how the ratio can vary within a quarter-hour period as given by Rice (1975). The loadings on the Victoria line not shown in Figure 7 are characterised by less variation about the mean than with the Northern line. But the maximum mean ratio rises to in excess of 2.5 as against a little under 2.0 for the Northern line. The amount of aggregation present in even 15 minutes is sufficient to smooth out the extremes in loading which do occur. However, the data do not warrant any finer division than 15 minutes. This approach is somewhat crude and the next step is to allow the passenger flows F_i also to vary. This is not attempted here.

PROGRAMMING MODEL FOR ANALYSIS OF CHANGES IN SERVICE AND LOADING

Stochastic programming incorporates statistical variations into a programming model typically given by:-

Minimise $z = c'x$, subject to $Ax \geq b$ and $x \geq 0$ (5)
where x is a vector which is chosen to minimise z.

Full stochastic programming allows vectors b and c and the matrix A to possess statistical distributions. However, this

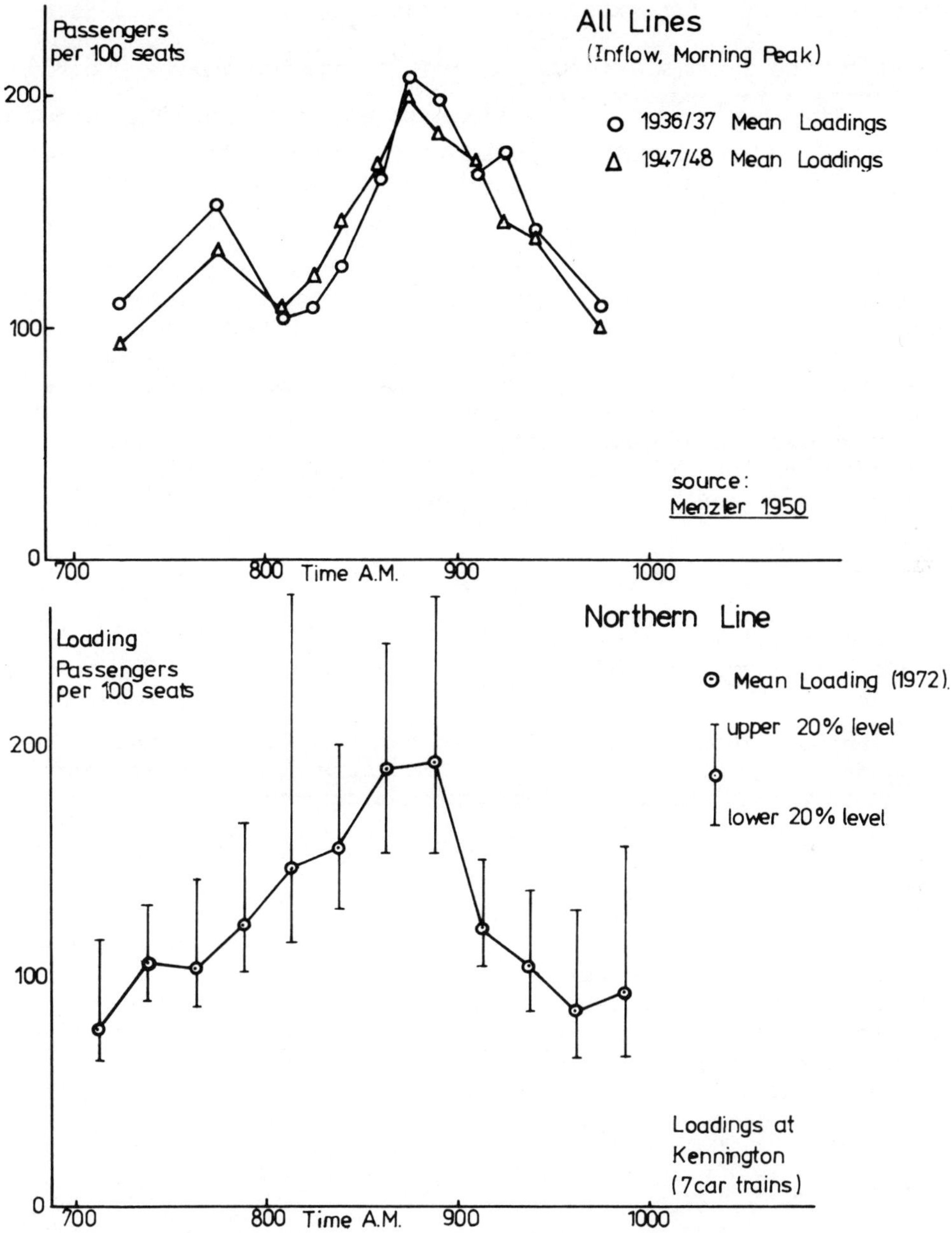

FIGURE 8. LONDON UNDERGROUND PASSENGERS PER 100 SEATS PROVIDED

degree of complexity makes such models intractable. If random variation is only present in the objective function so that the distribution of z has expectation $w'x$ and dispersion $x'Qx$, then

$$\text{Minimise } z = w'x + tx'Qx, \text{ subject to } Ax \geq b \text{ and } x \geq 0 \qquad (6)$$
where t is a scalar parameter,

can be solved by quadratic programming, assuming the multivariate normal distribution is a sufficient approximation (Sengupta, 1972, p214).

A further application is to interpret systems reliability programming to a transport service. The advantage is that a service is conceived as having probabilistic levels rather than being deterministic. We may require that at the 95% tolerance level the actual service headway at a particular time in the peak is less than three minutes. Equivalently we may specify that the system should operate such that the crush-level of loading is not exceeded more than 1% of the time. It is possible to include such chance constraints in equation 6 (Sengupta, 1972, p265).

Stochastic Model of Train Service with Reliability Constraints

For simplicity we may imagine a situation where the components of cost of an underground railway service are given by the waiting time and overcrowding costs. This is U_3y_3 and U_5y_5 of equation 2. A fuller model would incorporate the other costs of travel. The purpose is to derive a model and to indicate how sensitive cost may be to variations in waiting time and overcrowding.

We use the framework of equation 6 and divide the peak-period that we are considering into periods T_i. We are seeking to minimise the generalised travel cost across the peak-period subject to a budget constraint, which is interpreted as a maximum number of trains available. We assume that all the parameters of equation 2, valuation of passenger waiting time and the treatment of overcrowding costs are known, together with the parameters of the headway distribution, p_i and k_i. Let us generate the model in tableau form:-

Time Period	T_1	T_2	T_3	T_4
Planned mean headway	x_1	x_2	x_3	x_4
Headway actually run	p_1x_1	p_2x_2	p_3x_3	p_4x_4
Service Tolerance	$Pr(p_ix_i > S_i) < c_i, \quad i = 1...4$			
Transformed to constraint	$x_i(p_i + k_iG_i) > S_i, \quad i = 1...4 \qquad (7)$			

Where G_i is the $100e_i$ percentile ordinate of the headway.

The waiting time components and the overcrowding components of cost specify the objective function. Let us assume passengers arrive at a rate B_i which is fixed. In Period T_i

Expected waiting time $= 0.5B_i p_i x_i (1 + k_i^2/p_i^2)$, $i = 1 \ldots 4$

This is converted into a cost by multiplying by U_3.

Similarly the cost of the overcrowding in period T_i is derived for the mean loading consequent on the level of service. We require:

$$\text{Expected overcrowding cost} = E[h(x_i B_i/C)]$$
$$= h(p_i x_i B_i/C) + 0.5v.h''(p_i x_i B_i/C)$$

Where h" is the 2nd differential of h, and v is the variance of the mean loading.

If there is correlation between the headways in periods T_i, as there certainly will be in practice, we define v_{jk} as the matrix and use a simplification of Whittle (1971, p. 123) to obtain an equivalent form of the objective function. Then minimise z:

$$z = 0.5\sum_i \left[p_i x_i B_i \ ((1+k_i^2/p_i^2)U_3 + 2/C) + f\sum_j v_{ij} x_i x_j\right]$$

Where f is an expression involving h" and h', which we assume is constant, so that we obtain an expression of the form of equation 6, which is a simple quadratic expression. The constraints are given by equation 7 for each of the four time periods:

$x_i(p_i + k_i G_i) > S_i$, $i = 1 \ldots 4$

The final constraint is on the number of trains available:

$\sum_i T_i/(p_i x_i) < N$ (The budget constraint)

This model may be evaluated by the methods of Sengupta (1972) to find optimum x_i for various given inputs.

Final Comments on the Need for Headway Control

The programming approach to minimising the generalised cost of travel appears to offer a useful tool for evaluating the performance of computer devices to regulate headways. The problems associated with irregular headways, excessive waiting times and uneven loadings can have substantial social costs. An estimate of £60,000 was placed on possible savings annually for only a crude system of headway control on a short stretch of underground line. Clearly the time is now opportune for train signalling to be orientated to social cost.

ACKNOWLEDGEMENT

It is a pleasure to be able to thank the staff of London Transport for their help with data and for access to their reports. I am also grateful to the Science Research Council for their financial support.

REFERENCES

Barker,T.C., & Robbins,M. (1974) A History of London Transport, 2. London: George Allen.

Havers,G.C. (1973) Setting up a strategic transportation model for the 70's. PTRC Conference. London.

L.T. OR Memo 244, (1972). London Transport.

McIntosh,P.T., & Quarmby,D.A. (1970) Generalised costs and benefits in transport planning. MAU Note 179. London: Department of the Environment.

Menzler,F.A.A. (1950) London and its passenger transport system. Journal of the Royal Statistical Society, 113, 299-345.

Rice,P. (1975) An application of counting distributions to the analysis of a transport facility. Seventh UTSG Conference. University of Surrey.

Sengupta,J.K. (1972) Stochastic Programming. Amsterdam: North-Holland.

Whittle,P. (1971) Optimisation Under Constraints. London: Wiley.

ANALYTIC REFLECTIONS ON THE PLANNING OF PUBLIC TRANSPORT IN METROPOLITAN AREAS

Peter H.C.Engelbrecht

Municipal Transit System

Munich

The transport operator is supposed to make the system go and to keep it going (and not to talk about it) with maximum safety, outstanding reliability and fair economy. His task is to work on the lines of industrial engineering, combining labour, fixed and moving installations and funds to serve varying transportation demand and to overcome rather unflexible capacities and resources by using flexible methods.

Transport has three parties: passengers (customers, patrons), personel (of varying skill and of all ranks), and the general public. Interdependence between disciplines (e.g. engineering, planning, management), locations (e.g. township, region, province), political powers and objectives, and other tasks and influences causes a large variety of controversial issues. Clear understanding amongst all parties is essential to avoid nuisances in human, political, technical, operational and other fields. Numerical approaches are indispensable to decision-making.

Running a transport system involves nine major tasks: defining volumes of traffic, devising operations, funding, providing installations, providing labour, handling operations, maintenance, renewal, and monitoring. Each of these tasks is more or less done in five stages: planning, deciding, executing, implementing and checking. Even these stages might be sub-divided into further stages. All the elements are influenced not only from inside the transport organisation but are exposed to many external influences: for example, city development and planning, public funding, legal procedures. In some important fields external influences now prevail, as shown in Figure 1.

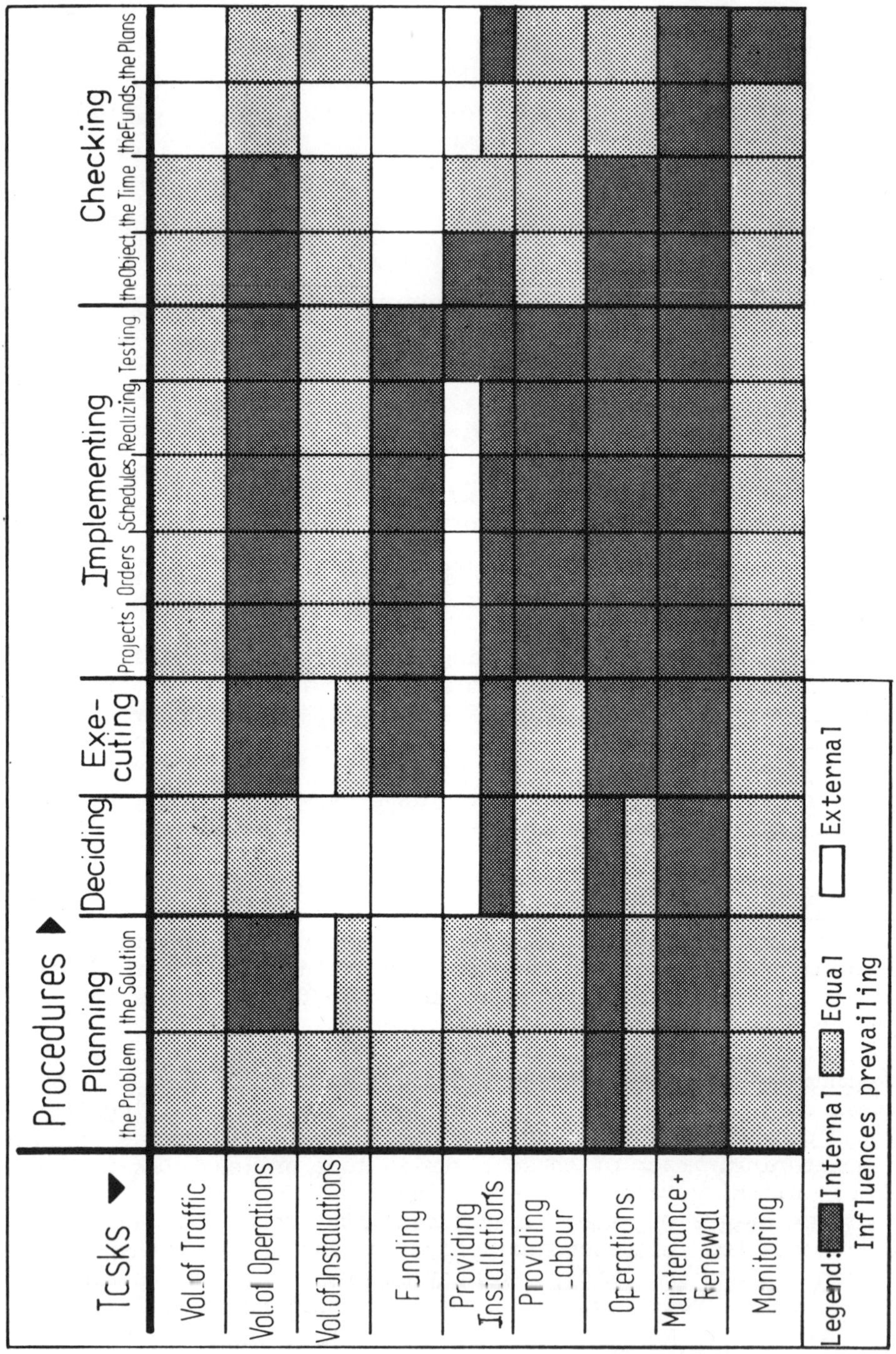

FIGURE 1. INTERNAL AND EXTERNAL INFLUENCES ON URBAN PASSENGER TRANSPORT

Human Spheres ▼	influenced by ▶ in Relation to ▼	Capacities (Installations)				Methods (Operations)				Resources		
		Vibration Noise	Pollution	Phys. Danger	Visual Intrusion	Vibration Noise	Pollution	Phys. Danger	Visual Intrusion	Inst. Funds	Operat. Funds	Labour
Passenger	Safety	↵	↵	↵		↵	↵	↵		↵	↵	↵
	Reliability			↵				↵		↵	↵	↵
	Economy									↵	↵	↵
Personell	Safety	↵	↵	↵	↵	↵	↵	↵	↵	↵	↵	↵
	Reliability	•	•	•								
	Economy									↵	↵	
Public	Safety	↵	↵	↵	↵	↵	↵	↵	↵	↵	↵	↵
	Reliability							↵		↵	↵	↵
	Economy	•	•	•	•	•	•	•	•	↵	↵	↵

Legend: • Mutual Interdependence

↵ Directed Influence

FIGURE 2. HUMAN SPHERES AND ENVIRONMENTAL INFLUENCES

The three human spheres (passengers, personel, public) may be affected in respect to safety, reliability and economy, through capacities, methods and resources, either mutually or in one direction. The initial installation and regular operation of a system will affect them in four fields of environmental influence: vibration and noise, pollution, physical danger and visual intrusion. Even resource questions such as funding or labour recruitment might do so (in Figure 2).

Many of the controversial issues are likely to occur in pairs and seem to be connected in a somewhat hyperbolic relation (Figure 3).

A batch of controversial pairings picked at random from fields such as decision-making, planning, management, operations or technology, may demonstrate the dangerous gaps in mutual understanding:

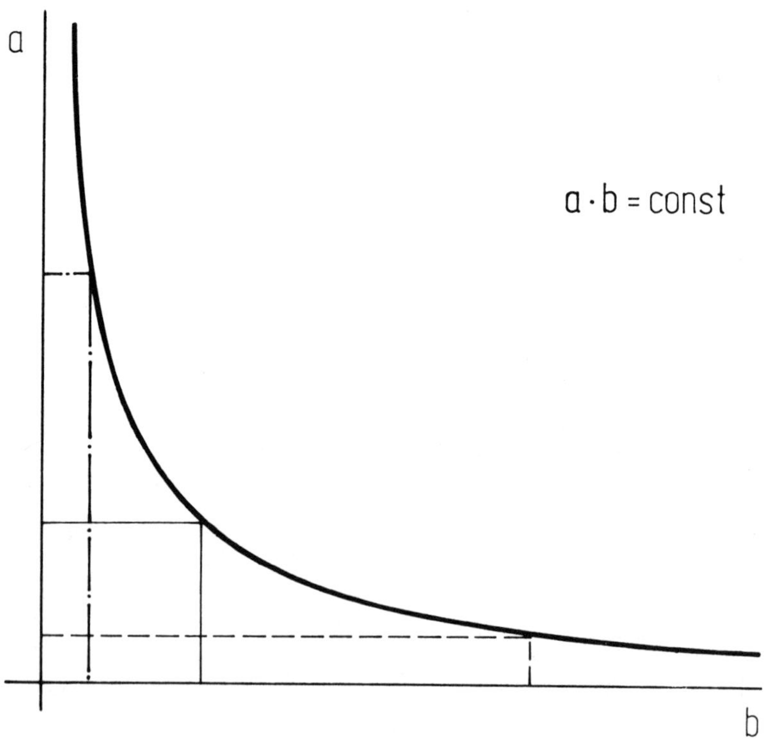

FIGURE 3. HYPERBOLIC FORMULA

(1)	Qualification	-	Competence
(2)	Facts	-	Views
(3)	Responsibility	-	Influence
(4)	Municipality	-	Region

Decision-making and advisory bodies tend to draw their competence from political powers and thus need to receive comprehensive information from qualified experts. Transportation generally is still intra-community financed, whilst transportation needs increasingly to extend beyond community borderlines. Citizen participation looks for influence without the burden of responsibility.

(5)	Function	-	Form
(6)	Serving masses	-	Serving marginal groups
(7)	Free movement for pedestrians	-	Transport for pedestrians
(8)	Preventing vulnerability	-	Mitigating vulnerability
(9)	Problem approach	-	Solution approach
(10)	Software approach	-	Hardware approach
(11)	Automation	-	Manpower
(12)	Centralisation	-	Decentralisation

Transport installations, designed to fashion and not to function, prove under heavy duty (e.g. the 1972 Olympics) to be inadequate and even dangerous. Special adjustments for better service to the handicapped must not combine the advantages aimed at with disadvantages to other users. Free movement in pedestrian precincts might be prejudicial to internal accessibility where public transport uses buses or trams. Where surface transport is to be introduced the dangers from and to other road-users have to be investigated thoroughly. New modes may offer interesting solutions, but not always for the problem in question. With automation and centralisation the information software might be of great significance for safety and reliability as well as for the relation of visible to non-visible manpower.

(13)	Varying demand	-	Non-flexible installations
(14)	Continuous flow	-	Intermittent operations
(15)	Renewing	-	Maintaining

(16) Implementing by contractors - Implementing within the organisation

(17) Components from mass production - Components from individual production

(18) Fare revenue - Non-fare revenue

Transportation needs vary by time of day, week and year, and call for flexibility in the capacity, frequency and occupancy factors which are likely to affect attractivity at times and locations of extremely high or low demand. Operating rolling stock of short-life expectancy allows one to follow up with new developments and to reduce to a minimum non-operative activities which normally have high running costs. So does employing contractors. All these policies are more likely to be maintained with buses embedded nationwide and worldwide into the motor industry. Rail transport components need to be assembled in smaller series; which will also be the case when initiating new modes. Passenger revenue will allow one to cover neither investment nor operation needs. Side-revenue from political as well as from commercial sources will increasingly be indispensable. Until the gap in these controversial issues is closed by means of substantial knowledge, the syndromes of 'Decision Evasion' have to be watched out for:

(1) Parochialism: "... certainly correct throughout the world, except here..."

(2) Revolving door: "... irrefutable but rejected..."

(3) New method: "... now let's try another method..."

(4) Precise calculations: "... we need the third digit after the decimal point..."

(5) Inactive activity: "... do something but leave things unchanged..."

To some extent the prototype syndrome of 'Human Attitude' may influence decisions as well. Any new element will gain special attention from passengers, personel and public; and will lose it as soon as it becomes part of everyday experience. Thus experience of a prototype cannot be fully representative of the future.

Industrial engineering has, therefore, to provide analytic numerical approaches leading from qualification to quantification, and aim at comprehensive integration of various interests to make all parties concerned better understand their mutual concerns. In the author's experience it is essential to define the relation between specific factors for incorporating experts' experiences in various fields and to allow calculations on alternative figures. Using graphs and formulae based on preferred numbers, as per ISO

$$\frac{\text{seats}}{\text{coach}} \cdot \frac{\text{coaches}}{\text{train}} = \frac{\text{seats}}{\text{train}}$$

$$\frac{\text{seats}}{\text{train}} \cdot \frac{\text{passengers}}{\text{seat}} = \frac{\text{passengers}}{\text{train}}$$

$$\frac{\text{passengers}}{\text{train}} \cdot \frac{\text{trains}}{\text{hour} \cdot \text{direction}} = \frac{\text{passengers}}{\text{hour} \cdot \text{direction}}$$

$$\frac{\text{passengers}}{\text{hour} \cdot \text{direction}} : \frac{\text{pass/hour} \cdot \text{direction}}{\text{pass/weekday} \cdot \text{direction}} = \frac{\text{passengers}}{\text{weekday} \cdot \text{direction}}$$

$$\frac{\text{passengers}}{\text{weekday} \cdot \text{direction}} : \frac{\text{pass/weekday} \cdot \text{direction}}{\text{pass/year} \cdot \text{direction}} = \frac{\text{passengers}}{\text{year} \cdot \text{direction}}$$

$$\frac{\text{passengers}}{\text{year} \cdot \text{direction}} : \frac{\text{passengers}}{\text{Inh} \cdot \text{year} \cdot \text{direction}} = \text{Inhabitants}$$

$$\text{Inhabitants} : \frac{\text{Inhabitants}}{\text{Area (ha)}} = \text{Area (ha)}$$

$$\text{Area (ha)} : \text{Length (m)} = \text{Width (m)}$$

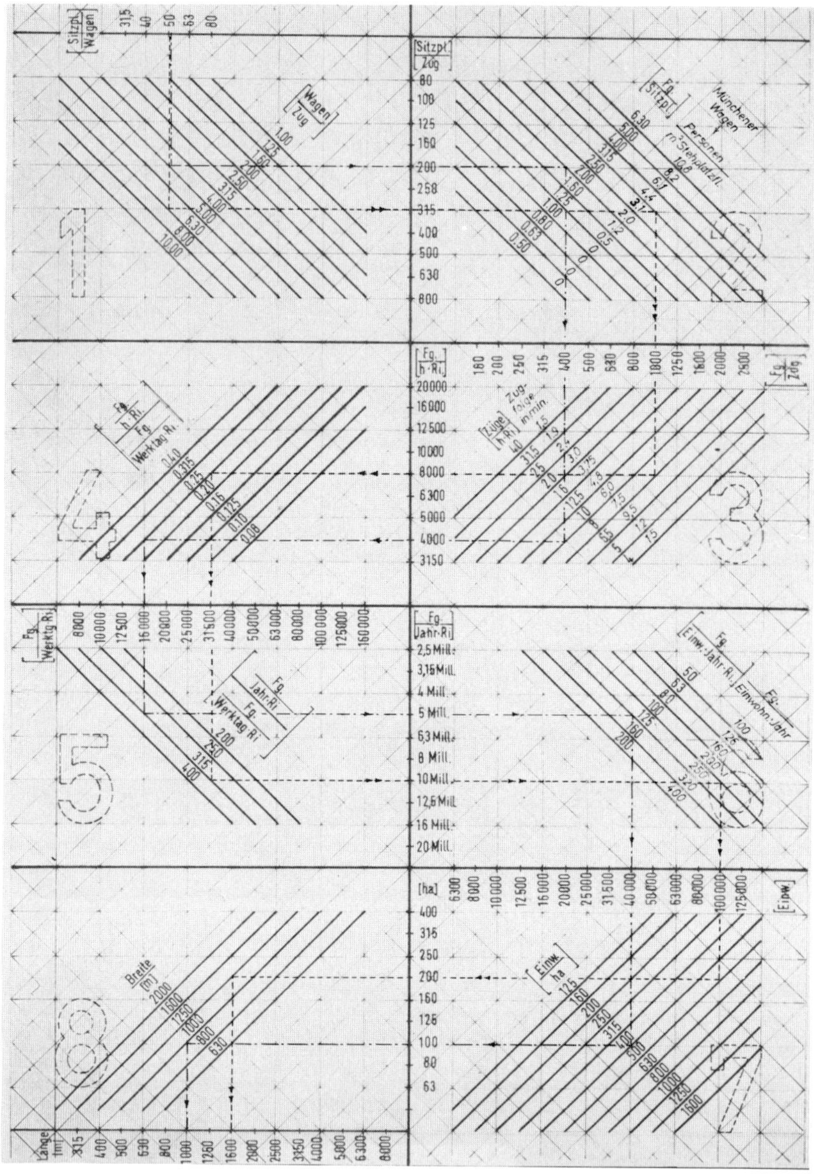

1. Seating capacity, train length. 2. Loading factor.
3. Train intervals, hourly service. 4. Maximum daily capacity.
5. Maximum annual capacity. 6. Frequency of journeys.
7. Catchment area - inhabitants/hectare. 8. Spatial extent.

FIGURE 5. TRANSPORT CAPACITY OF ONE RAILWAY SECTION

$\frac{\text{trains}}{\text{hour}} \cdot \frac{\text{hours}}{\text{day}} = \frac{\text{trains}}{\text{day}}$	trains daily circulated
$\frac{\text{trains}}{\text{day}} \cdot \frac{\text{days}}{\text{year}} = \frac{\text{trains}}{\text{year}}$	trains annually circulated
$\frac{\text{trains}}{\text{year}} \cdot \frac{\text{coacnes}}{\text{train}} = \frac{\text{coaches}}{\text{year}}$	coaches annually circulated
$\frac{\text{coaches}}{\text{year}} \cdot \text{kms} = \frac{\text{coach} \cdot \text{kms}}{\text{year}}$	annual volume of operations
$\frac{\text{coach} \cdot \text{kms}}{\text{year}} \cdot \frac{\text{DM}}{\text{coach} \cdot \text{km}} = \frac{\text{DM}}{\text{year}}$	annual operating expenses
$\frac{\text{DM}}{\text{year}} : \frac{\text{DM}}{\text{DM} \cdot \text{year}} = \text{DM}$	annual operations capitalized to be compared against investment capital

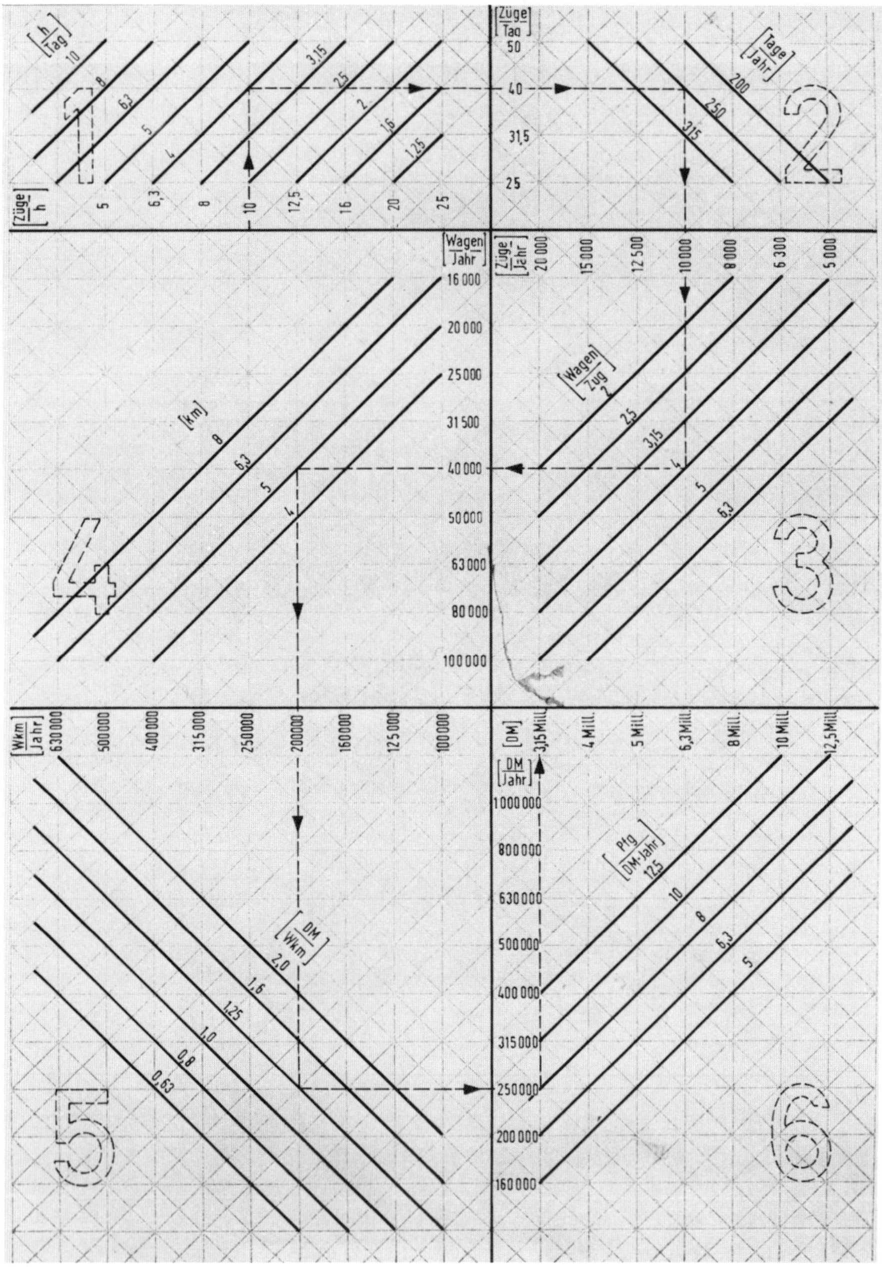

FIGURE 7. ECONOMIC ANALYSIS BY TRACK
(Rail capacity, track rationalisation, investment costs, annual outlay, overall and capital costs.)

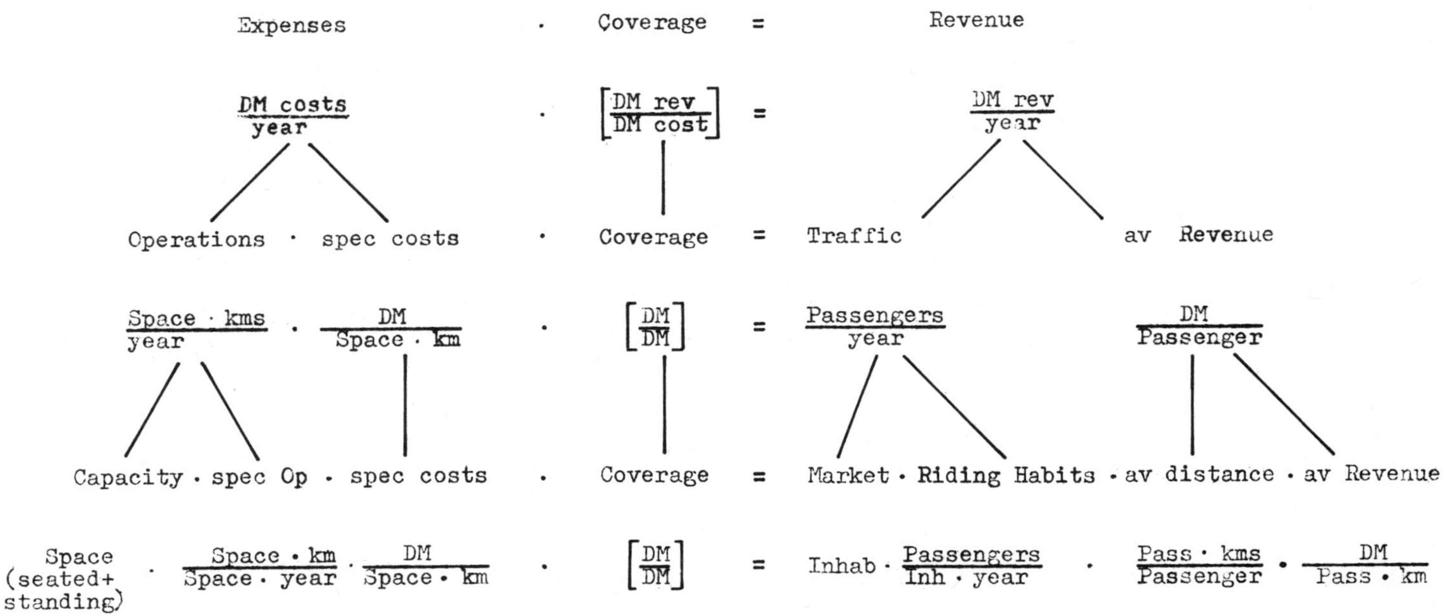

FIGURE 8. CALCULATION FORMULA : OPERATIONS

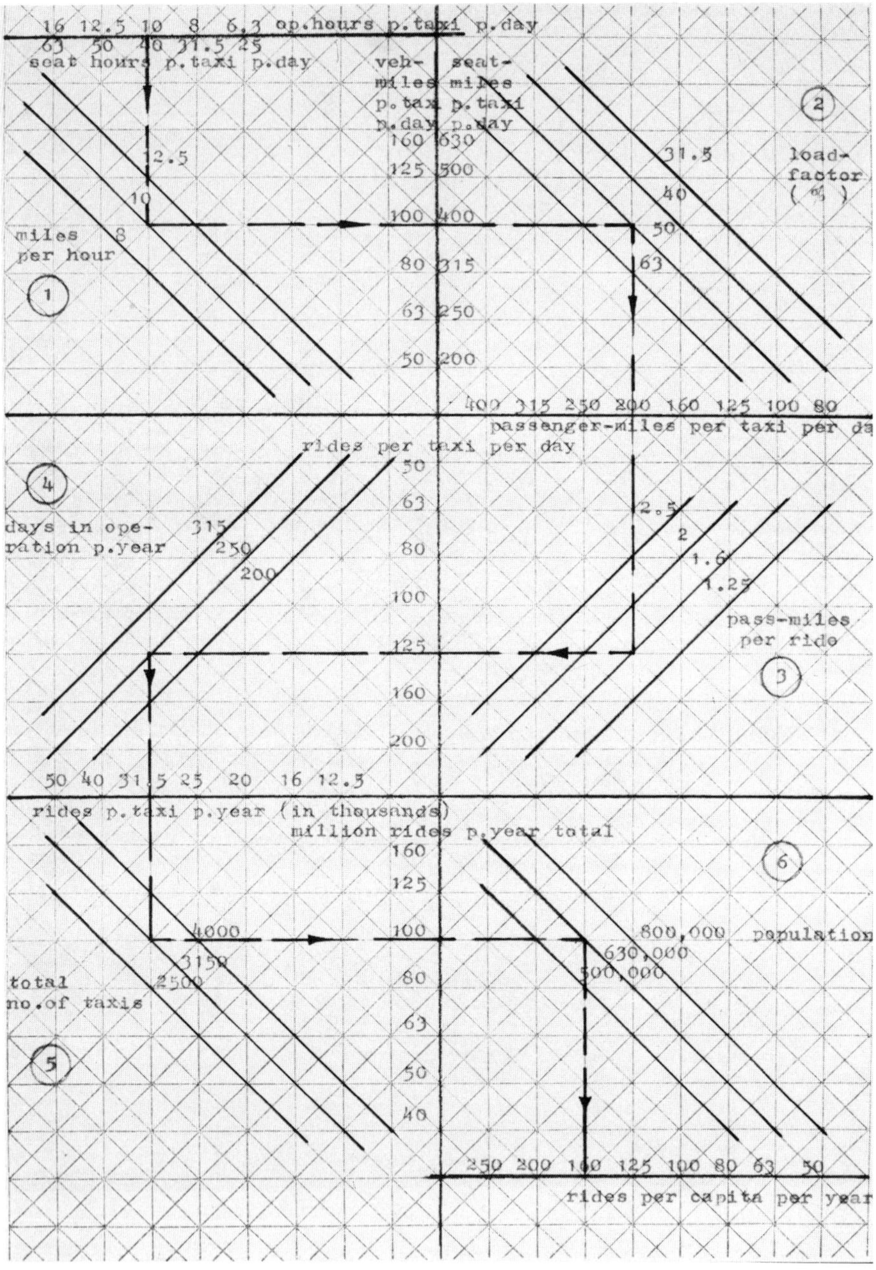

FIGURE 9. IBADAN CITY COUNCIL BUS SERVICE. ESTIMATED TRAFFIC BY TAXICABS

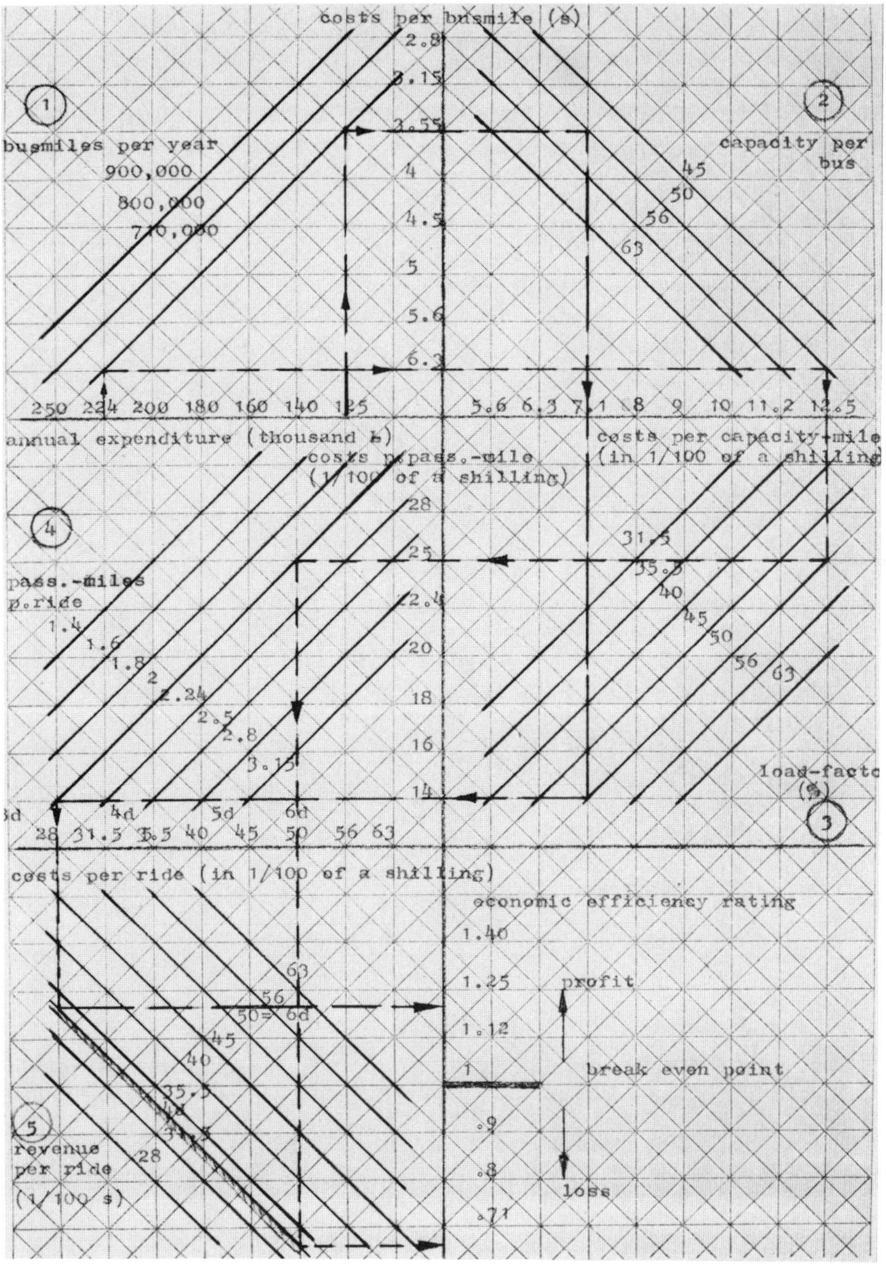

FIGURE 10. IBADAN CITY COUNCIL BUS SERVICE. COSTS PER RIDE, BREAK-EVEN POINT

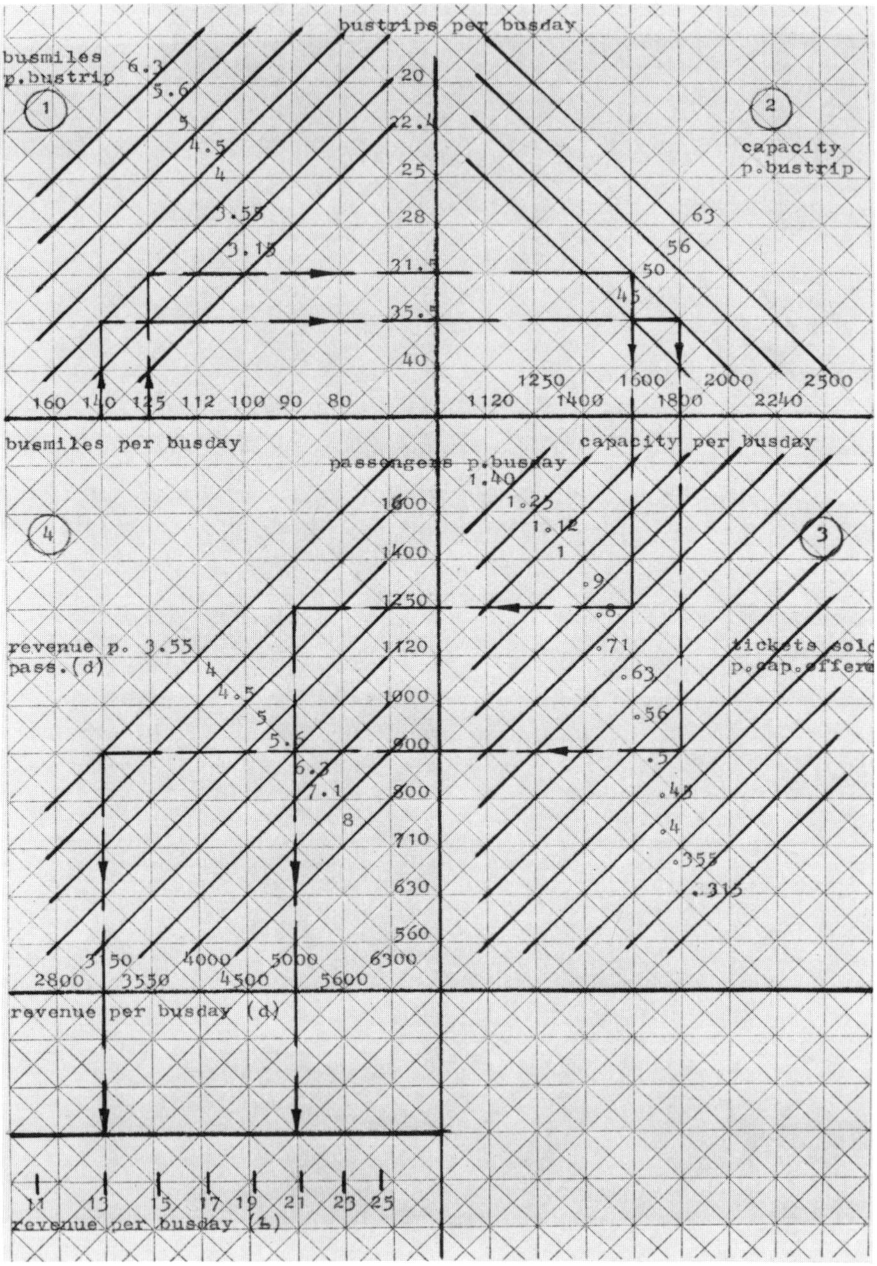

FIGURE 11. IBADAN CITY COUNCIL BUS SERVICE. PERFORMANCE AND REVENUE PER BUS DAY

Recommendation R 3 and German Standards DIN 323, the so-called AAI method* has proved simple and practical.

Planning formulae and graphs demonstrate interdependence among capacity, occupancy, frequency and peak factors and riding habits, population density and area size (Figures 4 and 5) and help to plan new lines in new settlement quarters. The calculation of a formula for operations against investment helps one to decide between working unnecessary services and constructing turn-back facilities instead (Figures 6 and 7). The calculation of a formula for operations (the graph is still to be developed) analyses basic cost factors in connection with revenue and expense readjustments (Figure 8). Checking taxi movements on the road might induce a better knowledge about the respective transport market (Figure 9). Operational characteristics set against commercial details might allow a rough but quick judgement about working at a fair or poor economy (Figures 10 and 11).

Analytic approaches might possibly be found for any multi-disciplinary problem and for its stages of planning, deciding, executing, implementing and monitoring. They are a successful means of supplying industrial engineering's three C's: Co-operation, Combination and Confidence.

---

*AAI stands for Analogous Analysis on Idealised figures or for Addis Ababa and Ibadan, where the method was developed and tested.

SECTION 5. HUMAN AND SOCIAL FACTORS IN TRANSPORTATION AND PLANNING

In the second half of this volume the human and social aspects of transportation which have often been touched on in the previous papers become the central focus of concern. All four papers in Section 5 argue in different ways that psychology and sociology have essential contributions to make to transportation planning.

Appleyard's paper is policy-oriented. Every transport decision is assumed to have a social impact. Traditional, technical variables need to be tied into their social context. Possible future policies for transportation are discussed in the light of how far they satisfy the needs and aspirations both of travellers and of those who are otherwise affected by transport systems. Attention is given to specifying priorities not only for meeting deficiencies in transportation, but also for building on its existing advantages. In addition to new policies, however, the basic concepts and values of transportation planning will need to be reformed. Transport should be considered as a user- and neighbour-oriented service, rather than a facility whose problems are defined by the supplier. Appleyard's final recommendations for a move to multi-disciplinary and open, participative planning are taken up in greater detail in subsequent papers.

The introduction of social policy as a major concern in transportation is complemented by Michon's review of the place of behavioural research and theory. Environmental psychology and behavioural modelling are beginning to meet the increasing demand for behavioural science theory and data in the field of transportation. The former has admittedly not yet found clearly delineated tasks when applied to traffic and transport problems. But it has the advantage of strong theoretical sources in main-stream general psychology. On the other hand, although behavioural modelling is a much more clearly defined approach, it lacks sound roots in behavioural theory. The two approaches should be seen as complementing one another.

The plea for inputs from psychology and sociology is repeated by Warman. But both he and Latchford and Pickering present the case from the less academic viewpoint of transportation consultants. He sees the role of the social scientist as one of mediator between the public and other professionals in the planning team. The programme and techniques that might be used are outlined in the context of their application to a pedestrianisation project.

Environmental assessments which compare predicted environmental impacts with the existing situation may sometimes need to be carried out without the relatively slow and expensive direct approaches to the public advocated by Warman. Aids are now being developed, as Latchford and Pickering explain, which enable technicians to measure and predict environmental factors rapidly and cheaply. They briefly describe some studies of the impacts of noise, visual intrusion and severance as they arise from road traffic, in cases where environmental assessment made an important and positive contribution to the search for a future land-use transportation strategy. We may expect such assessments to become routine features of transportation planning. Some more detailed examples of them are given in the next section.

TRANSPORTATION AS A SOCIAL ENVIRONMENT: CAN WE CHANGE A TRADITION?

Donald Appleyard

Department of City Planning and Landscape Architecture

University of California, Berkeley

Transportation planning inexorably tends to screen out the richness and relevance of everyday life in favour of a restricted set of given quantifiable variables which are seen as the essence of the transportation problem. These variables reduce transportation quality to a limited number of terms like 'accident rates', 'traffic volumes', 'ridership', 'costs' and 'decibels', while issues like comfort, privacy, amenity, politics and other less tangible, but often more important, variables are ignored or termed 'residuals'. These limited variables are used to set standards for design and then the 'building' syndrome sets in. Another set of technical variables such as 'right-of-way width', 'alignment', 'concrete mixes', 'curvatures', 'signaling' come into play, and we get even further away from the people whom transportation systems are meant to serve.

Transportation systems, once in operation, are primarily social environments. But their social, environmental and political meanings are suppressed in the 'technicalization' of the planning process. Transport systems have many more impacts on the quality of life than the few variables now being used acknowledge.

We not only have to make the intangibles tangible, we have to change the style of transportation thought, by tying variables already used into their social context. Every transport decision, whether it be about modal split, a right of way width or the height of a railing, has some social impact.

The fundamental questions to be asked in every transport decision, however small and 'technical' it may appear to be are: who will gain and who will lose? and which qualities of life will it encourage or discourage? Then, and only then, should we ask which systems, structures or policies can best serve these people and these values.

Such questions depend on some understanding of social life, human values and what the quality of life means. The philosophical nature of these questions makes them difficult to link to everyday decisions, but there have been efforts to operationalize these values into transport criteria (Appleyard & Okamoto, 1968). This paper will concentrate more on priorities and potential policies for transportation in the immediate future.

Two basic population groups are affected in transport systems- existing and potential travellers, existing and potential neighbours. Each group has urgent and essential needs and equally essential aspirations. Life is only worth living if both are satisfied. We shall therefore consider transport environments under four headings:

A1 Spreading transportation service to deprived population groups.

A2 Increasing the choice and quality of travel for the whole population.

B1 Reducing the negative impacts of transportation systems on the social and physical environment.

B2 Contributing to environmental quality through transportation planning.

SPREADING TRANSPORTATION SERVICE TO DEPRIVED POPULATION GROUPS

Transport planning has concentrated too much on the majority and on average travellers, neglecting those least able to cope with the modern city. The poor and the aged have become the subject of some attention in recent years. The most serious problem for the poor is lack of access. Suburbanizing industry has moved away from their inner city residential areas; infrequent, slow, radially-oriented transit service can take up to two-and-a-half hours to move between home and work place; they can only afford secondhand cars.

The problems of these groups reach into all aspects of transportation: the lack of a particular transit connection, long distances to be travelled daily under hot and crowded conditions, the fear of waiting at a bus stop, a feeling of 'lostness' in the transportation system, the height of steps on buses (Carp, 1970).

Some can be resolved by the provision of new systems. Much can be done by the modification of existing systems and changes in land-use patterns (Owen, 1972).

But there are other neglected groups besides the poor and aged: the physically handicapped and poor in health, children and teenagers in suburban areas who cannot drive cars, housewives who must drive all the time, strangers to the city, - especially non-native speaking immigrants and tourists - those who by personality live in fear, need privacy, or are esthetically sensitive, and those who travel in families or large groups. These and others may rightly demand equal travel rights in the coming years. We are still relatively ignorant of their transportation needs.

Suggested Policies

a. A Social Impact Report, similar to the environmental impact reports required by the U.S. National Environmental Policy Act 1969, should be required of every major transportation project. Such a report should include an assessment of whether the project will aid or further deprive various designated populations. It should be reviewed by representatives of deprived groups.

b. Ways of ensuring that there be a systematic concern for the 'left-outs' must be found. Reviews of the characteristic transportation needs and problems of all potentially deprived traveller groups should identify gaps in knowledge and develop criteria for the successful transportation of deprived groups. A survey we carried out in the Bay Area examined the travel perceptions and needs of low income, aged, teenagers, housewives and various handicapped groups (Carp & Appleyard, 1973). We found, for instance, that working women are more transportation-deprived than housewives. Such reviews should be summarized in reports on each deprived group, should be updated as new research findings come in, and should be used to evaluate future investments.

c. Transportation planners should acknowledge and cooperate more with the consumer and citizen organizations who represent various deprived groups, encouraging them to identify their transportation needs and participate in system management and design.

d. Systematic efforts to identify transportation programmes, components and systems that would help the poor, handicapped, and otherwise deprived should be undertaken and adequately funded. While the problems of the deprived are multifarious, the solutions to their problems are likely to be even more varied, ranging from the redesign of buses to accommodate the handicapped to guidelines for sign systems, from various forms of fare subsidy to priority bus lanes and dual-mode personal transit systems.

INCREASING THE CHOICE AND QUALITY OF TRAVEL FOR ALL

Transportation planning has habitually taken a narrow view of what travel is all about. The expressed concerns of transportation planning have been geared to moving the traveller from origin to destination in the cheapest and most convenient way. The numbers of travellers on any system have been taken as the sole measure of its success.

Transportation surveys conduct thousands of interviews, in order to record travel patterns. Typically, no questions on traveller satisfaction are asked. It is assumed. So long as he travels on the system the traveller has done his bit. Yet surveys which ask questions about satisfaction find that people are frequently dissatisfied with the systems they must travel on.

Commuter travel is seldom an enjoyable activity. It can be unsafe, frightening, stressful, tiring, uncomfortable, disorienting, monotonous, dirty, overstimulating, trivial, dominating, depressing, and more. I can think of only a few urban freeways, and almost no transit systems or arterial streets in any country which possess pleasurable qualities during rush-hour. A few exotic forms of transport like ferries, cable-cars, and minitrains are still enjoyable. Urban freeways are nerve-racking, subways are gloomy, buses are noisy and jerky, city streets are crowded and unsafe. No profound assessment of the quality of a metropolitan transportation system has to my knowledge ever been carried out. It is clear, however, that we have a long way to go before the present transportation systems are safe, convenient, comfortable, clean, pleasant, interesting, civilized forms of travel.

Travel plays many roles beyond moving a person from one place to another, although we know little about those roles. Travel is an escape from the world of home and work; travel can be a rest from more disciplined activities; travel can be the time to meet people or the time for privacy. It can be a time to plan the day's work or a time to sleep it off. Travel can be an exciting adventure or a relaxing hobby. It can be an education. For many, the commute is the only time they see other people or the city or have contact with the natural environment. On one U.S. commuter system people can study for professional degrees. But little research has been carried out on these potential values of travel.

Travel can also make a contribution to the problems and aspirations of a democratic society. Equality of travel opportunity should clearly be a transportation goal, but there are other societal functions that transportation systems might perform. Transportation systems can affect the physical and perhaps mental health of the population. The diminution of walking is an obvious threat to physical health. Transportation can perform a social function, bringing together or further fragmenting the various groups in society. Before the automobile, the street was a meeting place. It was common ground and neutral territory, a place where major events were celebrated. Since the automobile, friendships have spread, but they have become more selective. In the automobile society, people live more with those of similar interest and status. The private and mobile nature of the automobile allows people to escape meeting other social groups - an isolation that emphasizes, at its worst, intolerance or paranoia about other social groups, at its best, a posture of unreal romanticism. It would be revealing to compare the social structure of cities which have dominantly transit and dominantly automobile systems.

The educational value of transportation systems has been neglected, yet the degree to which travellers comprehend the structure and meaning of cities and society, and to which transport systems encourage democratic values and responses - such as tolerance, freedom, and feelings of personal individuality may be more significant than we realize. The impersonality of relations between people travelling in automobiles is symptomatic of the growing impersonality of our society.

Finally, in this age of energy shortage and inflation we must seriously reconsider the values of mobility and choice, which have been the gospel and justification for transportation projects over the past twenty years. While mobility allows us to have conferences like this one, mobility has also contributed to the eclipse of local communities, the destruction of many beautiful places, and the possibility of global war. Do we need it so much? Just as some are arguing for a return to smaller-scale enterprise and society, so there are many arguments for restricting mobility on a selective basis. This, of course, will be easier to contemplate for us who already have the choice, than for those who are still waiting to purchase their first automobile or fly their first aeroplane.

Suggested Policies

a. Criteria other than economics, accessibility, function, safety, or ridership must be included in system design and evaluation. The identification of travel values, many of them latent or emergent, is the task of more careful research than the usual attitudinal survey. Research efforts to uncover the whole array of present and emergent traveller values and their incidence among various population groups should be launched. These could then be converted into quality criteria for transportation systems. Such 'civilized' criteria should also be built into the environmental and social evaluations of transportation proposals (Appleyard & Okamoto, 1968).

b. The restoration of neglected systems should be seriously considered. Systems that have not met efficiency criteria have fallen by the wayside in our headlong rush towards mobility. For example, in the interests of health alone, national needs for walking and cycling systems should be reassessed.

The lot of the pedestrian in the American city is not an enviable one. Yet a recent study in San Francisco found that between two-thirds and three-quarters of those over sixty must walk to get their groceries. By night the pedestrian travels at his peril along poorly lit streets. By day, in suburban areas, he is suspect. He trips over decaying sidewalks, he risks his life to cross the street and parking lot, he picks his way through the concrete caverns of parking garages, he is attacked by fumes and noise, he freezes at bus stops, he bakes on asphalt wastes, there are no benches on which to rest. It is more dangerous for the children, for they are less aware of the hazards. And still the environment continues to decline, the sidewalks are narrowed or eliminated to accommodate more traffic, garages and parking lots grow larger, trees are cut down.

Outside the relatively small areas of 'pedestrianization', the European city dweller fares little better. In Rome's narrow streets with no sidewalks, the pedestrian must creep along between the parked cars and fast-moving traffic. The development of safe, well-lit, continuously surfaced, ample, pleasant and agreeable pedestrian path systems with adequate resting places and assistance vehicles should be a much more important transportation goal than it is at present. Turning one commercial street over to pedestrians in a whole city is hardly enough.

Cycling, a transportation system which is available to youngsters of all ages who cannot drive cars, is not yet being considered as a serious commuter transport system in the United States. Safe and continuous cycling networks can be created, especially in cities which have river and valley systems, and in new extensions to cities. But they will cost more than most cities contemplate. If there were more continuous auto-free path systems, other forms of recreational travel such as horse-back riding (as in San Diego) and cross-city skiing (as occurs in Oslo) could be contemplated.

Other older systems - tramcars, cable cars, jitneys are ripe for resurrection, though in many cities at this moment they are about to be eliminated. Where cities are on water, ferry systems might become as popular again as they are now in San Francisco and Seattle.

c. The rehabilitation of transportation systems to create more dignified travel environments should be another priority. Perhaps the most seriously neglected system in the American city is that of the arterial streets. These streets have been in the hands of impoverished city departments for years, and everywhere they have created a blight on travel and on the urban environment. They are poorly maintained, lined with indiscriminate curb cuts, multifarious and disjunctive uses, parking lots, signs, and disorderly arrays of street furniture. They have been the prime target of conservationists. Although some have a brash excitement of their own, most are dull, tedious, depressing.

The same can be said for many of the urban freeways constructed over the last ten years. During the massive freeway programs of the 1950's and '60's, many mistakes were made, but few attempts to remedy them have come forth. Highway designers rarely go back over their old projects. They have been too preoccupied with building new ones. A freeway renewal programme would evaluate the quality of existing urban freeways, with the intent of eliminating danger spots, improving orientation, and the appearance of signs and street furniture, of landscapping, guard-rails, and side walls. Views which have been unwittingly blocked, for instance on the older bridges, should be opened up.

Neither should we forget country roads. The U.S. Scenic Roads programme was oriented to the building of new scenic roads, rather than to the care of existing roads, - a likely way of destroying the very scenic qualities that were to be revealed.

The rehabilitation of subway systems, many of which in the U.S. are a national disgrace, has scarcely begun. It has been left to enterprising youngsters ('vandals' they are called) to decorate the drab subway trains and stations of New York and Philadelphia.

d. New technologies which offer multi-functional qualities of service are of course still worth exploring. Demand-actuated mini-bus systems, people-movers of various kinds, designed to reduce absolute reliance on automobile travel, look promising. Responsive systems that allow a degree of personal control, choice and convenience, without deleterious impacts on the environment, should be encouraged. But I emphasize the renovation and resurrection of old systems, because too much attention is paid to new systems, too little to that which we have. The 'build-at-all-costs' syndrome is one of the main reasons why transportation planning is unpopular today.

REDUCING THE NEGATIVE IMPACTS OF TRANSPORTATION SYSTEMS

During the last ten years protests from the public have halted transportation projects in many countries. Transportation planners at first were defensive and bewildered. Then conferences were called, and research projects were financed. Concerns broadened from the narrow right-of-way to the transportation-corridor, and joint-design teams were initiated. Later, some cities like Boston embraced massive programmes of citizen participation.

Since the passing of the National Environmental Policy Act in 1969, the environmental impact of transportation systems has become a national concern. Environmental impacts have become a matter for more or less systematic assessment and a whole new industry of impact assessors has sprung into existence. While this has made transportation planners more sensitive to environmental issues, the social impacts have often been overridden or forgotten. The natural environment has more powerful defenders than have low-income neighbourhoods, and in many cases it is easier to assess impacts on the natural environment than the social environment. The scars of erosion, the demise of a rare species or the pollution of streams are easier to discern than disruptions in community life.

There has also been a tendency to concentrate on the impacts of large projects like freeways and transit systems while the incremental but vastly more pervasive effects of traffic on city streets have been neglected. In the San Francisco metropolitan area we have found that 40 to 55 percent of the population sample were annoyed at the noise produced by local traffic, whereas only 24 percent were disturbed by noise from freeways and major roads (Carp et al, 1975). In a current study we are discovering just how deeply traffic has affected neighbourhood life. Besides the more obvious dangers to children and old people of noise, fumes, dirt and vibration, we find that traffic forces a withdrawal from the street, a reduction in the sense of home territory and therefore responsibility for what goes on in the streets. Local friendships are reduced, children must be accompanied to school, forbidden to go beyond the block, nearly all activities cease. Our residential streets are more dead than alive (Appleyard et al, 1975).

Transportation agencies should consider not only social impacts such as losses in housing, jobs, job opportunities, but also various dangers and discomforts from noise, fumes, dirt, visual intrusion, parking, friendship patterns, neighbourhood functioning, enforced adaptations, and changes in density, heterogeneity, and mobility.

Neglected, too, are the more indirect secondary and system impacts of transportation facilities. The primary impacts of transportation channels and facilities are often immediate, visible and identifiable. The secondary impacts - the traffic and development generated by their presence - and system-wide impacts - on the form and growth of metropolitan areas over the long term - are complex and elusive, but usually more significant. For instance, those who live around the BART stations in San Francisco are more concerned about future development which will raise rents or evict them from their homes than about the environmental impact of the system itself.

Suggested Policies

a. Proper attention to social and environmental factors requires radical changes in the whole process of transportation planning. These factors should be considered right from the beginning of the planning process - from the system level down - if they are to be adequately recognized. They cannot be considered in the last phases of the process, as the 'frosting' on the engineering plans. This requires, for example, working out the detailed physical and social implications of regional transport decisions before they are fixed. The Boston transportation study proposals were displayed to neighbourhood groups at scales at which they could see how future routes would affect individual houses and neighbourhoods. In the past communities were unaware of the character of such projects until they were built.

b. The 'opening-up' of the planning process to include representatives of impacted populations and environments is the most direct way of understanding the impacts of existing transportation systems on everyday life. Many ways of involving citizens in the planning process are now being experimented with in North America, the Boston and Toronto transportation studies being good examples.

c. But citizen participation alone is not the solution to all social and environmental problems. Substantive knowledge in the methods of social and environmental assessment is needed. A system of guidelines for impact studies should be developed (Dickert & Domeny, 1973). Guidelines might include:

i. a description of community, social and environmental values, as they are held by different population groups.

ii. a checklist of possible social, psychological, physical and natural impacts and how they could impact different groups.

iii. the setting of social and environmental performance standards which transportation facilities should meet.

iv. methods of obtaining social and environmental information by which to assess and predict the consequences of impacts, including data sources, field survey methods and sample questionnaires.

v. methods of recording, simulating and communicating such information for public understanding.

vi. the identification of transportation system components and actions which create impacts.

vii. methods of controlling and mitigating such impacts.

These guidelines should emphasize specificity and should aim for quantifiable indicators of environmental and social quality. But they must be tied back to people, to which groups gain and which groups lose. Most cost-benefit analyses shy away from identifying these social and political consequences of projects. It is too tempting only to talk of the gains. Even at the recent OECD Conference on "Better Towns with Less Traffic" (O.E.C.D., 1975), the possible social costs of traffic restraint schemes were suppressed in the general enthusiasm for a new cause. Guidelines should be couched at different levels of sophistication, and should be well-illustrated and comprehensible to the professional and the layman.

d. Efforts should be made to plan the primary impact zone of major routes more comprehensively and sufficient allocations to suport adequate relocation and rehabilitation in those areas should be mandatory. Such plans must be worked out in close collaboration with city and local agencies and with local community groups who have control of funding.

CONTRIBUTING TO THE QUALITY OF THE SURROUNDING ENVIRONMENT

'Impact', the most commonly used term to describe the relation between transportation and its environment, implies a negative consequence. Transportation systems should be viewed also as potential benefactors to their surroundings. A transportation system should be like a Christmas tree on which many good things can be hung. Transportation facilities can be designed to include other functions than the travel function. Air rights developments and broader corridor planning in the United States have allowed housing, schools, public buildings and parks to be built over noxious freeways. Linear parks have been designed along the BART elevated tracks in the Bay Area and streets over the covered parts of the route have been landscaped with benches, fountains and other amenities. Tokyo has a shopping street under one of its freeways. By spreading investment in these ways, there is a chance that major transportation routes could change the tradition that has created corridors of blight through our cities.

Suggested Policies

a. Multi-modal corridor planning in close collaboration with cities and citizen groups should be initiated in all areas where transportation facilities are to be built. Multi-modal planning would start with the transportation needs of the corridor without any preconceptions about modal solutions. Every modal alternative including the 'no-construction' alternative would be considered.

b. The use of air rights both under and over the right-of-way, for parallel transportation systems, for viewpoints, and for various uses that can serve the surrounding community should be encouraged to prevent blighted transportation corridors from forming. These improvements will depend a great deal on design ingenuity.

c. Transportation agencies should participate in broader regional and metropolitan planning. Such agencies should include regional highway, transit, airport, and port agencies. They should be oriented towards region-wide transportation planning which optimizes the social, environmental and land-use patterns of metropolitan areas.

CONCLUSIONS

A revolution is taking place in transportation planning due to public disenchantment with traditional planning methods and the closed nature of the old transport planning process. The revolution demands not merely the adoption of new policies, but a restructuring of the basic concepts and values of transportation planning; ways of thinking that it may be impossible for many to adopt without re-education.

1. Transportation must be seen as a <u>service</u> rather than a <u>facility</u>. Too many in transportation planning are 'locked-in' conceptually to a system or a technology - whether it be freeway, bus, transit or automated highway - as the solution to the transportation problem. Too many have been trained only for the planning and construction of one product. Their viewpoint dominates transportation planning. Yet transportation should be a public service. The criteria for the success of transportation system should be <u>user</u>- and <u>neighbour</u>-oriented. Success should be measured in degrees of satisfaction and numbers satisfied rather than in amount built. There should be as much emphasis on management operations and incremental improvement as on the invention of new

technological systems.

2. Transportation systems should be seen as complementary rather than competitive. Yet transportation planning is characterized by rival systems aiming for permanence and/or profit. It is depressing and frustrating to find the highway, BART, and bus transit agencies in the Bay Area spurred on by press and public vying for ridership without any view of the overall public benefit. Multi-modal metropolitan and local transportation agencies appear to be essential to the resolution of such problems.

3. Transportation planning must be multi-functional and multi-disciplinary. Transportation decisions have multiple and simultaneous effects. A channel designed only for safety and traffic capacity may destroy a neighbourhood or create an eyesore. Good, safe street lighting can be a source of uncomfortable glare for neighbouring residents, a curve designed for 40 m.p.h. may take more property than one designed for 30 m.p.h. The clearing of trees from around the right-of-way in the cause of safety makes highway travel arid and monotonous. The engineering profession has a style of operating that focusses only on priority problems, often scorning other values. It performs extremely well in the planning and construction of transportation facilities and in the more quantifiable indicators of transportation success. Social values in a pluralistic society, environmental values based on the human psyche, or the natural ecology require a more complex, subtle, less precise view of the world. Transportation agencies have only begun to consider these factors.

4. The transportation planning process has to be more open to public participation than the present closed professional system of planning. The myth that we are value-free technicians able to make decisions on behalf of the public has surely been exposed by now. Every transportation standard or decision involves values and usually benefits some of the public while hurting others. Every decision is therefore a social and political decision - even if it be so apparently technical as the design speed of a highway or the creation of a one-way street system.

The transportation establishment has talked a great deal about public participation in transportation planning, and some progress has been made. But there is all the difference in the world between token participation aimed at selling the public a product, and a genuine effort to find out the public's needs and desires. Allowing the public access to the information on which decisions are based and participation in decisions, even to the point of veto, involves a sacrifice of power that cannot be made with good feelings unless the public is genuinely viewed as the clientele of the transportation agency.

Public participation is the most eagerly discussed issue in U.S. planning at this time. Debates on the nature and composition of citizen committees; or even on whether to have committees at all; discussions on how and when to have public hearings, on how much of a veto to give citizens groups and on the representativeness of citizens groups are intense. Yet professional dominance of planning seems inexorable, given the erratic nature of citizen interest and their understandable apathy when faced with obscure remote decisions which do not affect their own neighbourhoods. Planners often fail to realize that they have the responsibility of selflessly catalyzing citizen involvement.

The need for changes in the planning process suggest a number of changes in the transport planning profession.

1. A major re-education programme for engineers, planners, and administrators in transportation agencies should be undertaken. In-house training programmes to educate all personnel in environmental and social matters are a priority. Such a programme is presently being attempted by the California Transportation Agency (formerly the State Division of Highways) with help from the University of California. It is not easy. A sabbatical system whereby every seven years personnel could take a leave of absence for refresher courses might also be instituted. This could allow more upward mobility within agencies, since returnees could be moved up to different jobs and others would take their place. Such a programme would also require that universities and colleges reorient their academic or extension programmes towards the older professional.

2. The hiring of environmental and social professionals, environmental planners, landscape planners, urban designers, sociologists, community organizers, political scientists, psychologists, ecologists and others as full-time staff members in transportation agencies should be actively pursued.

3. A closer collaboration between schools of engineering, planning, architecture, environmental design, natural resources and the social sciences should begin. Service courses in environmental and social planning, science and design for engineering students should be provided, and they should be encouraged to take them. The new emphasis on the behavioural and natural sciences in the design and planning fields should spread to transportation engineering. Such programmes should also encourage those related professions and disciplines to become involved in transportation problems.

It is not that other professions and disciplines will be any better at transportation planning than those traditionally in the field. Every discipline suffers from isolated specialization, a phenomenon that is destructive to the quality of life in our cities.

REFERENCES

Appleyard,D., Gerson, M.S., & Lintell,M. (1975) Living on Streets: the effects of traffic on street life. Unpublished report to the U.S. Department of Transportation. Institute of Urban and Regional Development. University of California, Berkeley.

Appleyard,D., & Okamoto,R. (1968) Environmental criteria for ideal transportation systems. In Barton-Aschman Associates Guidelines for Future Transportation Systems, for U.S. Department of Housing and Urban Development.

Carp,F.M. (1970) Correlates of mobility among retired persons. E.D.R.A. Conference Proceedings. Pittsburgh.

Carp,F.M., & Appleyard,D. (1973) Pre-Bart traveller attitudes and perceptions. Institute of Urban and Regional Development. University of California, Berkeley.

Carp,F.M., Appleyard,D., & Zawadski,R. (1975) Noise in the urban residential environment. Unpublished paper.

Dickert,T., & Domeny,K. (1973) Improving the Environmental Impact Assessment Process. University of California Press.

O.E.C.D. (1975) Better Towns with Less Traffic. Conference Proceedings. Paris.

Owen,W. (1972) The Accessible City. Washington: The Brookings Institution.

THE MUTUAL IMPACTS OF TRANSPORTATION AND HUMAN BEHAVIOUR

John A. Michon

Sub-faculty of Psychology

University of Groningen

Recent years have seen a considerable change in opinions and actions pertinent to transportation and traffic. As a result there is a considerable movement which may lead eventually to transportation policies that are based on social and psychological arguments pertinent to the user, rather than on the usual technological starting points. It is gradually becoming acceptable to admit that the development of transportation systems should not be guided by its own intrinsic possibilities, but by the needs and values of individuals and society.

In this light it is not surprising that by implication a heavy responsibility is being laid on the shoulders of the behavioural sciences. It is not difficult to quote long lists of Transport Ministers, high and low ranking civil servants, journalists, politicians and others, who subscribe to the idea that:

> "Our overall transportation planning process will be broadened to encompass even more of the community goals and objectives, and to bring citizens into the process, if means can be found to make their contributions useful and effective. (.......) Accordingly we will shift our emphasis toward the intangible factors related to the humanities". (Turner, 1970)

At the same time many citizens are convinced that "the future is too important to leave it in the hands of the politicians" (NRC, 1974).

All such remarks - and there are many - point to the important role society is trying to impose on the Sciences of Man as a contributor to more responsible and humane transportation planning.

Of course there is the legitimate question: Can the behavioural and social sciences meet this demand, in particular in view of the present doubts about transportation planning practice? In recent years decision-makers and users have shown an increasing dissatisfaction with conventional transportation models. The doubts expressed by many with respect to the usefulness of the Integral Traffic and Transportation Study carried out in the Netherlands a few years ago (NEI, 1972) could serve as an example; but many other countries could provide their own, analogous cases (U.K. : House of Commons, 1972; U.S.A. : Binder, 1973). Some authors, in fact, are afraid that "Unless models can soon be developed that can be demonstrated to assist decision-making to a noticeable extent, the opportunity to regain respect for travel demand modelling may be lost for all time" (Stopher & Meyburg, 1974, p. 3). Some - presumably psychologists - are highly optimistic about the potentiality of behavioural science in this field (e.g. Michaels, 1974, p. 53). Conversely there are sceptics - presumably engineers (e.g. Shuldiner, 1974, p. 45). Apparently there is little unanimity about the value you get when you hire a sociologist or a psychologist for your transportation planning team. Even though it may be difficult, or even impossible to judge the value of what you will get, it is quite well possible to outline what you will get; and this I want to do in rough outline.

The slow penetration of the behavioural sciences into the area of transportation and traffic appears to be taking place along two rather different routes. One is emerging from the psychological sedimentation of the growing concern for the relation between man and his environment: environmental psychology. The second is a derivative of traffic systems analysis and consumer psychology and has become known as disaggregate behavioural demand modelling. While the two should (and could) be intimately connected, their development has followed quite distinct lines.

ENVIRONMENTAL PSYCHOLOGY

Environmental psychology is one of the responses of psychology to the growing concern of society for its habitat. Of course, psychology has always dealt with the interaction between the human organism and its environment, but only since some forty years do we find systematic study of the way man performs integrated tasks under realistic circumstances. (For an introduction and many recent references, see Ittelson et al., 1974.)

Environmental psychology still lacks a solid body of theory, and in some respects quite deliberately so; but it is, on the other hand, able to draw on the large body of perceptual, cognitive, and motivational theory and fact*. In some - rare - cases it does so. I shall briefly discuss some of the more plausible attempts to throw light on the problems of human 'life-space' in general, and transportation and urban life in particular. Each is related to a quite different type of psychological theory, reflecting the multiformity of contemporary psychology. At present any attempt at a unified theory would be futile.

Information Processing

Man can be considered as an information-processing system that interacts adaptively with its environment. As such, his capacities are limited; in fact, estimates are that the rate at which information is presented is of the order of 10,000 times that at which the human brain is capable of handling it. Adaptive interaction with the environment is possible by means of processes in which hypotheses about the real world are tested by computations carried out on the internal representations of that world (Miller et al., 1960; Newell & Simon, 1972).

Such computations must necessarily be limited, in depth and extent, and the allocation of this limited capacity is the core problem of our attentional mechanisms. Performance is usually best if the task to be performed is of intermediate load. Underload may have adverse effects on performance, such as loss of vigilance; but overload is more serious. In the first place the subject will fend off the load by selective filtering and an increasing amount of low level analysis of incoming information. At the same time irritation will increase, and attention will become unstable (Kahneman, 1973). If this condition lasts the so-called 'general adaptation syndrome' (GAS) will become manifest. After a transient state of alarm (high arousal), a state of resistance will develop, with feelings of depressiveness and psychosomatic symptoms. A prolonged period of persistent stress will be followed by a state of mental exhaustion (Selye, 1950).

This view of man as an information-processing system has found wide application in the area of road-user research - in particular in relation to the perception, orientation and performance

---

*This is not necessarily so in the case of "behavioural modelling" as it developed in connection with disaggregate traffic models. The format of data and theoretical relations in that case must conform with the requirements of traffic systems analysis, rather than with psychological ways of thinking.

of drivers. There it does lead to well-defined and quantitative results (summarized in e.g.: SWOV, 1971; Forbes, 1972). We may conclude that this area of psychology has thoroughly established itself in traffic operations. A special field of research which has somewhat more general implications for the transportation process is formed by the studies on "cognitive mapping". The work of Lynch (1960) has had an extensive follow-up (see e.g. Pailhous, 1970; or Downs & Stea, 1973). I think this is a very important and promising entry for the behavioural sciences into the area of traffic and transportation studies. It is discussed elsewhere in this volume by Lee & Tagg (1976).

Information, Stress, and Social Behaviour

The information-processing view has been put forward by Milgram (1970) in relation to some other aspects of the transport process. In his view the urban environment offers such an overwhelming amount of information - not only perceptually, but also by way of emotional stress factors and pollution - that the city-dweller runs the risk of permanent mental overload. As a result he will, sooner or later, start showing symptoms of defense. In their milder form these will be restricted to filtering the flood of information, for instance by greater selectivity in social contacts or distrust of strangers; but in its more extreme forms it may among other things lead to the decay of social responsibility. This theory certainly has the attractiveness of being "plausible", but it is based on precious little empirical data, and it may be too strictly representative of the big North American city. In many Oriental and European cities the urban processes seem to show quite different characteristics.

It seems that we need many more data about the size and extent of contacts between strangers and residents in towns of various sizes. Appleyard and Lintell (1971) have performed an interesting analysis of local interaction patterns before and after relatively big changes in traffic-density in a particular street. Increasing traffic results in a sharp decline of "across the street" contacts and a compensatory increase in "this side" contacts. The latter tend to show a rather transient character, and apparently the final result is a reduction in the felt need for contact.

At the descriptive level such information may help policy decisions about the seriousness of severance of social communication. However, theoretically one does want to make a further explanatory step, and preferably one which enables us to quantify the interrelations between the explanatory concepts as time-varying functions. In the case of the example I discussed, we seem to have such a framework in a theory of interaction in social groups put forward by Homans (1950), and given a mathematical basis by Simon (1957), and Coleman (1963).

The theory describes the interrelation between four variables: the intensity of contact (interaction) between group members; the level of friendliness among the members; the amount of activity carried on by members in the group; and the amount of external force exerted on the group - in our case represented by the traffic density. Simon (1957) translated the qualitative assumptions regarding the relation between those variables in terms of differential equations.

To my knowledge this model has not been used in the context of the effect of physical constraints on the patterns of activity in residential neighbourhoods. However, the approach may turn out to be fruitful even if relatively few empirical data are available. One of its positive points is its dynamic character. The second advantage is the unequivocality of the relations between the theoretical constructs in the model.

Motivation and Scaling

In a recent study Mehrabian and Russell (1974) have tried to describe behaviour motivation in the living environment in a theoretically consistent way. Their basic concept is the "approach vs. avoidance tendency" which people show with respect to objects and situations in their environment. The interaction between environmental information and personality characteristics (motives) determines the level of three "primary emotional responses": pleasure, activation and dominance, which in turn determine the observable approach-avoidance behaviour. The special merit of the line followed by Mehrabian and Russell is that they obtain a quantification of the primary emotional responses by means of scaling procedures and then make a sucessful attempt to relate these results to their model, describing the observed approach-avoidance behaviour. It would be feasible to express mode-choice or route-assignment in terms of approach-avoidance tendencies.

Conditioning of Behaviour

Behaviourism has been an important trend in psychology. It has left us with a system of "behavioural technology", based on the theories of Skinner (1972), which may have implications for social policies that we do not yet fully appreciate. Skinner's "operant conditioning" technique starts from the observation that a desirable consequence of a behavioural activity will reinforce the tendency toward that activity, while a negative outcome (e.g. pain, fine) will reduce that tendency. Manipulation of positive and negative consequences provide the opportunity to manipulate behaviour. By this means - Skinner believes - it will be possible to bring about fundamental changes in society. Apart from criticism regarding the ethical implications, there are grave doubts as to the applicability of the technique on a global scale (Carpenter, 1974; Brewer, 1974).

In a more limited context, however, such as traffic, the technique seems a forceful argument for those who feel that one should expect more effects on behaviour from situational modifications than from persuasive techniques or instruction. This has been the subject of a discussion by Klein (1971) about the relations between traffic and asocial behaviour. In Klein's opinion the present cultural and educational system over-reinforces behaviour patterns that we might call "brave", "sporty", "strong", or even straightforwardly "aggressive". Unlike the past, such values today rarely have a role in everyday life, such as in the work situation. One of the realms where one can still conform to them is traffic, and other situations in urban life which guarantee a high degree of anonymity. At least that is what advertising, movies, and maybe the example of father or mother, make us believe. Behaviour modification could be achieved if we succeeded in changing the values attached to this so called "sporty" behaviour. On a large scale such a programme would demand a considerable effort from authorities and citizens.

On a much smaller scale some applications of conditioning theory in the area of enforcement may be mentioned. With respect to urban processes there seem to be attractive possibilities for the behaviour of smaller children. Another interesting application might be a study of the influence of traffic congestion on trip-generation, distribution, mode-choice and assignment. Until recently congestion was followed by immediate action from the authorities to remove the cause of the congestion, a very reinforcing situation for the road-user. Introducing congestion (e.g. road works) frequently leads to extinction. It is my conviction that a scientific manipulation of congestion may do some essential things for us when we try to cope with the present level of mobility.

Conclusion

While there are various individual and group models of the sorts described here, environmental psychology has by and large not yet reached the stage where such models are put to test in the urban context. If this branch of psychology is to make any progress in meeting the demands of society it should become more serious about finding and formalizing behavioural variables that, when controlled, do indeed affect urban behaviour, and in particular the pattern of mobility. Only then we may expect that it will really become the cornerstone of transportation planning that it ought to be.

DISAGGREGATE BEHAVIOURAL MODELLING

One of the basic problems of transportation planning is indeed the relation between the travel behaviour (and its underlying motivation) of the individual, and the aggregate behaviour that is the concern of the authorities: the total pattern of all travel in a certain area. Until recently all traffic models were essentially aggregate models describing travel in a rather deterministic way related to geographical zoning and without any reference to personal characteristics. This has changed considerably over the past few years in favour of so-called "disaggregate" models, based on data from individuals or households (see Stopher & Meyburg, 1974; Richards & Ben-Akiva, 1975). This leads, however, to the delicate question of how to aggregate the individual data. As one very pessimistic expert in the planning field laments: "We have, it seems, in forsaking areal aggregations for more discrete analysis gone from the unsupportable to the intractable" (Shuldiner, 1974, p.43).

In brief the problem is as follows. Instead of determining the trip-generation, distribution, choice and assignment on the basis of geographical areas, various socio-economic and behavioural variables are determined for representative samples of individuals. With regression or multidimensional scaling techniques "behavioural groupings" are constructed on the basis of similarity of scores on the socio-economic and behavioural variables. On the basis of such behavioural groupings the travel patterns in the region studied may be determined for different types of trips, different sub-samples of the population, for different (new) transportation systems, and for different policies aimed at influencing travel-demand.

One of the few examples of the latter is the analysis in the Netherlands of travel patterns between the suburbs of the city of Eindhoven and the centre (Richards and Ben-Akiva, 1975). The authors have calculated the effects of various changes in travel-time on the modal split, on the basis of a disaggregate multinomial logit model. Their results are in Table 1.

From this table one can read the effects of single or combined influences. The most effective policy from the point of view of pushing back private car-usage, for instance, consists of increasing parking problems and congestion, plus an increased service (frequency and stopping distance).

Percentage change in					Percentage travellers choosing			
in-vehicle travel time		out-of-vehicle travel time (parking etc.)	walking time to or from bus-stop or station	waiting and transfer time	bicycle	moped	car	bus + train
car	bus + train	car	bus + train	bus + train				
base situations					5.2	17.1	77.0	0.7
-10					4.5	15.0	79.8	0.6
+10					5.9	19.4	74.0	0.8
	-10				5.2	17.1	76.8	1.0
		+30			8.9	29.3	60.6	1.2
	-10	+30			8.8	29.2	60.4	1.6
		+30	-10		8.8	29.0	59.9	2.3
		+30	-20		8.7	28.7	59.4	3.2
	-10	+30	+20		8.9	29.5	61.0	0.6

TABLE 1. EFFECTS ON MODAL SPLIT FOR CAR-OWNERS BETWEEN PARTS OF THE SUBURBS AND THE CENTRE OF EINDHOVEN (from Richards & Ben-Akiva, 1975).

This approach to transportation planning introduces a number of tasks for the behavioural scientist. The first problem in this approach is to find the relevant behaviour variables and to measure them. Measuring them, by observation or scaling techniques, may be the lesser of this double task. Defining the relevant variables will be much more difficult. The planning process currently is not carried out any more under the assumption that unlimited means are available, for whatever the mobility needs of the community area. Planning now means controlling and influencing these needs in an effective way. If the behavioural variables that enter into the planning model are really the variables that determine the travel pattern, they are by implication the variables one wants to influence. Therefore the task of finding relevant behavioural variables may turn out to be much more complicated, and require more theoretical wit than has been invested thus far. I feel that considerations of this kind might put some useful constraints on some parts of environmental psychology which tend to be carried away by academic arguments, notwithstanding the fervent claims of social relevance.

The second main problem is that of aggregation of whatever data will enter into the travel-demand model. The usual procedure is to establish - on the basis of some statistical procedure - a "model" or "average" member of a behavioural grouping. Because of the nature of the procedure such models will necessarily be stochastic, which makes them computationally more difficult than the classical, deterministic, aggregate models. The psychologically interesting part, however, is not so much the technical problem of aggregation, but the rather fundamental, and thus far burning, question of how far there is a psychological basis for the conventional sequential treatment of trip-generation, distribution, mode-choice and assignment. It seems most unlikely that there is such a basis. Certainly there are decisions in parallel or even in reversed order, as for instance in the case of multi-purpose trips.

This typically requires an analysis of the cognitive procedures used by travellers planning and realizing their trips. The outcome would offer a tremendous aid to the modelling process. Such an analysis is, however, rare - again with few exceptions. The strategies determining route-choice by drivers have attracted some attention (e.g. Pailhous, 1970; Janssen, 1975).

Extending the analysis to other aspects of the transportation process does seem extremely important, in particular if it would be carried out separately for behavioural groupings that have specific travel problems: children, the aged, the blind, etc. Also one should consider that people are not only transport consumers, but always enter into a complicated multi-level interaction with the transportation process. To this point I will turn now.

FOUR LEVELS OF TRANSPORTATION IMPACT

While we have two lines for approaching the intricate problems of the relation between man and his environment, we still appear to lack a descriptive framework in which to specify our research goals and results. The following proposal is based on some assumptions about Man as a processor of information and a solver of the problems the environment offers, derived from what is commonly called cognitive psychology. My inclination towards these assumptions will already have become clear in the preceding sections. In this view human activity may be summarized in the following schematic way (Fodor, 1975, p.28):

1. The subject is in a certain situation (task environment).
2. By experience or instruction the subject believes that in this situation he can do several things to reach a particular goal.
3. The probable effects of each of the alternatives are estimated (computed), on the basis of the internal representation of the situation available to the subject (his knowledge, beliefs), and constrained by his computational power.
4. A preference ordering is assigned to the consequences.
5. The actual behaviour chosen is a function of the preferences, and probabilities assigned to the alternatives.

	behavioral level			
	I	II	III	IV
human quality as a problem solver	road user	transportation consumer	social being	psycho-biological organism
problem to be solved	vehicle control	trip making	social activities (communication)	satisfaction of basic needs
task environment	road	road network (geographical structure)	socio-economic structure, land-use	nature (environment)
task aids	vehicle, signs, etc.	transport mode	transport system	"culture"

FIGURE 1. TRANSPORTATION AS PROBLEM-SOLVING AT FOUR LEVELS OF HUMAN ACTIVITY.

Whenever the situation is such that there are no acceptable behavioural alternatives available to them, human beings will create task-aids, technological or mental artifacts which make it possible to negotiate the difficulties in the task-environment (Simon, 1969). Alternatively they may abandon the task situation.

Human interaction with transportation and traffic can be specified at four different levels of "problem solving", as in Figure 1 (Michon, 1975). This division indicates four types of interaction between man and traffic of an increasingly abstract and complex kind. It should be pointed out that the task-environment and the task-aids at one level combine into the task-aid at the next complex level.

Man as a Road-User

At the first level the problem-solving task at hand is to change one's position with respect to the environment, in such a way that one remains on the road surface and does not collide with any obstacle. The problem is usually tackled with the aid of a vehicle and various "roadside means" such as traffic signs, lane markers, etc. Many studies have been carried out at this level, with such considerable success that we may consider the role of the behavioural sciences in this area well established.

Man as a Transportation Consumer

Here the task-environment consists of the locations (destinations) in three-dimensional space. Problem-solving consists of finding a connection between two points, under probabilistic constraints pertaining to time and route. The available task-aids - the transportation modes - may vary and determine these constraints to some extent. The problem here is essentially one of deciding between alternatives, but the seemingly simple structure of these decisions is highly contorted by the limited processing capacity of the decision-maker. We have already seen that it is this level which contributes most to the chronic headache of the transportation planner.

Man as a Social Being

At this level, the person must solve the problem of spatio-temporally coordinating an intricate pattern of social activities (in large measure: communicating with others). The task-environment is determined by the societal structure and the prevailing land-use. The transportation system in toto is the aid to help solve the problem that different activities must take place in different locations. While we know a fair amount about the actual trip-distributions as they relate to various social activities, we still know very little about the motivations underlying mobility. This is where a principal task of the behavioural sciences is waiting, and where very little progress has been made.

Man as a Psycho-Biological Organism

At this fourth level we are dealing with what may be called the "traffic culture": the whole of potential mobility as a part of the activities aimed at satisfying one's fundamental needs. This level becomes problematic as soon as the boundaries of the "traffic culture" as a regulative mechanism are reached. The harm done to the environment by traffic is one of the direct results of such a transgression of the boundary conditions of the system. At this level of discourse considerable progress has been made in recent years with the quantification of 'imponderable' aspects of economic value and risk. We may also expect that the rather intensive efforts of environmental psychologists to establish models for the relations between man and his physical environments will end in some success.

Figure 1 pointed out some important fields of study better accessible to the behavioural scientist than to the engineer. Much will depend on the ability of the former to understand the needs of the latter. While there is much theory and fact available about the interaction the traveller and the transportation environment, much of it does not meet the conceptual, dynamic, and manipulation standards required in transportation modelling. Research programmes should first of all try to change this. The next few paragraphs suggest some formal requirements for future studies of the relation between man and the transportation system.

Static versus Dynamic Models

The first shortcoming of most behavioural data is their static character. The essence of demand modelling, or other planning procedures in prediction, requires over-time data. The usefulness of static data for a systematic approach of transportation problems or transportation decision-making is close to zero.

The lack of dynamic designs in psychological studies is very persistent, at all four levels defined above. It is true that dynamic models of the driver have reached a high degree of sophistication. The time-dependent description of perceptual and decision processes is much less well developed, however (Forbes, 1972; Riemersma et al., 1972). At the higher levels the dynamic analysis of behavioural and social variables remains rudimentary at best. Only when we succeed in predicting the marginal changes in travel behaviour in response to marginal changes in the relevant behavioural variables, will the efforts to isolate such variables be rewarded. We are badly in need of longitudinal data, rather than the cross-sectional data usually available. At any of the four behavioural levels we need to follow individuals in their travel-behaviour over quite long periods of time. Only this will enable us to gain detailed insight into the differential effects of changes in the system on the travel patterns of specific groupings, as well as of the total population.

Simultaneous versus Sequential Models

The practice of transportation modelling has been to treat trip-generation, distribution, mode-choice and assignment separately and sequentially. The odds are that this does not reflect the cognitive processes of the traveller. The analysis of problem-solving and decision-making has led to rather impressive results with respect to some problem areas, such as chess playing, propositional logic, or blast-furnace operating. A similar, necessarily extensive and difficult, analysis of the cognitive strategies by which people plan their travel, is needed before a behaviourally adequate type of (partly?) simultaneous model can be constructed.

The Effectiveness of Policy

The evaluation of planning proposals, as well as of effects of changes in the system, or regulations and campaigns, is generally lacking. Not infrequently, in fact, a change is not introduced with the intention that it be evaluated: political import can outweigh the long-term consequences on mobility. Much more attention should be paid to evaluation procedures, first of all for real-world circumstances, and for planning models, and for the relevance of behavioural variables which enter into such models. Part of this task will involve a better analysis of the decision-making process at the transportation policy level. Decision-making is intrinsically restricted by the processing limitations of the human mind, irrespective of the aids offered by computer and planning experts. A timely topic for study would be the structure of policy-making in the light of the present-day requirements of citizen participation.

REFERENCES

Appleyard, D., & Lintell, M. (1971) Environmental quality of city streets: the residents' viewpoint. Highway Research Record, 356, 69-84.

Binder, R.H. (1973) Major issues in travel demand forecasting. Highway Research Board. Special Report No.143. Washington DC.

Brewer, W.F. (1974) There is no convincing evidence for operant or classical conditioning in adult humans. In W.B. Weimer & D.S. Palermo (eds.), Cognition and the Symbolic Processes. Hillsdale, N.J.: Erlbaum.

Carpenter, F. (1974) The Skinner Primer: Behind Freedom and Dignity. New York: Free Press.

Coleman, J.S. (1964) Introduction to Mathematical Psychology. London: Collier-McMillan.

Downs, R.M. & Stea, D. (1973) Image and Environment. Chicago: Aldine.

Fodor, J.A. (1975) The Language of Thought. New York: Crowell.

Forbes, T.W. (1972) Human Factors in Highway Safety Research. New York: Wiley.

Homans, G.C. (1950) The Human Group. New York: Harper.

House of Commons (1972). Second report from the expenditure committee. Urban Transport Planning, Volume 1. London: HMSO.

Ittelson,W.H., Proshansky,H., Rivlin,L.G., & Winkel,G.H. (1974) An Introduction to Environmental Psychology. New York: Holt, Rinehart & Winston.

Janssen,W.H. (1975) Denken, voorstellen en plannen: de bestuurder als redelijk wezen. In J.A.Michon & H.H.van der Molen (eds.), Verkeer en Samenleving. Assen: Van Gorcum.

Kahneman,D. (1973) Attention and Effort. Englewood Cliffs, N.J.: Prentice-Hall.

Klein,D. (1971) The influence of social values on rates of death and injury. Journal of Safety Research, 3, 2-8.

Lee,T.R., & Tagg, S. (1976) The social severance effects of major roads. In P. Stringer & H. Wenzel (eds.) Transportation Planning for a Better Environment. New York: Plenum.

Lynch,K. (1960) The Image of the City. Cambridge, Mass.: MIT Press.

Mehrabian,A., & Russell,J.A. (1974) An Approach to Environmental Psychology. Cambridge, Mass.: MIT Press.

Michaels,R.M. (1974) Behavioral measurement: an approach to predicting transport demand. Transportation Research Board. Special Report No. 149. Washington D.C.

Michon,J.A. (1975) Een structuurschets van de "Sociale Verkeerskunde" In J.A.Michon & H.H. van der Molen (eds.), Verkeer en Samenleving. Assen: Van Gorcum.

Milgram,S. (1970) The experience of living in cities. Science, 167, 1461-1468.

Miller,G.A., Galanter,E., & Pribram,K.H. (1960) Plans and the Structure of Behavior. New York: Holt.

Nederlands Economisch Instituut (NEI) (1972) Integrale verkeers - en vervoersstudie. (Main Report, plus 6 annexes). The Hague: Staatsuitgeverij.

Nieuwe Rotterdamsche Courant (NRC) (1974) Zaterdags Bijvoegsel, p.1, 10 October.

Newell,A. & Simon,H.A. (1972) Human Problem Solving. Englewood Cliffs,N.J.: Prentice Hall.

Pailhous, J. (1970) La Representation de l'Espace Urbain. Paris: Presses Universitaires de France.

Richards, M.G., & Ben-Akiva, M.E. (1975) A Disaggregate Travel Demand Model. London: Saxon House Studies/Heath.

Riemersma, J.B., Mess, K.W., & Michon, J.A., (1972) Perception and maintenance of speed and course of a moving vehicle: a review of the literature. Institute for Perception. Technical Report 1972-C7. Soesterberg.

Selye, H. (1950) The Physiology and Pathology of Exposure to Stress. Montreal: Acta.

Shuldiner, P. (1974) Extension of present methodology. Transportation Research Board. Special Report No. 149. Washington D.C.

Simon, H.A. (1957) Models of Man. New York: Wiley.

Simon, H.A. (1969) The Sciences of the Artificial. Cambridge, Mass.: MIT Press.

Skinner, B.F. (1972) Beyond Freedom and Dignity. New York: Knopf.

Stopher, P.R. & Meyburg, A.H. (1974) Introduction and summary of research recommendations. Transportation Research Board. Special Report No. 149. Washington D.C.

SWOV, Netherlands Institute for Road Safety Research (1971) Psychological Aspects of Driver Behavior, Vols 1 and 11. Voorburg: Institute for Road Safety Research SWOV.

Turner, F.C. (1970) In Panel Discussion on Highway and Urban Transportation in the 1970's and 1980's. Highway Research Board. Special Report No. 122. Washington D.C.

CITY CENTRE AND TRANSPORT PLANNING - HUMAN FACTORS AND SYSTEMS SCIENCE. A PROBLEM OF IMBALANCE.

Peter Warman

Alan M. Voorhees and Associates

London

It is a familiar catch-phrase that "planning is for people". Is it not therefore sensible to make the most of those professional disciplines which illuminate the behaviour and attitudes of people? Is it not necessary to extend our understanding so that those developing, evaluating and implementing policies do not make unrealistic assumptions about the probable effects of policies on the community?

It is becoming recognised that the psychologist and the sociologist can make a useful contribution to planning. However, the natures of their disciplines do differ significantly from the disciplines of physical town planning, engineering and economics. The latter group has had a tendency to dominate the planned development of cities in recent years because those who have the practical responsibility for planning are most often skilled in these technical disciplines. The process of planning has already become complex, lengthy and expensive. It is also highly numerically-oriented, whereas much (although not all) of the psychological/sociological contribution tends to be verbal. Whilst the need for a sociological and psychological input is recognised nowadays, by no means has it been further recognised how these additional disciplines can be effectively integrated into the planning team.

This paper is based upon the views which I, as a transport planner, have formed as a result of some recent British studies. It is also based on experience gained as a member of planning teams in which both psychology and sociology have been represented.

TRANSPORT PLANNING TRENDS IN CITY CENTRES

To discuss the subject I have chosen the planning of transport associated with the central area of cities. Over the past twenty years policies have evolved around the need to reduce traffic congestion and environmental conflicts. In the early years attempts were made to accommodate the growth of private vehicle usage through the planning of major highway schemes. Additionally more technically sophisticated public transport systems were proposed. However, in Britain and elsewhere, such schemes have often generated feelings of disillusionment among the community. Their contribution, believed by the planner to be important, does not appear to many of the public to advance the interests of living in cities. This has resulted in a change in emphasis in transport planning. In recent years, although the trend continues of providing for the private motorist in the central areas of cities, it is not done at the expense of inhibiting the efficient operation of conventional public transport. Where possible there has also been an accelerating development of the concept of more streets for pedestrians only.

The development of traffic-free zones in the central areas of cities in Western Europe has become a trend with increasing momentum. Analysis of the 445 cities with a population greater than 50,000 showed that, by 1973, at least one hundred had created pedestrian precincts in the previous six years. From the study of future plans it is estimated that an additional 250 cities will have pedestrian zones by the early 1980's and many of the existing zones will be extended.

The widespread adoption of similar solutions is derived from a problem analysis technique which indicates that similar types of transport problem exist in varying degrees in most Western European cities. The transport planner has devised various possible approaches to provide transport policies for city centres. The present success and acceptance of traffic-free areas encourages their adoption in other cities. Indeed, the criterion for the acceptability or suitability of a transport policy in one city often seems to be the application of similar policies elsewhere, irrespective of the compatibility and salience of the particular transport planning issues.

Together the planner, the engineer and the economist are able to identify the factors which affect present and future transport operation in a city. In other words, these are the people who recognise the implications of present transport problems, can anticipate future ones and are often the people who have the responsibility for resolving them.

I wonder, however, when one looks at the marvellous variety

which provides so many cities with their special attributes, whether these common solutions to transport problems may not be generating an unnecessary trend towards uniformity; whether, also, quality-of-life attributes should not play a greater part in influencing transport policies.

It is here I think that the sociologist and psychologist might make their contributions. If so, it is reasonable to ask what could be the nature of the contribution.

SOCIOLOGICAL/PSYCHOLOGICAL CONTRIBUTION

Often nothing more ambitious is envisaged than an attitude or opinion poll, but in my view there are four fundamental areas in which the sociologist or psychologist can provide a contribution of deep significance. These are:

1. To inform the planner how people, in the different roles that make up their lives, perceive transport-related problems and the nature of the impact that transport issues have upon human life-styles.

2. To indicate how important to people are these transport issues relative to other planning issues.

3. To reveal which values and criteria might additionally be relevant for evaluating the adequacy of proposed policies.

4. To suggest which behavioural changes are most likely to occur as a consequence of implementating particular policies.

For different disciplines to work effectively together it is necessary that the professional skills of each discipline are understood by the others. In this respect, it is as important that the psychologist and sociologist are as aware of the nature of the planning processes as it is that the planner appreciates the skills of the psychologist and sociologist. Much more is required than a perusal of text-books and technical journals. Contact is necessary, because the combination of academic discipline and experience has a profound influence upon the way an individual perceives and understands issues.

In providing a contribution to the four subject-areas given above, the sociologist and psychologist become mediators between the public and the other professional disciplines in the planning team: a role which involves a two-way process of communication and interpretation.

PRACTICAL APPLICATIONS TO TRANSPORT PLANNING ISSUES

Within the transport field, psychological methods of investigation have produced useful results in a number of specific areas. These include: the design of railway rolling stock (British Rail have commissioned numerous studies for local and Inter-City carriages); bus design (see 'Motorbus & Man' by O.W. Schultz. Paper to the 41st U.I.T.P. Congress, 1975); marketing of public transport (numerous studies for British Rail and UMTA); modal choice; community involvement and public participation; rural public transport (the author conducted a study in Bedfordshire which revealed the interaction between the physical provision of transport and the development of communities).

DEVELOPING THE CONTRIBUTION

The results of these types of study, where psychological techniques of research have been employed in the study of transport, suggest that methods are available to test the major assumptions made about people in a particular city, regarding their probable responses to changes resulting from the transport planning processes. So far the techniques have been used for transport issues that are fairly self-contained and specific. However, I suggest that their application would be equally beneficial in research relating to the broader and probably more significant transport planning issues.

Often a statement is made about the need to improve public transport. How is this view arrived at? What systematic methods of research are generally employed which illuminate the nature of any real dissatisfaction with the services amongst particular sections of the community? Does the exploration go beyond the rather stereotyped answers and give definite leads to how the services should be improved? It should be acknowledged that a person can put forward conflicting objectives about what is required and a more in-depth investigation can better reveal the nature and salience of problems which in turn should lead to more effective policy solutions. An example where more appreciation of the passenger's view is required is the presentation of information about the services. Often this is projected in the form of 'what the operator provides' and not necessarily how the passenger can make use of it to plan and carry out a journey.

Many policies are discussed and implemented on the issue of traffic restraint. A policy of minimum or maximum traffic restraint could have enormous impacts on transport opportunities. What conclusive evidence exists to favour a bias towards one policy or the other in any given city?

Assumptions are made in transport planning concerning the effect of walking distances and waiting times on choice of transport mode and travel patterns. Indeed this whole question of what factors can influence the perception of time and distance in the context of travel requires more understanding by those involved in transport planning.

PUBLIC TRANSPORT ACCESS TO PEDESTRIANISED SHOPPING STREETS

Described above are some of the issues that show the need for a greater input from the social sciences. To discuss an example of how this process can work in practice I return to an issue associated with the central area of cities - the access of public transport to pedestrianised shopping streets.

As already mentioned there is a significant trend towards the creation of traffic-free zones. From an examination of such schemes in a hundred cities throughout Western Europe, the dominant reasons given by planners for establishing the pedestrian areas are to improve safety and the quality of the environment, particularly in the shopping areas of the city centre. An equally important trend is the desire to increase patronage of the public transport services in preference to use of private transport in the central area of cities. As pedestrian areas become more extensive, conflicts arise between these two desired trends. Bus and taxi services are usually the dominant modes providing public transport in the central area and their existence in otherwise relatively traffic-free streets can be thought incompatible with the reasons given for creating pedestrian areas. Certainly the design possibilities for the pedestrian environment are severely restricted by the existence of these services. However, the public transport operator's view is usually that the services must provide good access to the intended central-area destinations of passengers, reducing the need for interchanging and keeping walking distances to a minimum.

The transport planner will probably suggest a variety of feasible schemes for transport operation in combination with the development of pedestrianised streets in a particular city centre. The costs and changes required to implement each of the schemes can be assessed. However, to what extent can the community's preferences be incorporated into the evaluation of the schemes? In the initial generation of the schemes many assumptions have to be made about people's attitudes and behaviour, to predict the use of the facilities. How can these assumptions be checked systematically so that we may be more certain that the assumptions are realistic?

The techniques likely to be used for answering these questions have been well tried by psychologists in other contexts.

Apart from observing behaviour, there are a number of methods used for interviewing people better to understand their attitudes and behaviour. These lie along a continuum ranging from a predetermined set of questions involving fixed responses, to the less structured approach where the topics are explored with the respondents at wider and deeper levels. Interviews in this less structured form can be with individuals or in the form of a group discussion. The range of techniques and their uses are described in greater detail in Appendix A. A combination of the techniques would probably be used by psychologists to illuminate community preferences for a scheme which essentially is attempting to resolve a conflict between public transport access and more extensive pedestrian areas.

Applying such techniques to this type of study, a suggested work programme for a psychologist to examine community preferences for public transport in or around pedestrian areas is outlined below.

Transport Planning Input

(i) An exploratory statement by the planning team of the primary issues that the proposed schemes are attempting to resolve.

(ii) A description by the planning team of the feasibility of each scheme, including its assessment of how effective each scheme would be in dealing with each of the identified issues.

(iii) A description of the expected differences between the present situation and that created by each of the proposed schemes, identifying significant variables which appear to differ between the schemes.

Exploratory Research by the Psychologist

Some exploratory research is required to identify, by their roles, the potential users of the area to be changed and to find ways of communicating with these people about the nature of the proposed development. The objectives of this stage can be stated as:-

(i) generating language - ways of describing the various schemes and the effects which they might have on the users in terms which have real meaning to those users;

(ii) determining the 'issues' which appear to be important and are associated with each alternative;

(iii) eliciting the salient attitudes towards the schemes and opinions on the general impact of each on life in the city;

(iv) aiding the identification of those who seem most likely to use the future facilities under each of the proposed schemes.

Besides discussions with those responsible for transport planning and operation in the central area, group discussions would be held with people who currently use the affected area of the city centre for various purposes.

These group discussions would enable some testing of the way in which schemes and concepts are communicated (i.e. the actual language used) and also allow probing of important attitudes towards each of the different schemes. In order to explore the attitudinal area more fully a number of depth interviews could also be undertaken with potential users of the schemes.

Selected 'Role Playing' Panel

On the basis of the exploratory qualitative research, a panel would be selected which typifies the projected target population of central area users, i.e. all who are likely to have an interest or viewpoint. The aim of this phase is to enable a group of people to state their preferences using a technique which leads them to consider how other members of the community might react. Basically, it involves role playing - adopting the standpoint of another in viewing the situation, and attempting to react in the way that it is thought that the other would react.

The process is iterative in that the panel would have a second chance to state their preferences, but in the light of having seen the actual preferences given by other members of the sample. In this situation the respondent is able to update his choice in the light of new information, although he is not obliged to take the views of others into account.

The technique, therefore, aims at familiarising a respondent with the elements of the schemes, and then asking him to react for himself, and in the role of others. In effect, the person is being encouraged to view the problem from a number of angles and weigh up the relative 'pros and cons' which he believes would weigh most heavily with different groups of the target population.

The panel would consist of a minimum of fifty respondents, chosen to represent various sectors of the potential user population, e.g. shopper, office worker, shop-keeper, shop manager, etc.

Initially, each respondent would receive a brief description of the schemes and be asked: (i) his own beliefs about and attitudes towards the schemes; (ii) to adopt a number of other roles and to state assumed preferences for each role, with reasons. This technique would uncover the many issues which people associate with the alternatives, under a number of role conditions.

On-Street Study of Preferences

Stage 2 and 3 of the research would generate information on the issues involved, or seen to be involved in the proposed schemes. This stage is aimed at structuring the important issues and obtaining reactions from a large number of citizens.

A list of statements concerning the schemes and their effects would be drawn up and respondents contacted on the street would be asked to indicate their level of agreement. The material would include items on such issues as pedestrianisation, environmental features, convenience, shopping, working in an area and transport parking provision.

The results and conclusions from this kind of work programme can provide useful indications and insight into the factors which affect a particular community's preferences for different transport planning 'solutions'. It can be undertaken within a period of three or four months and become an integral part of the evaluation process.

CONCLUSION

I called this paper City Centre and Transport Planning - Human Factors and System Science. A Problem of Imbalance. The system-oriented approach grew from the need, recognised by all planners, to incorporate a range of factors into a logical objective process. In most cases the replacement of hunch or intuition by an evaluation process derived from a scientific system analysis has improved the quality of planning. It has also made planning into a multi-disciplinary team activity.

My concern is not that this evolution is undesirable, but that the process can become too greatly oriented to technology and economics and too little oriented towards those less easily defined factors which we generally describe as 'the quality of life'. I do not suggest that the disciplines of psychology and sociology should dominate the planning process. But I do believe that they can very helpfully contribute to it. Often in my view they do not contribute sufficiently and it is here that the imbalance lies.

Planning has long since ceased to be an easy task. Psychologists and sociologists are not always the easiest of people with whom to deal. They have their own jargons and they can be masters of equivocation. Nevertheless, they are the people who, better than others, can understand the scales of values operating within communities. As planners we need to develop an understanding of human desires and aspirations so that we can tailor our plans to accommodate them. The first stage is for it to be recognised that these other disciplines have practical use. Secondly I think we should understand what techniques they have available and how they use them. Without this understanding we are in danger of confining the contribution we ask from them. I set out therefore in an appendix a resume of some of the techniques.

APPENDIX. SOME BASIC PSYCHOLOGICAL RESEARCH TECHNIQUES USED IN TRANSPORT RESEARCH

Unstructured Interview

An unstructured (or depth) interview does not involve the use of a questionnaire. The interviewer is very fully briefed so that he or she understands the problem to be solved. It is important to ensure comprehension of the information required within the total study context. The interviewer may construct or be supplied with a checklist of topics to be covered within the interview, but he does not have an exhaustive list of specific questions. All the questions are formulated within the interview situation, and their content is dependent on the response to previous questions. On each topic the interviewer probes until he is satisfied that he has obtained all the available information. Good rapport, between the interviewer and the respondent is essential, so that there is a limited number of physical environments in which this type of interview can be conducted. We have found it possible for example to use the technique in the vehicle (train) in the inter-urban situation; but it is desirable to reserve a block of accommodation to ensure privacy, since respondents are inhibited by being overheard. It is also undesirable to interview a passenger who has already overheard a previous interview, since responses will not be spontaneous.

Unstructured interviews are appropriate for exploring attitudes in depth, and for gaining information about travel motivation and the travel and modal decision processes. The technique is costly, requiring a trained psychologist or sociologist for both application and analysis. Furthermore the extent to which data can be quantified is limited; although this does not mean that such data lack validity.

We have found it preferable to record the interview on tape, as this is less inhibiting to the respondent than long pauses while the interviewer notes his responses. Furthermore the use of tape eliminates the temptation to edit in the interview situation, and allows meaningful key words to be identified which might have been edited out or altered in written notes. It has been found possible to use a tape-recorder in modern rolling-stock.

The technique is frequently used as one element in a study. It is particularly useful in the development of a structured questionnaire and subsequently to probe in depth areas of particular interest revealed by structured data.

Group Discussion

About eight respondents constitute an ideal group. Fewer than six or more than ten give limited information. The discussion is led by an experienced psychologist or sociologist and is recorded on tape. The usefulness of a group discussion is entirely dependent on the experience and skill of the group leader. In a successful group discussion it is possible to simulate and observe the attitude-forming process in microcosm.

We have found that there are certain mixtures of people which lead to less successful group discussions. For example, it is not generally wise to mix social classes within one group, as this can inhibit one section. We have found that, in relation to the journey to work, it is not useful to mix the users of different modes, as attitudes tend to polarise, and a series of rationalisations and justifications emerges rather than underlying attitudes and images. Usually in transport studies it is quite possible to mix sexes and ages within a group; with the reservation that it might be unwise to mix age-groups in a discussion of short-distance leisure-travel, since offspring, for example, will often not readily discuss their evening activities in the presence of their seniors.

In any group discussion stereotyped attitudes will emerge at the outset. It is necessary that this should happen, as the stereotypes must be established in order to examine in greater depth the attitudes beyond them.

The group leader must make judgements about how and when to steer the discussion towards another aspect of the topic, and whether it is fruitful to return to earlier topics at a later point in the discussion. Here he must also take care to involve all the members and to avoid allowing the group to be dominated by any section of it. Both of these tasks must be performed discreetly, as the group leader himself must not perceptibily dominate the group or attract any hostility from its members. He must in fact be a neutral figure rather than a catalyst. (We have had more success with female group leaders, as women respondents will talk more freely and men are less competitive towards women.)

Respondents are not usually told in advance the precise topic for discussion. A small financial incentive to attend is usually offered.

We have conducted small-scale attitudinal studies entirely by use of the group discussion technique. It is almost always an element in large-scale studies. The data are not susceptible to quantification. The technique yields purely qualitative data about the action and interaction of members of a segment of society in the attitude-forming process.

Semi-Structured Interview

A semi-structured interview employs some structured questions with pre-coded responses. Most of the questions, though, are open-ended, and a question printed on the questionnaire is not regarded by the interviewer as finite, but rather as a "topic heading". Other questions may be asked within the topic-heading until the interviewer is satisfied that he or she has obtained sufficient information. These secondary questions are not pre-formulated, as their content will depend on the response to a previous question. This technique allows attitudes to be probed in some depth. It is also useful in the assessment of knowledge and awareness of a transport facility.

The administration of semi-structured interviews requires relatively high-level staff with some training in the technique. There is a particular danger of the introduction of interviewer-bias in questions formulated in the field situation if these questions are any less simple than, for example, "In what way?" and "Could you tell me more about that?" Furthermore, the interviewer must be carefully and thoroughly briefed to ensure an understanding of the type of information required.

Semi-structured interviews can be subjected to limited quantitative analysis; but this must be conducted by qualified staff, and the analysis is time-consuming. A sample of the interview reports can be used to identify the important response-themes. All the questionnaires can then be coded accordingly.

Because of the relatively high costs involved in the use of this technique it is not usually employed on a large mass of respondents, but rather on a limited scale, often in support of a mass structured survey. It is essentially a qualitative rather than a quantitative technique.

Structured Interview

A structured interview has pre-coded responses. Strictly it has no open-ended questions; but where these are employed there are few, and they are very simple in the nature of the response. Responses are indicated by ticking in the appropriate box or encircling the appropriate answer.

Structured interviews may be either self- or interviewer-applied and are subjected to statistical analysis. Interviewers are quickly and cheaply trained in their use and do not require any major skills or qualification. Attitudinal data can be elicited by a structured interview, but not probed in depth. Filter questions can be used, but the flexibility of this type of interview is severely limited. In the transport context structured interviews are used for mass surveys, either of households or in-transit. They are particularly useful for recording travel patterns and travel history.

ENVIRONMENTAL ASSESSMENTS FOR UNITED KINGDOM URBAN TRANSPORTATION STUDIES

J.C.R. Latchford and D. Pickering

Jamieson Mackay & Partners

London

Background

Medium-term land-use transportation studies, considering the ten to fifteen year period, are the currently accepted context within which transportation plans for urban areas are formulated. Such studies evaluate a range of alternative strategies, developed within a defined budget, comparing the strategies in operational and economic terms within an established modelling and cost-benefit framework. However, with the changing emphasis of public opinion towards consideration of environmental aspects, the environment has also become an important component of the evaluation process. This paper outlines the environmental evaluation procedures which have been developed and applied in many recent studies, and describes in particular some of the techniques for the quantification of environmental impact.

Context

The strategic evaluation of transport plans, considering the regional or townscale implications of a range of transport policies, is generally the first stage in the planning process. Subsequent studies focus attention on particular aspects within the preferred strategy, and require more detailed evaluation of environmental effects. It is important to recognise the different stages of the planning process and to apply evaluation techniques appropriate to each level. This paper considers the procedures for environmental assessment appropriate specifically for strategic evaluation.

Section 2 describes the framework for evaluation; Section 3 considers the procedures at present available for the quantification of environmental impact, and Section 4 briefly describes some areas for future research and development.

EVALUATION FRAMEWORK

In the past environmental evaluation has relied largely upon the subjective assessment of environmental impact, but with the increasing complexity of the evaluation procedures there is a need to base the evaluation on some quantification of the impact of alternative strategies. The objective of such a quantified evaluation is to provide a clear basis for judgment.

It is suggested that the framework for evaluation should meet three basic requirements:

(a) The procedures should be able to encompass all environmental factors, whether quantified or based on subjective assessment; and be flexible enough to incorporate new knowledge and techniques as they become available.

(b) The procedures should be relatively simple in application, demanding a level of manpower and expenditure appropriate to the overall strategic evaluation process.

(c) The procedures should be capable of differentiating the environmental effects of a wide range of transportation policies, presenting a meaningful assessment of the environmental implications of each strategy.

Various approaches have been adopted in developing an overall framework for evaluation (UMPT, 1973), some attempting to translate environmental impact into cost-benefit terms, others presenting environmental indices. At the present state of knowledge there are, however, problems in translating the full range of environmental effects into reaction or monetary terms.

An individual's valuation of travelling time can be derived from observations of choices available in the market for trading time against money. Although property prices partially reflect environmental factors such as noise, and some individuals install double glazing to prevent it, there is no reliable market mechanism from which individual valuations of environmental impacts can be inferred in a comparable way to inferences from travelling time. Environmental benefits are usually valued by the recipient's willingness to pay for such benefits. Where environmental costs are imposed by transportation improvements the imputed cost might be considered as the compensation required by those affected to be willing to accept the environmental deterioration. The former is a compensating surplus and the latter an equivalent surplus, which in most cases is expected to be considerably higher. However, in each case the problem arises that environmental costs and benefits may not be fully perceived (e.g. health effects), and valuations based on individual responses may underestimate the full social costs and benefits. The integration of environmental externalities into the formal cost/benefit framework requires considerable further research.

A framework has been developed working towards an overall summation of impact, but undertaken in stages allowing a meaningful evaluation of the 'disaggregated' environmental factors to be undertaken at the present time. Figure 1 illustrates the evaluation framework which may be summarised in three stages:

(a) The prediction of the physical level for each environmental factor; presenting a simple assessment in terms of the number of people affected by degree of effect.

(b) The transition of the physical level into a reaction score, for each factor; presenting an evaluation of people affected by 'reaction' level.

(c) The ranking of the relative importance of the environmental factors; providing a 'weighted' reaction for each factor, and enabling the factors to be aggregated to provide an overall impact index.

Environmental Factors

The report of the Urban Motorways Committee of the Department of the Environment (DOE, 1972) identified the major adverse consequences of new highways as noise, severance and visual effects, together with nuisance during the construction period. Recent measurements and the results of ongoing research (OECD, 1973) suggest that, while no danger to health due to air pollution from traffic has been positively identified, there is a need to include this aspect as a significant component of the evaluation procedure.

The prediction methods for some factors are now well established; whereas further research and development is required for other factors. It will not be possible to treat all factors in equal depth.

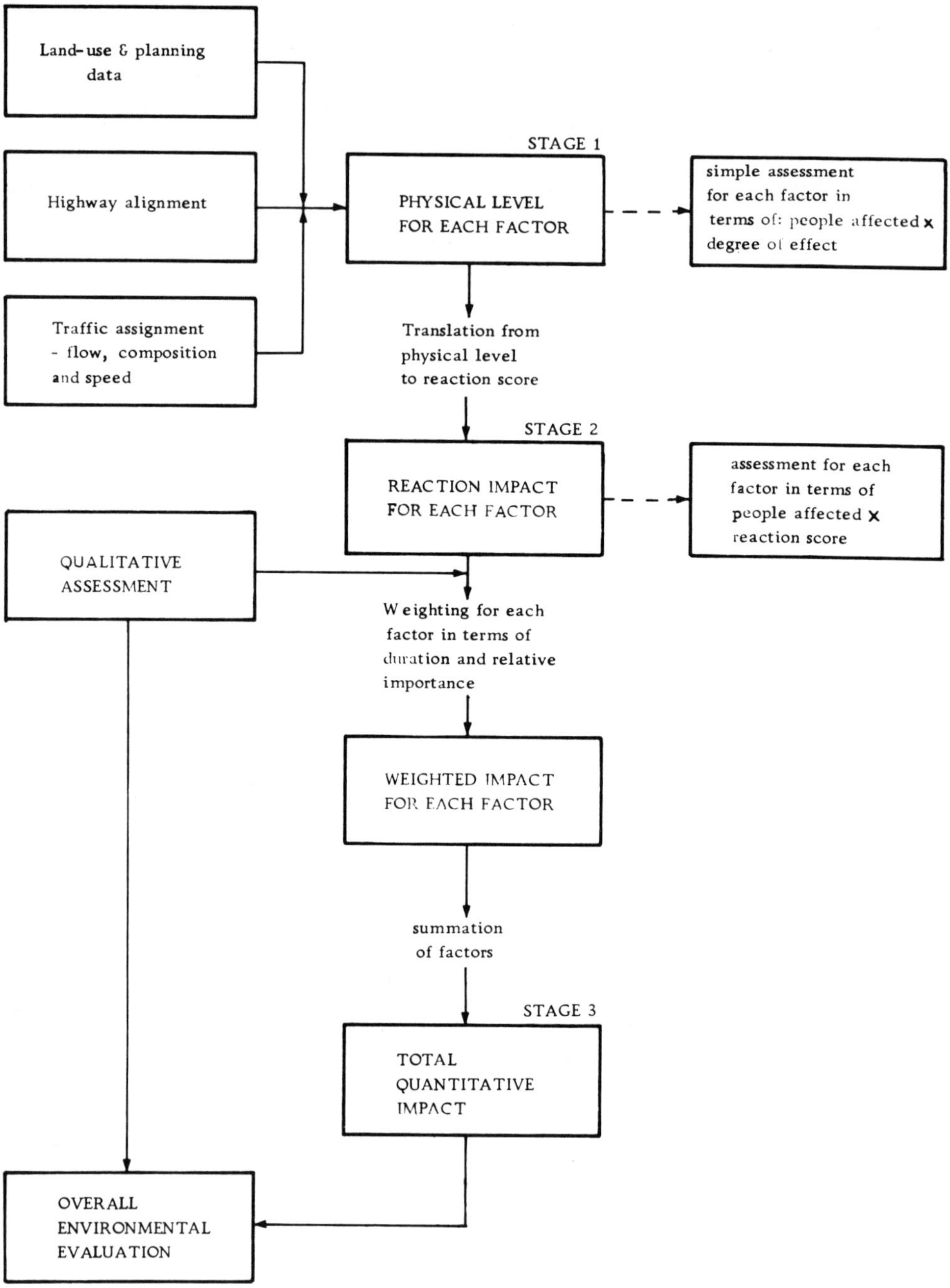

FIGURE 1. EVALUATION FRAMEWORK

Locational Variations

The predictive models discussed in this paper were developed for application in the United Kingdom. It is recognised that variations in vehicle size and differing driving habits may lead to differences in the levels of environmental impact and almost certainly in reaction, requiring some re-calibration of the predictive models in application outside the United Kingdom. The framework of evaluation is however generally applicable.

PREDICTION TECHNIQUES

Noise

The strategic evaluation of noise-impact records for each strategy the noise levels or reactions for the sensitive land-uses on the highway network. For each strategy noise-impact may be expressed in terms of the number of households (or schools or shops) by degree of effect.

The noise prediction method is based on the procedure developed by the Department of the Environment (DOE, 1975). The key to its simplicity in strategic application is that traffic-source noise and attenuation may be regarded as independent. Traffic-source noise is dependent on traffic flow, composition and speed and on road gradient. The source L_{10} noise level may readily be calculated for each link of the network from simple charts and from assignment data.

It is generally found that in current transportation studies only some 5 to 10% of the network links will be physically altered between options, and for the majority of links the distance between the traffic and the building facades will remain constant for all options. Thus attenuation is constant for all options and the change in noise-impact is due to variation in traffic flow. It is found that in most urban areas the development along the main highway network is seldom more than 15m from the edge of kerb. At that distance the fall in noise level due to attenuation is balanced by the increase due to facade reflection, and there is no significant difference between source and facade noise levels. The source noise and facade noise level for the sensitive land-uses for some 90% of the network may be calculated by computer from the traffic data derived from the assignment.

For the remainder of the links where new construction is proposed, the noise attentuation may readily be calculated for each option by using the noise attenuation charts which have been developed for a wide range of highway and barrier configurations (Figure

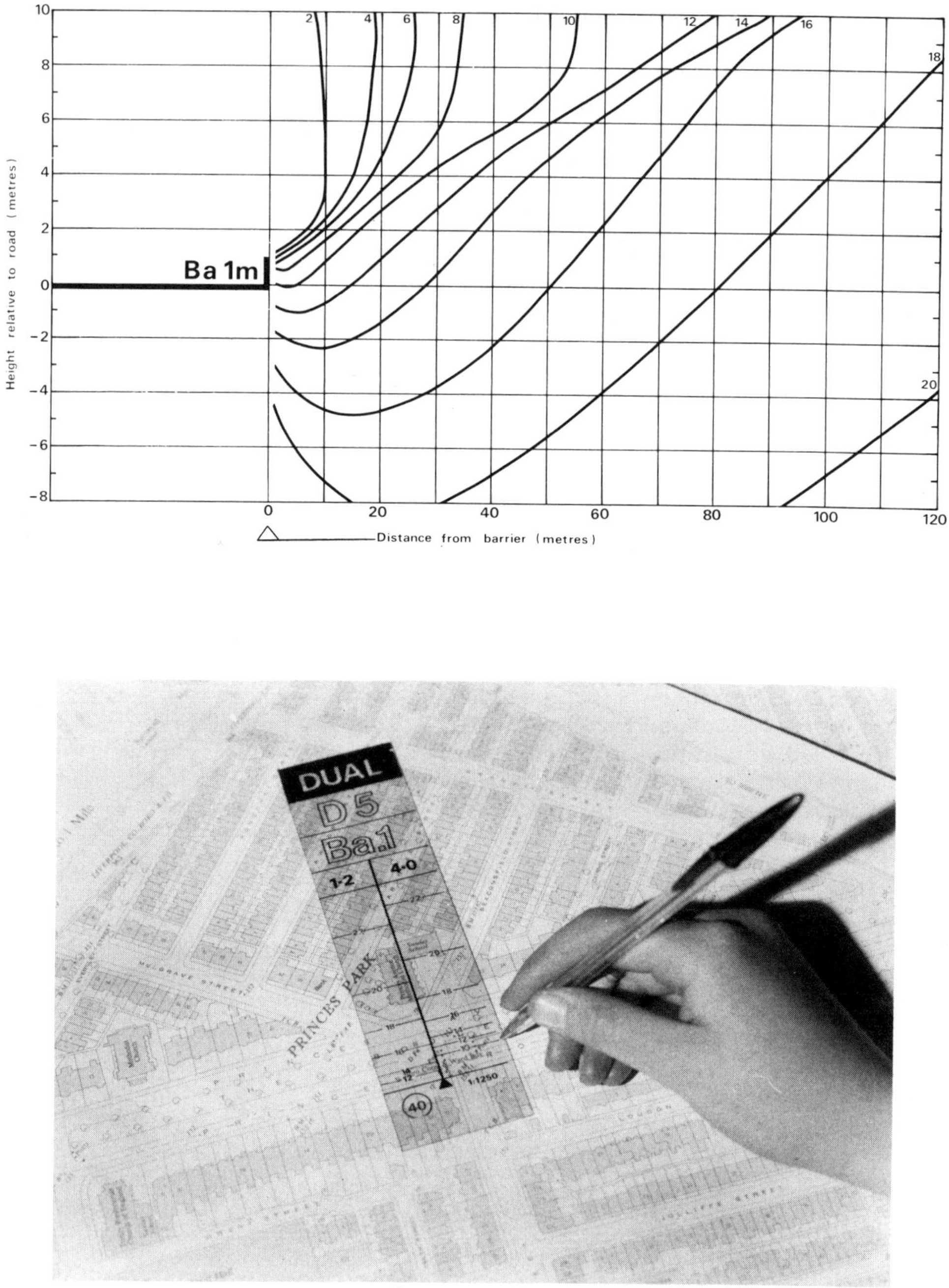

FIGURE 2. NOISE ATTENUATION CHART AND SCALE

2). In addition, sets of noise scales have been developed to overlay 1/1250th and 1/2500th scale plans (Figure 2) allowing the work to be undertaken by relatively unskilled technicians.

The noise impact to dwellings and other sensitive land-uses may be presented for each option in the form of cumulative distributions, providing a basis for comparative assessment. A further assessment can be derived from graphs of noise difference for each option relative to the economic base situation.

Recent research has provided information on people's reaction to noise (Lassiere, undated). It is possible to translate the network noise level into reaction scores (on a seven-point satisfaction score) for comparative assessment of the strategies.

Visual Effects

Recent research on behalf of the Department of the Environment (Hopkinson & Watson, 1974) has assessed people's visual reaction to highways and traffic. It has been found that visual judgment can be determined in terms of a number of predictive equations. The predictors relate either the size and location of the road structure or the volume and composition of the traffic, and an index for the visual context of the location, to a visual satisfaction scale.

The visual reaction to elevated structures is determined by the measurement of the solid angle subtense (steradians). The solid angle may be predicted in a number of ways, but is most easily measured by use of a visual protractor developed for use by unskilled staff.

In the majority of cases where there are no elevated structures, the visual impact will be due to moving vehicles and the traffic parameters are the only change between strategies. For these links a simple prediction equation allows rapid assessment of visual satisfaction to be calculated by computer, based on the total number of households affected by degree of satisfaction, and presented as a cumulative distribution in similar manner to the noise evaluation. It is important that the quantified data predicted by the above techniques should not stand alone, but should be complemented by a qualitative assessment of visual amenity for each strategy.

Severance and Pedestrian Effects

Severance may be defined as the sum of the divisive effects

of an urban highway. The effects may be both psychological and physical. While it is not presently possible to quantify severance in the same precise terms as noise, it is possible to identify the areas of impact due to severance, and provide an initial comparative assessment of the strategies.

The assessment of the alternative strategies would consider the relationship of the highway network to the land-use pattern and to areas of conservation and amenity. In addition, the evaluation would assess the extent of community severance and the diversion of movement, particularly pedestrian movement. Some identification of communities may be achieved in terms of the catchment areas for walk-trips to local shops and primary schools. The evaluation would relate the network flows for each strategy to the pattern of catchment areas, to identify the relative impact of each strategy. Pedestrian counts may be undertaken in sensitive areas to determine the strength of pedestrian movement, and to provide a basis for the prediction of pedestrian vehicular conflict in the areas of high pedestrian concentration.

Traffic Air Pollution

Evidence available in the United Kingdom suggests that present concentrations of traffic air pollution, though often unpleasant, do not generally constitute a significant risk to health (Sherwood & Bowers, 1970). However, as congestion in urban areas increases so does awareness that the only long-term solution to air-pollution lies in emission controls. Concern for possible health hazards has mainly centred around carbon monoxide and lead, with the result that, for the former, European (ECE) standards for emission levels have been adopted. The lead content of petrol is also being progressively reduced.

In the short and medium terms, however, roadside concentrations of most pollutants will not be reduced dramatically by emission controls. In the near future the growth in vehicle stock will largely offset the effects of improvements to individual vehicles (Thompson, 1973). Strategic transportation planning is the most immediate means of achieving desirable air-quality standards. To this end predictive models are being developed (HRB, 1973) which reflect both the emission and diffusion of pollutants and which would be applicable for strategic evaluation. Roadside concentrations are normally predicted as varying directly with traffic flow and inversely with speed.

Although further research is required in the areas of predictive models and health effects, a framework for evaluating pollution impact is clearly emerging. Whilst the benefits and costs of various pollution levels cannot at present be valued it is desirable that concentrations predicted by strategic studies be compared against both community and health standards.

APPLICATION OF PROCEDURES

The prediction techniques discussed in the previous section have been applied in a number of strategic studies. The scope and content of these studies varied widely in terms of the elements considered and the factors quantified, ranging from simple quantification of one or two major factors to a complete summation of environmental impact (Mackay, 1975). Three examples taken from recent studies are briefly discussed to illustrate the application and interpretation of the prediction techniques.

Noise

The quantification of noise-impact has formed a major component of the environmental evaluation in all the studies undertaken. The case study illustrated in Figure 3 related to the transportation study undertaken for a medium-sized coastal town which is both a major seaport and a holiday resort. The town is divided by a river which provides shipping access to the docks.

The study considered two basic alternative strategies within a common fixed budget for the Town. Strategy A would provide a high-level bridge across the river, linking the two halves of the town and connecting to the main coastal road. The second, strategy B, would provide a new route skirting the inland boundary of the urban area and providing a lorry route via a bascule bridge to the docks.

The cumulative distribution of noise-impact to dwellings demonstrates that the general environmental effect of adopting a "do nothing" policy is a deterioration in the noise climate, particularly in the higher bands. In comparing the noise-impact of the options it is found that worse noise conditions occur for the first option (A), based on the high-level bridge. Although the traffic on the bridge itself does not create significant noise impact, the option leaves the opportunity for only minor changes to the remainder of the highway network, creating heavy traffic flows along the existing urban streets. The use of the coastal route leads to high noise-impact to holiday accommodation, showing a poorer environment than the economic base.

By comparison, option B, based on the outer ring, channels the traffic away from the existing urban streets, relieving the noise-impact. The distribution shows less noise-impact to both dwellings and holiday accommodation. Option B was preferred in terms of

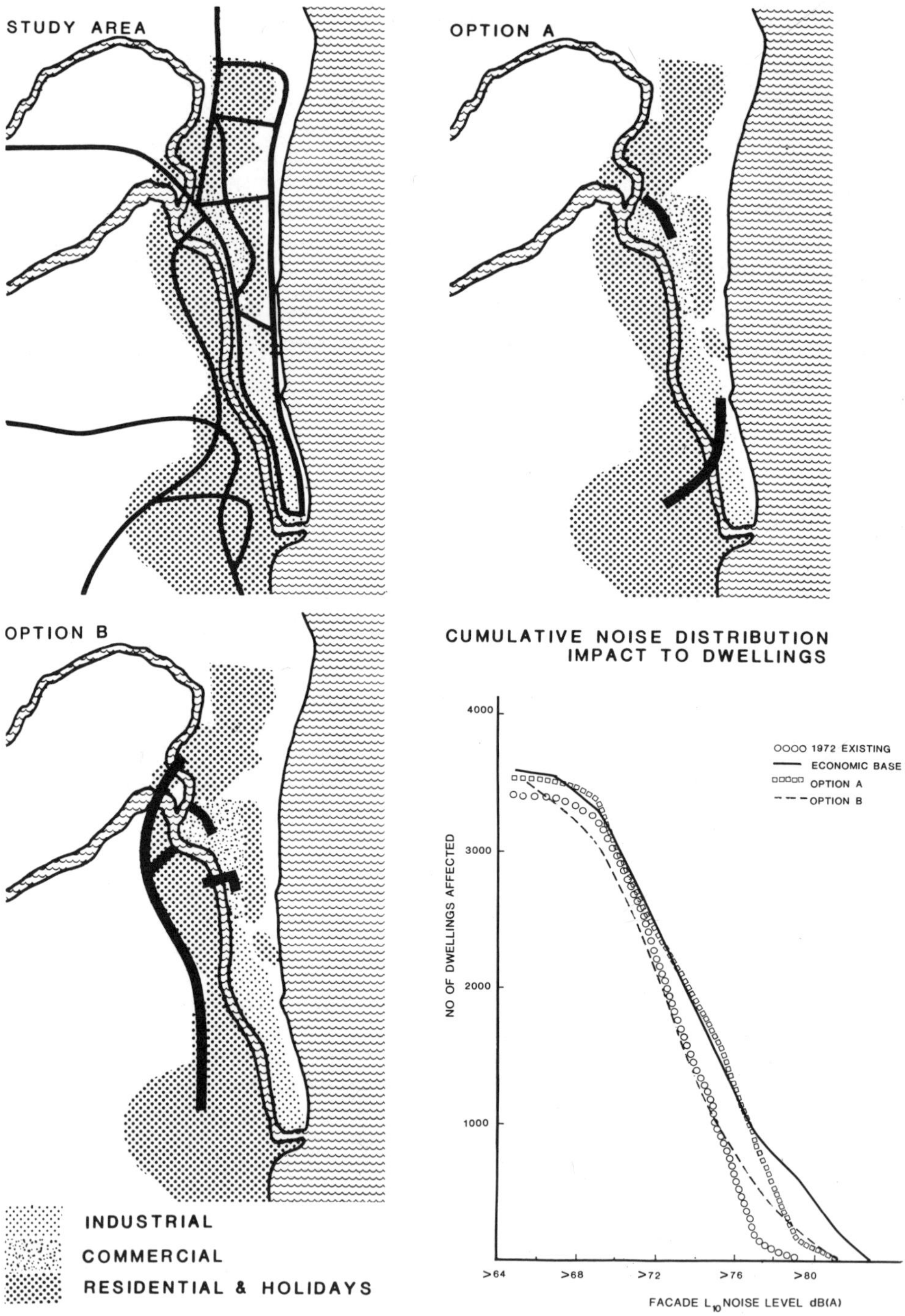

FIGURE 3. CASE STUDY-NOISE.

Option	Reaction Score
1972 Existing	1770
1986 Do-nothing	2040
1986 Option A	1990
1986 Option B	1810

TABLE 1. REACTION SCORES

noise-impact and produced a noise climate which showed only slight deterioration from existing conditions. The evaluation produced reaction scores for residences (Table 1) confirming the results of the distribution.

Visual Effects

The introduction of an elevated road structure into a neighbourhood will create significant visual impact. However, as nearly all the roads in a strategic network are at grade the evaluation of visual effects must also take into account the visual impact of moving vehicles. The second example considers a sector of a major urban area for which a transportation study was recently undertaken. A number of strategies were tested including both highway and public transport orientated tests. Their visual effects relative to the base situation were evaluated by visual satisfaction scores.

Table 2 shows the number of residences according to their respective levels of visual impact on a satisfaction scale from 0 to 7. The scale covers reactions to both road structures and

Visual Impact	0-1	1-2	2-3	3-4	4-5	5-6	6-7	Total
Base	472	601	1439	591	337	242	50	9974
Highway Option	454	911	1395	389	291	292	0	9356
Public Transport Option	600	603	139	464	370	206	50	9548

TABLE 2. LEVELS OF VISUAL IMPACT

traffic, from 'indifferent' to 'highly dissatisfied'. The overall visual score is the sum of the number of residences multiplied by the median of the range within which they fall. The figures illustrate the quantified improvement in visual effects for the highway option resulting from the attraction of traffic from local roads onto a purpose built primary distributor, which in itself causes only minimal visual impact. The public transport option, by restraining traffic movements along the main bus corridors, also shows an improvement over the base, though not as marked as the highway option.

Severance

The severing effects of new roads should not be considered in isolation from attendant improvements in pedestrian-vehicular conflict elsewhere in the strategic network. However, the disruption of pedestrian walk patterns and community/social interaction are probably the most pervasive consequences of the construction of high capacity roads. Figure 4 shows an example of two alternatives for improving the strategic network, each with widely varying severance effects. The figures are taken from a sector of a medium-sized town, and illustrate how the simple identification of land-use and pedestrian movement patterns can provide a preliminary basis for comparing the severance effects of alternative strategies.

The sector, which is largely residential, contains four radial routes (A, B, C and D) and a north/south railway on embankment, which forms a boundary to the community areas. School and shopping catchment areas, bus routes and conservation areas, all of which influence the pattern of pedestrian movements, are plotted.

The introduction of route D is common to both options. Option 1 also proposes a major highway along route A, which would create some severance of walk trips to primary schools. In addition the dwellings to the north of route A would be isolated from the bus service along route B. The major route would sever a conservation area.

By comparison option 2 proposes the widening of route B, which is a long-standing divisive element, with the tendency for separate communities to develop on either side. The widened route does not sever the school catchment areas. There is, however, some severance of school trips due to the link from route B to D, but crossings over the depressed route will allow segregated pedestrian movement. The bus services along route A and C would serve the neighbourhoods on either side of the main route.

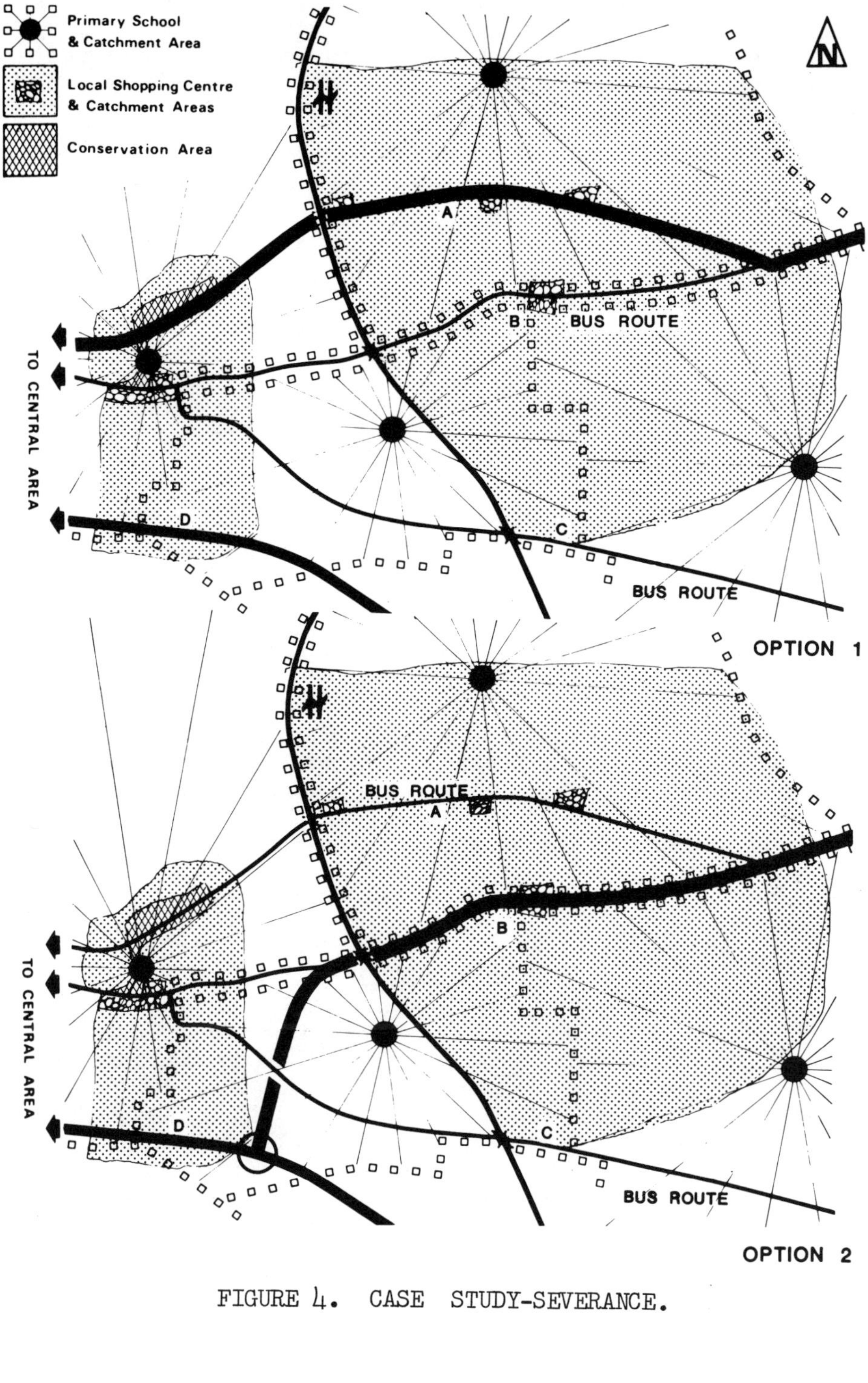

FIGURE 4. CASE STUDY-SEVERANCE.

Local shopping centres have been established on both routes A and B, and thus both options will present problems of severance at the local centres. The sum total of the physical and psychological divisive effects of option 2 are likely to be less severe than those of option 1.

CONCLUSIONS

In recent years the procedures for the quantification of environmental impact have been applied in many transportation studies, for towns and cities varying in size from some 100,000 to over 600,000 persons. The quantifications have provided a basis for environmental judgment, and have been undertaken within a limited budget. Although the number of factors quantified and the extent of the evaluation has varied, it has been found possible to use the procedures to provide a meaningful basis for strategic environmental evalution.

In order to extend the evaluation procedure to its full potential further research and development are required, fulfilling three basic requirements:

(a) A refinement of the prediction techniques and an extension of the range of environmental factors for which quantification is possible, in particular severance and pollution.

(b) Further knowledge regarding the translation from physical impact to reaction; and the determination of the relative importance of each factor, to allow a summation of impacts.

(c) The development of the relationship between environmental impact and monetary valuation, to allow environmental assessment to be incorporated in a formal cost/benefit framework.

ACKNOWLEDGEMENTS

Some of the evaluation techniques described in this paper were developed as part of a research contract with the Department of the Environment and the authors are grateful to the Department for permission to include these items. The authors also wish to note their thanks for the cooperation of Local Authorities in undertaking the case-studies. Thanks are recorded to Mr M. Goodge and Mr K. Sowerby for assistance given in the preparation of this paper.

REFERENCES

Department of the Environment (1972) New Roads in Towns. Report of the Urban Motorways Committee. London: HMSO.

Department of the Environment (1975) Calculation of Road Traffic Noise. London: HMSO.

Highway Research Board (1973) Air pollution controls for urban transportation. Report No.465. Washington,D.C.

Hopkinson,R.G., & Watson,N. (1974) Visual effects of roads and traffic. Report to the Department of the Environment.

Lassiere,A. (undated). The environmental evaluation of transport plans at the strategy level. Research report. Department of the Environment.

Mackay,J. & Partners (1975) Strategic environmental evaluation techniques: Bristol study.

O.E.C.D. (1973) Environmental implications of options in urban mobility. Environment Directorate. Paris.

Sherwood,P.T., & Bowers,P.H. (1970) Air pollution from road traffic. LR 352.Transport & Road Research Laboratory. Crowthorne.

Thompson,D.N. (1973) The Economics of Environmental Protection. Winthrop.

Urban Motorways Project Team (1973) Report to the Urban Motorways Committee. London: HMSO.

SECTION 6. ENVIRONMENTAL ASSESSMENT

The effects of noise, severance and pedestrianisation resulting from road traffic are the main topics of this section. In each case the effects are empirically determined. Severance is conceived by Lee and his colleagues as a complex social response to the presence of a physical barrier, and is comprehensively measured at the levels of perception, cognition and behaviour. Although they found that major urban roads did act as strong barriers to free movement, both cognitively and behaviourally, there was also sufficient accomodation to the barrier for people to shift their 'neighbourhood' back from the road and allow it to expand. Lee is less sanguine, however, than Latchford and Pickering, believing that further work needs to be done before one has a predictive index which may be used as a practical design tool.

Even conventional physical indices may be unsatisfactory for some kinds of environmental evaluation. In a before-and-after study of a pedestrianisation scheme, Hills found a lack of correspondence between the public's perception of the benefits of banning road traffic and the factors conventionally used in evaluation. Even where there was a correspondence, in relation to noise, safety, and access by transport, the usual formulation of these factors could not be guaranteed to match the way in which they were perceived. This is yet another example of the need to relate the professional and public viewpoints, to which the final section is devoted.

Although the next two papers both examine aspects of noise from urban traffic, they differ in their approach. Brown and Saur concentrate on establishing associations between noise annoyance and certain socio-economic, cultural and physical variables. The results of their study in Vancouver suggest that socio-economic status is a major factor in noise annoyance. They conclude that traffic management must more often include as criteria the socio-economic and cultural environments as well as traditional traffic planning criteria.

Ó'Catháin's attempt to evolve a systematic and manipulable theoretical model of housing layout takes research a step nearer to practical, environmental planning application. Traffic noise is related directly to housing configuration and density. Variables such as layout geometry, orientation, access and proximity to highways are taken into account. The aim is to derive feasible, optimal development patterns for given background environmental conditions, with clear implication for highway planning strategy. A process of evolving preferred built forms is demonstrated.

Mobility deprivation may appear to be an unlikely effect of road traffic. In common with a number of recent writers, however, Doubleday points to this as one of the most significant results in the rapid growth in motorisation. Its assessment is as important as the attention given to noise and severance. Doubleday offers a conceptual framework within which various uses of the term 'mobility' can be considered. He describes a simple approach to measuring the effects of different levels of mobility in the community. He concludes that a radically different approach to travel description and modelling is needed if mobility deprivation is to become a priority for social policy.

THE SOCIAL SEVERANCE EFFECTS OF MAJOR URBAN ROADS

Terence Lee & Stephen Tagg

Department of Psychology

University of Surrey

It is by now generally known that careful consideration is being given in this country to possible procedures for assessing the impact upon the environment of large scale and complex developments, such as airports, power stations, chemical works or gravel pits. Such procedures exist already in France, Sweden, New South Wales and, notably, in the United States. In the latter case, the National Environmental Policy Act 1969 requires that whenever developments at the Federal level are likely to significantly affect the quality of the human environment, there shall be prepared an Environmental Impact Statement which sets out in detail the nature of the action, any unavoidable adverse consequences, the relationship between short term and long term exploitation of the environment, any irretrievable commitments of resources involved and, finally, possible alternatives to the proposed action.

In Britain there is already a considerable and sophisticated machinery for the review of development proposals by the planning authority. This includes the arrangements for planning permission under the Town and Country Planning Act 1971, procedures by which proposals initiated by Government Departments must be considered by local planning authorities and a range of special legislation for controlling various forms of industrial pollution.

Hence there is understandably strong opposition in some quarters to the establishment of additional mandatory procedures, on the grounds that they will be costly to operate, cause aggravation and delay to projects that are in the national economic interest and because they will duplicate existing safeguards. The opposite view in the debate is that existing planning authorities have neither the time nor the necessary skills to carry out scrutiny of the detailed kind that is needed, that they possess no machinery for post hoc monitoring of cumulative or delayed 'impacts', that they lack the wider perspective that is required to consider alternatives to the proposed actions and this process is likely, anyway, to be limited by the 'permit/refuse' orientation that they are required to adopt at present. Finally, there is concern that the public at large combines an increasing concern for environmental conservation with a lack of confidence in the present decision making process. National and local groups are expressing themselves strongly on these lines by all means at their disposal, but particularly at public enquiries where they raise issues about the wider or related impacts of proposed developments that have sometimes patently not been considered at the design stage.

The desirability and feasibility of E.I.A. is likely to be studied and discussed for some time yet and if finally approved, its implementation would probably be contingent on more favourable economic conditions than obtain just now. It is by no means the purpose of the present paper to argue further the pros and cons, but to offer an indirect contribution by discussing the feasibility in one limited but important area - the assessment of the impact of major road proposals upon the environment.

Application of E.I.A. to Road Proposals.

Of course, we do not know whether road proposals would be subject to E.I.A., even if it were introduced. But it seems likely that there would be considerable public pressure in support of their inclusion because it has been an extremely prominent argument in some recent public enquiries that the system of public participation based on the examination of three alternative routes does not allow for consideration of the wider and long term impact of the new trunk road or motorway or, more particularly, of the possible alternative of having no road at all. Government Departments and other Crown bodies do not have to seek planning permission but are required to 'seek the views' of local planning authorities on their proposals. The process is presumably one that might be covered by the ubiquitous term 'negotiation'. It seems inescapable that if E.I.A. is introduced, for say, power stations, there will be strong pressure for its application also to roads.

If environmental impact statements become an integral part of the planning machinery, it seems most probable, for obvious reasons of expedience, that they will be required at the very earliest point in development, prior to the stage of outline planning permission in the private sector cases and at the first design stages for public sector roads etc. The task of identifying a stage that is late enough for the impact consequences to be realistically predicted but early enough that the various major options have not already been pre-empted is a daunting prospect and, indeed, a rock on which such an enterprise might well founder.

As with other developments, E.I.A. of roads would need to include, firstly, consideration of the physical impact - the effects on the adjacent environment and the 'intrusion' upon the view or outlook; secondly, the consequences for the natural ecology of plant and animal life and thirdly, the effect on the lives of people.

It is possible that all of these could be subjected to measurement using modern statistical scaling techniques. What is inconceivable, however, is that agreed procedures could be devised for combining quantitative measures of variables that are totally different in kind to yield a single index of impact. How can the estimated disturbance to the habitat of a marsh harrier be set off against the improved access to the Welsh holiday resorts for car workers from the Midlands? The conversion of the two into standard monetary terms, as sometimes suggested, might have a virtue in facilitating arithmetic, but the sum would be meaningless. The weighting of each variable must depend on subjective values and attitudes and while these values may sometimes be subject to widespread norms they will just as often be subject to conflicting sectional norms. The attitude towards improved access to Wales is likely to be sharply different for the indigenous population than for the would-be holiday makers. Ultimately then, the weighted combination could be made only by 'reasonable and just men', attempting to act fairly in the light of the social values of our society, which include a due regard for minorities.

This does not mean that the separate parameters should not be scientifically measured. On the contrary, the more information that is available about the impact of the proposed road, relative to other situations, upon the various aspects of the physical and ecological environment and upon peoples' lives, the more judicious the ultimate assessment can be and the more realistic the consideration of alternatives.

As an example, the impact of roads upon one aspect of peoples' lives - disturbance to urban neighbourhoods by severance - is considered here. The evidence is drawn from a recently completed study carried out on behalf of the Department of the Environment.

Social Severance by Urban Roads

The Urban Motorways Committee (1972) identified severance as one of the main environmental problems affecting road design and recommended a more comprehensive approach in which "The planning of new major urban roads should form an integral part of planning urban areas as a whole". More specifically, "Decisions on road location and design should take full account of possible severance of communities, and the design of the road should, from the beginning, pay special attention to providing pedestrian routes which are easy, agreeable, safe and convenient. Consideration should be given to the relocation of facilities which pedestrians use".

The research to be reported here represents an attempt to elucidate the concept of severance and to provide methods for measuring its extent. The primary objective was to increase understanding at the design stage in order to alleviate some of the adverse effects; but it should also be possible to roughly index these effects (the 'impact') for any given section of road.

Attention in the past has focused on two types of severance, 'physical' and 'psychological'. Both imply barriers to free movement. The first appears to be operationally defined in terms of the extent of enforced route changes imposed by the barrier and the second in terms of a 'felt aversion' - the essence of which is its indefinability.

This distinction is not a useful one. We advocate that all forms of human response to (or interaction with) a barrier in the environment should be regarded as severance. Attention should then be focused on the nature of these responses and on their measurement at various alternative levels; through perceptions of the barrier, through cognitions, through the structured organisation of these cognitions with motives and feelings (i.e. with attitudes) and finally, through behaviour. As part of this approach, severance responses would be measured to different physical barrier ('deterrence') parameters. For example, we have begun to look at the effects of pedestrian and car crossing points but not so far at road width, traffic flow, etc.

The main hypothesis that we brought to the enquiry is simple. We predicted that the presence of major roads or motorways would impose a shrinkage or atrophy of people's neighbourhood by constraining them to one side of the road line and that there would be a corresponding reduction of their social activity and its spatial patterning. These effects were expected to diminish with distance back from the road line and to be relieved by crossings.

Overall Research Design and Sampling

The overall design of the study was a comparison between two control sites and five major road sites, scattered throughout the London area. Other variables governing the selection of sites were social class and the length of time for which the major roads had been in existence. This is illustrated in Table 1. The age of the roads is approximately five years for the 'recently constructed' and ten years for the older ones. The Kingston bypass was added to the original design in order to include a comparison with a much longer established road - in this case about 30 years old. Only one of the roads (Greenwich) is classified as a motorway, the others are six lane dual carriageways which present the same barrier properties as motorways. Since it is not possible to select control sites of an appropriate size without roads, each of these has a smaller road running through its length.

There is little alternative to a clustered design of this kind because of the economies of survey interviewing but it has the obvious limitation that such small samples to represent control sites, age and social class, may be quite untypical. In order partially to offset this difficulty, sampling was carried out at each site on axes at right angles to the road line and at successive 200m steps back from the road up to 800m. This allows not only the exploration of the effects of distance within sites but an intersite comparison of the distance slopes. The sampling layout is shown in Figure 1.

	TIME SINCE CONSTRUCTION			
Urban/ working class community	No Major road (East Ham) A 124	Major road recently constructed (Greenwich) A 102(M)	Major road less recently constructed (Greenford) A 40	
Suburban/ middle class community	No Major road (Muswell Hill) - B 550	Major road recently constructed (Barnet) A 1	Major road less recently constructed (Chiswick) A 4	Major road long established (Kingston) A 3

TABLE 1. OVERALL DESIGN

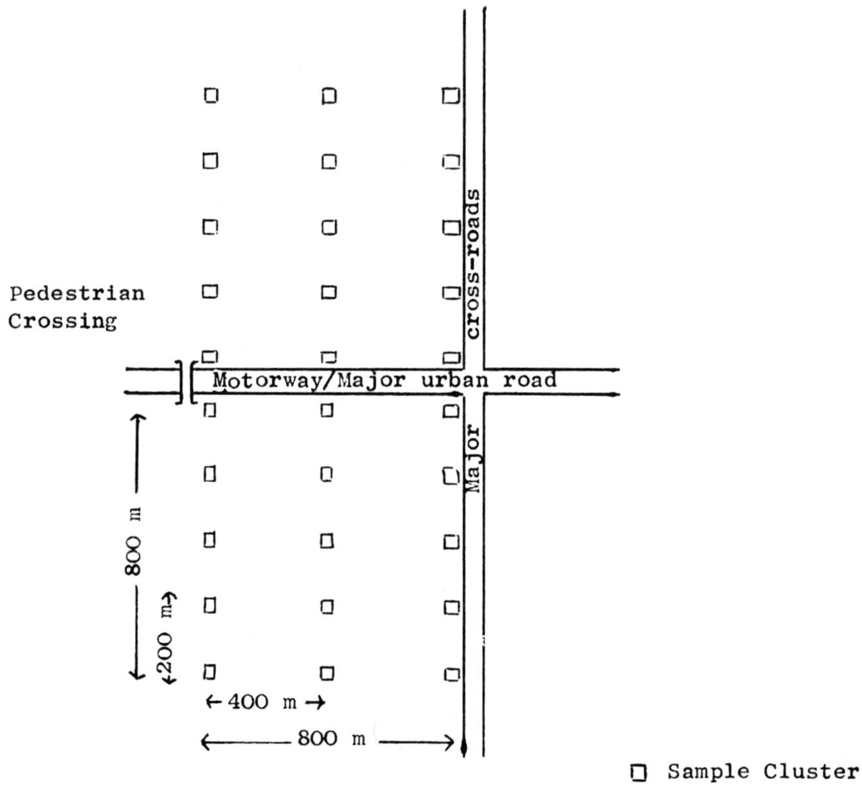

FIGURE 1. SAMPLE LAYOUT

Another important feature of the layout (and one that made it difficult to locate an adequate number of sites) is that development is approximately homogeneous on each side. The reason is that we wished to combine the results for each side of the road line in all cases to overcome the difficulty of including as a measured variable the density of different facilities. All severance responses are likely to be strongly influenced by the attractiveness of the facilities on the 'other side' but what constitutes a reason for crossing by people on one side is a reason for not crossing by people on the other. Hence by averaging these responses from each side we hoped to control for differences in the urban fabric.

The total sample (N = 960) was drawn to represent the adult population, using the Kish Procedure.

The Interview

The interview was conducted in the home and lasted 45 to 60 minutes. An advance letter referred to 'environmental planning' but there was no specific mention of roads at this stage or during the interview.

One section covered the usual biographical data. Another was concerned with a group of attitude questions, dealing with the respondents general liking for the area and any grounds of dislike. Social activity (trips) was probed in a series of specific items dealing with common activities, followed by open ended questioning on up to four activities nominated by the respondent. The frequency and the precise location of all of these were determined and the latter was converted to an exact grid reference using a transparent overlay at the coding stage.

A similar procedure (including grid reference location) was followed for determining social contacts, except that the initial listing by the respondent was followed by a procedure used by the interviewer to select a representative sample of these contacts for more detailed specification.

The respondents were given 24 cards bearing the names of prominent 'landmark points' in the wider locality and asked to sort these into two piles, those with which he was familiar, and others. In a second sorting, he was carefully instructed to distinguish those lying within his own neighbourhood. Later, he was asked to sort a sub-set of 10 cards in rank order of their distance from the home.

The landmark points for each site had been selected by dividing the area into a grid of 800m squares, 6 on each side, such that the two centre squares represented the sampling area. Two landmark points were selected from each grid square.

For a proportion of the sample, data were collected on street name identification, using a 'blank map' but this task was too time consuming and had to be discontinued.

Finally, the respondent was asked to define the boundaries of his neighbourhood and to draw it on an A4 sized map section of the locality.

Data Analysis

The neighbourhood maps were superimposed and their coincidence of boundaries provide useful visual inspection data. In addition, a number of indices were derived from the maps and the landmark points tasks. These may be seen as measures of three postulated processes which form part of the severance response; size adjustment of neighbourhood, bridging (across the road line) perceptions and behaviour, and shifting of neighbourhood and locality perceptions:-

Size Adjustment.

1. Area in hectares of neighbourhood map.
2. No. of points included in neighbourhood.
3. Total points known.

Bridging.

4. Crossing road line with neighbourhood boundary.
5. Proportion of landmark points included from the 'other side'.
6. Ditto known points.
7. Proportion of crossover trips.
8. Proportion of crossover social contacts.

Shifting.

9. Displacement of neighbourhood map 'centre of gravity' (digitiser) from home, in the direction away from the road line.
10. Disequilibrium of 'other side' landmarks in distance rank ordering.

The Presence and Age of the Motorway

The principal comparison drawn from the data is between the two control sites and five major road sites. In addition, the major road sites had been selected to enable us to look at the effects of age and possible adaptation effects. In each case the dependent variables are the various indices of severance. Initially these were analysed separately, and they are grouped here for purposes of discussion under the headings of 'bridging' 'size and adjustment' and 'shifting'.

The simplest index of <u>bridging</u> is the proportion of neighbourhood boundaries that cross the road line and Figure 2 shows minimum severance for the control sites and an age effect extending up to maximum severance for the 30 year old Kingston Bypass ($F = 8.62$; $p < .001$). The other indices of bridging behaviour are broadly confirming of the main effect (except that the measure of differential knowledge shows no effect at all) but they do not discriminate by the age of the road. A good example is the mean proportion of all trips which cross the road line. This is shown in Figure 3, where it will be seen that although the control sites are markedly different from the major road sites (in showing substantially more bridging behaviour) the latter are bunched together.

Turning to the indices of <u>size adjustment</u>, the presence of a major road does <u>not</u> appear to attenuate the area of neighbourhoods. In fact, the reverse. The mean areas at the major road sites are almost 50% higher than at control sites and the differences are highly significant ($F = 6.79$; $p < .001$). There is no consistent effect with the age of the major road site, but such an effect is strongly apparent with the second index of size adjustment, i.e. the number of landmark 'points' included in the neighbourhood. The pattern here is highly consistent, as will be seen from Figure 4. The control sites are different from all other site pairs and the means are 10.07, 11.83 and 12.75 respectively (t tests significant at $< .01$) with increasing age. The 'number points known' confirms the main effect but does not consistently support the age effect.

What is clear so far is that people at major road sites do respond to the road line as a barrier, substantially reducing their commitment to the other side (though not their knowledge of it) and their overt trip behaviour - but that they make a positive adjustment by enlarging their neighbourhood; incorporating 'additional' territory on the same side.

The evidence of <u>'shifting back'</u> is just sufficient to confirm this general pattern, but since the extent of the shift is not large overall and almost non-existent in working-class sites, it must be presumed that the size adjustment mainly occurs laterally, i.e. along the road line. This makes sense when it is considered that shifts in this direction would tend to include more crossing points.

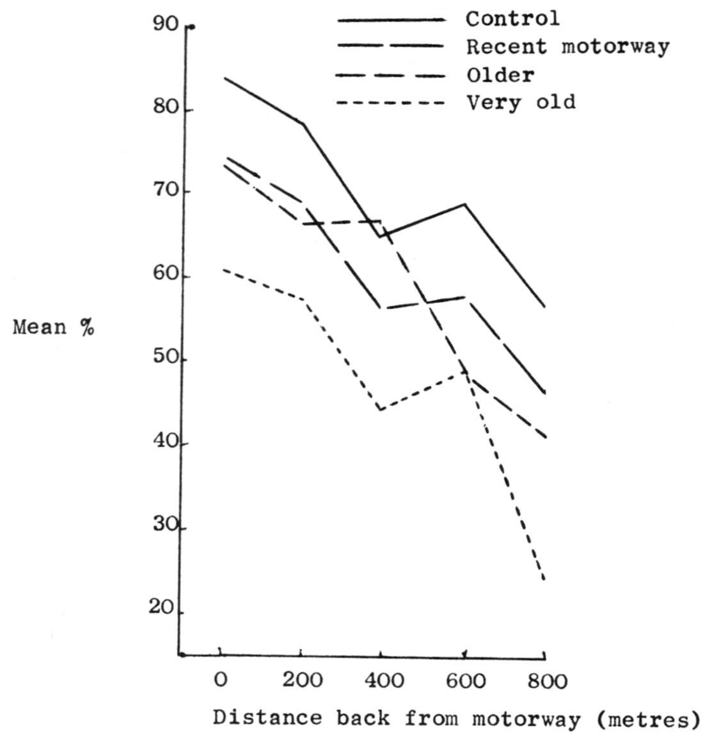

FIGURE 2. THE EFFECTS OF AGE OF MOTORWAY (Mean percent lines crossing)

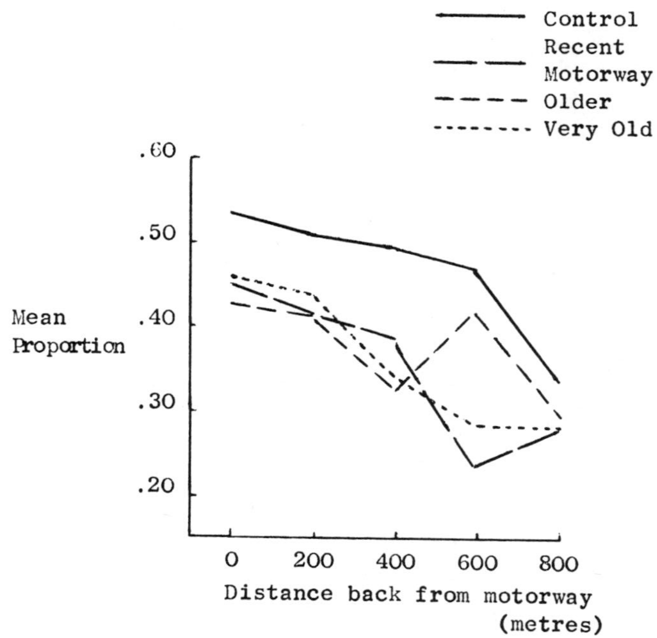

FIGURE 3. THE EFFECTS OF AGE OF MOTORWAY (Mean proportion all trips crossing)

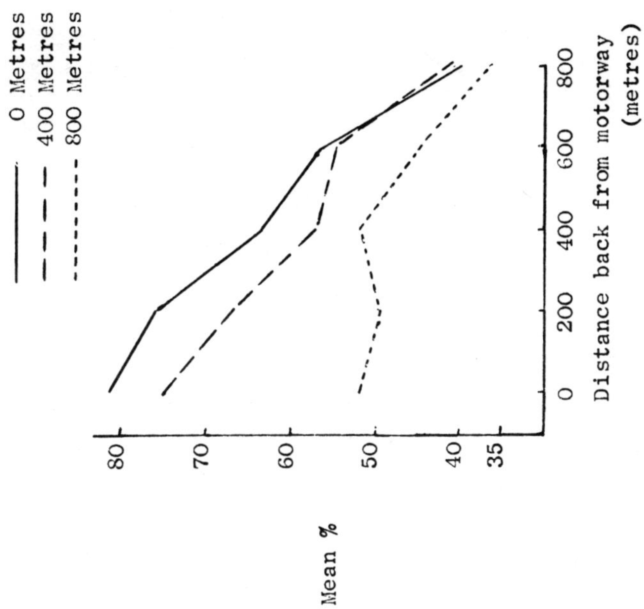

FIGURE 5. THE EFFECTS OF DISTANCE FROM MAJOR CROSSING-POINT (Mean neighbourboods crossing)

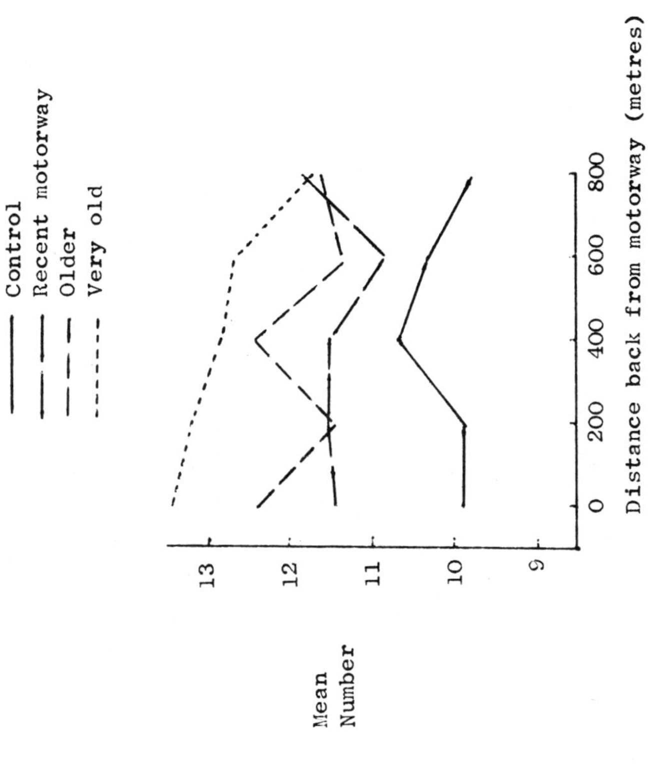

FIGURE 4. THE EFFECTS OF AGE OF MOTORWAY (Mean number of points included)

Most of the indices which show differences between control and major road sites indicate, surprisingly, that the effects extend back as far as 800m from the road line. For example, although the proportion of crossing trips naturally falls away with distance back from the road line, the disparity between the control and major road sites remains approximately the same (Figure 5).

Distance From a Major Crossing Axis

It will be recalled that the sample was drawn on three axes at right angles to the road line and one of these axes constitutes a major crossing road which gives easy access by car or on foot to the other side. In all cases except one (Chiswick) this axis is at one or the other end of the site and the other axes can therefore be regarded as laterally displaced by approximately 400m or 800m. Some of them occur at a crossing for pedestrians only, others at a minor road.

The same methods of analysis were used to compare the severance response and to see, in effect, if physical-environmental facilitators of this kind produce measurable differences in peoples perceptions and behaviour.

The answer is plainly 'yes', but since the type of evidence to be deployed has now, we hope, become familiar, the findings will be described more generally.

The neighbourhoods of people who live actually on the major axis are relatively smaller, at least for those up to 400m back from the road line, although this effect is only significant for the criterion of number of landmark points known ($F = 5.21$; $p < .01$). Again, this fits our emerging pattern of more barrier equals larger neighbourhood.

In the same vein, the bridging index of number of neighbourhood boundaries crossing the road line is a copy book effect. There are large differences between the three axes in the expected direction (means = 63.80, 59.44 and 46.93; $F = 5.74$; $p < .01$) but there is also a convergence with distance back. Notwithstanding, the convergence is not quite complete at 800m back. The other bridging indices show similar effects though with less consistency. The proportion of crossover trips is significantly related to lateral displacement ($F = 3.59$; $p < .05$) as is also the proportion of crossover landmarks included in the neighbourhood. In both cases the effect shows only uncertain signs of diminishing at 800m back and is at its maximum at 600m back from the road line.

In one sense (i.e. that greater ease in crossing the barrier increases the crossover activity) these findings are obvious. But the fact that adjustments of neighbourhood size take place where crossing is frustrated is not so predictable and it fits the findings already reported. The fact that the effects of a crossing persist up to 800m at least is also surprising.

Pedestrian Crossings

Comparisons were made between 15 axes which meet the road line at or near a pedestrian crossing and 6 axes where there is no crossing. The former included vehicle crossings.

There is virtually no consistent size adjustment for those living on a pedestrian axis, but bridging is substantially increased on every criterion. The differences in bridging persist up to and presumably beyond the full 800m back from the road line, except for the criterion of neighbourhood boundaries crossing the line, which shows a convergence of the means of respondents living at this point.

Shifting back is also greater for those who are apparently forced to make an adjustment because of the lack of a pedestrian crossing (t, one tailed = 1.74; $p < .05$).

Subject Variables

Age and length of residence are correlated with the severance indices, and could be included in a regression equation to predict the severance response for a given area, providing their distribution could be determined. It is unlikely that they would explain much of the variance. The general effect is predictable, i.e. an increase in bridging up to about age 45 and a decrease thereafter, and similar effects with length of residence, partly because the two are interrelated.

Social class effects were estimated in two ways, i.e. by individual and by site. The differences in this case are very important, with a greater tendency for middle-class sites to show the general pattern of adjustment to severance than is emerging - i.e. reduced bridging, enlarged neighbourhood and some 'shift back'. The implications of these differences for routeing policy are too large to discuss here.

The Distribution of Facilities

It is self evident that both bridging and size adjustment will be strongly influenced by the proportional distribution of essential (e.g. schools, workplaces, doctors' surgeries) and attractive (e.g. clubs, libraries) facilities on the two sides of a road barrier. The impact of severance is much greater for a residential area where such facilities are sparse than for one where they are prolific, with correspondingly less need to surmount the barrier.

In order to measure the effects of other variables, the main strategy of the present investigation was to select areas with a roughly equal distribution of facilities on each side of the barrier and, more important, this variable was controlled by averaging the responses of the two sides.

However, in order to demonstrate the possibilities, a crude measure of "side attractiveness" was made on a three point scale by an examination of map data and from local knowledge. This index was entered in a multiple regression equation with other variables.

The criterion variable was SUMSI - a sum of eight severance indices, combined after standardising so that each contributed equally to the score. Distribution of facilities made a useful contribution towards explaining the variance, but the detailed results are not presented here because SUMSI was a crude criterion and the results would be misleading. In particular, it was computed before it was appreciated that the area of neighbourhoods increases with severance and not the reverse - so its loading at present reduces the validity of SUMSI.

It is clear that the proportional distribution of facilities should be included in the calculation of severance effects. It should not create too much difficulty. For any given road line proposal it would first be necessary to specify areas of equal size on each side. On the basis of our evidence they should extend up to 1000m back from the road. Their lateral extent is more arbitrary. Basically, so long as the dimension is equivalent for both sides and large enough to cancel out idiosyncracy, the only other limitation is that it should roughly correspond to the average span of neighbourhoods. Also, since vehicle and pedestrian crossings clearly serve to channel bridging behaviour they should not be selected as the boundaries of the measurement areas - rather, these should be located midway between crossings.

The next step would be to make a simple count of facilities, which is usually possible from an existing Land Use Survey and to express each side as a simple proportion of the total.

It frequently happens (as for example with the M6 section through West Birmingham) that residential development on one side is complemented by wholly industrial (or semi-derelict) land on the other. In such cases the 'facility ratio' will be very high and the degree of severance correspondingly high.

The main difficulty that can be envisaged in the application of such a 'facility ratio' is where the selected area is itself subdivided by barriers such as canals, railways or further roads. Those which run athwart the main barrier are not likely to be the problem because they usually affect both sides equally; more serious are those which run parallel on one side only, with essential/attractive facilities on the further side. The only way of overcoming this obstacle would seem to be to enter a correction factor based on the apparent strength of the subsidiary barrier and its distance from the main one. The importance of this element in the equation may be seen within the context of the present study by pointing out that subsidiary barriers will tend to prevent the size and shifting-back adjustments.

Conclusion

It has been shown that a range of measures derived from the urban environment and from its population are related to the severance response. The identification of these measures for a given area, selected as described above, would make it possible to estimate the degree of severance by applying beta weights derived from an empirical study, i.e. the present one or, preferably, further confirmatory investigations.

The size of the task involved in deriving satisfactory Environmental Input Statements can be gauged from the fact that this would only be one input, i.e. that dealing with the effects on the lives of local residents.

PEOPLE'S RESPONSES TO PEDESTRIANISATION SCHEMES

P.J. Hills

Institute for Transport Studies

University of Leeds

Since the early 1960's there has been a growing awareness of the serious damage that road traffic can do, if unrestrained, to the urban environment (Buchanan Report, 1963). Although the recent reaction against urban motorway building stems from a concern for the fabric of cities through which roads may be constructed, there is a continuing concern about the threats to our living environment posed by vehicular traffic already using the ordinary street system. This is especially so in the dense central areas of old towns and cities. Inevitably, this generalised concern has been translated into political calls for action - action to ban traffic entirely, to confine it to certain types or times, or to restrain its volume, speed and composition in the interests of "the environment". Since banning traffic over wide areas tends to become impracticable (as well as impolitic), limits have to be imposed on the degree of restraint applied. In any case, the congestion and other resource-costs imposed upon vehicle-users, as restraint is applied, increasingly offsets the social benefits of the environmental improvements that are gained. All pedestrianisation schemes are involved in this classic trade-off situation.

Whilst acknowledging the difficulties of measuring the impacts of traffic upon the immediate environment, numerous attempts have been made in recent years. In Britain, the work of Crompton and Gilbert (1971) and of Joyce and Williams (1972) is probably the most advanced. OECD (1973) has published a comprehensive report on the effects of traffic and roads in urban areas in relation to its member countries. Efforts to incorporate environmental impacts into the transportation planning process have been fewer, although reports by Wigan (1975) and Manheim & Suhrbier (1974) have emerged.

More speculatively, Dawson (1974) is continuing with his experiments using a full-size simulator.

A fundamental weakness, however, lies in the assumption (frequently made) that to identify a source of disturbance to the environment and to measure it, is axiomatically part of an evaluation. It is nuisance that is reflected as environmental cost, not the physical disturbance itself. Thus, unless the changes in sound or vibration levels, fumes or smoke concentrations, visual or physical obstructions to movement are perceived as nuisances, they contribute nothing to the social costs imposed by traffic. Strictly speaking, we should except those sources of pollution which give rise to clinical hazards at levels which are unperceived by the subjects (lead and other trace metals may be in this category). For the most part, in relation to traffic, we are dealing with environmental conditions that may be uncomfortable or inconvenient but which are not inimical to health or life itself.

The realisation that we must seek not just to measure the objective change in conditions but also subjective response of people affected has provided the mainspring for our research programme in the Environmental Studies Unit at the University of Leeds (Hills, 1975). A project, launched in 1973 and using the shopping area in Central Liverpool as a case-study, is still in progress; but it has already demonstrated the practicability of applying psychometric techniques to the problem of environmental evaluation. The results achieved so far are outlined in this paper.

PERCEIVED ATTRIBUTES OF THE ENVIRONMENT

The aspect of environmental disturbance which has received the most study is, fortunately, the one most systematically investigated in terms of subjective response. Noise is, by definition, a subjective term (i.e. sound unwanted by the subject) and, in the evolution of noise measurement techniques, reliance upon pure sound pressure-level, as a proxy for the nuisance perceived by people exposed to it, has long ago been abandoned. Instead, more or less elaborate weighting systems and statistical conversions have been tried to make noise index measurements accord with human responses (Griffiths & Langdon, 1968). Arguably this has yet to be fully achieved.

Nevertheless, similar psycho-physical investigations have not been undertaken for most of the other measurable aspects of environmental disturbance. For those thought not to be measurable, their assessment is often dismissed as being "too subjective" or determined "merely by political whim". Our contention is that a particular aspect of the environment is only worth measuring if it is appropriate and that the method of measurement is bound to be psycho-physical in nature, since the test of appropriateness is whether or not that aspect is perceived as a nuisance.

The technique we chose to adopt in Liverpool, to determine which aspects of the environment were perceived, involved non-directive tape-recorded interviews of people in the street. The aim was to record and analyse a broad range of opinions given spontaneously in answer to the question "What are your impressions of Church Street?" The interviewer was strictly confined in the type of prompting that was allowed and the interview continued until, by specific probing, the required level of detail was obtained. The sites for the interviews were located in Church Street, the major shopping street in Central Liverpool, which was at that time (April 1973) carrying heavy traffic. The 3,900 impressions obtained from 1,825 respondents were then subjected to a content analysis, such that, with the minimum of semantic interpretation, a total of 66 groups of impressions were identified. No fewer than 23 of these groups were of favourable impressions: in all, nearly one third (1,267 responses) were expressions positively in favour of the street as it then was.

By rank-ordering the groups on the basis of their frequency of mention, some idea of their relative importance as perceived by users of the street was obtained. Little or no effect was caused to the rankings by weighting responses on their order of mention (it may have been invalid, anyway, to assume that an impression mentioned first was most "weighty" in perception terms). Of the grouped impressions, both favourable and unfavourable, eight clearly emerged as predominant. These are shown in Table 1 together with their constituent impressions. Predictably delay (18.4% of all adverse responses) and danger (12.8%) were most frequently mentioned; whereas, surprisingly, all aspects of air pollution combined to less than 2% of responses and vibration was scarcely mentioned at all. Of the favourable responses, 45% related to the shops, reflecting the main land-use activity in the area. By selecting the 8 attributes, about 78% of all responses were wholly included and a further 20% partially included (depending upon interpretation). Less than 3%, with both air pollution and vibration among them, were excluded. These eight attributes were then taken forward to the structured questionnaire which was designed to measure attitudes.

Attributes and Impressions	No. mentions per impression	No. mentions per attribute	% of all mentions
SHOPS		821	21.2
- Poor range of shops and goods	122		
- Lack of food shops and cafes	40		
- Shops too crowded	10		
- Bad service	9		
- Good range of shops and goods	569		
- Good cafes	8		
- Good service	63		
AMOUNT OF NOISE		174	4.5
- Noise from all traffic	120		
- Noise from buses and lorries	48		
- Excessive blaring music	6		
EFFECT OF CROWDS		376	9.7
- Footpath too narrow	329		
- Crowd density too great	47		
FREEDOM TO CROSS		658	16.9
- Delays in crossing	484		
- Insufficient crossings	133		
- Adequate crossings	41		
SAFETY FOR PEDESTRIANS YOUR FEELINGS OF SAFETY		513	13.2
- Danger from vehicles	336		
- Speed of traffic excessive	127		
- Intimidation by buses and lorries	49		
- Good driving	1		

TABLE 1.1 COVERAGE OF THE ATTRIBUTES SELECTED BY PERCENTAGE OF IMPRESSIONS INCORPORATED WITHIN THEM

Continued ....

TABLE 1.1 (cont.)

Attributes and Impressions	No. mentions per impression	No. mentions per attribute	% of all mentions
ACCESS BY TRANSPORT		110	2.8
- Difficulty of bus access	11		
- Car parking insufficient	47		
- Ease of bus and train access	46		
- Car parking handy	4		
- Closure of Church Lane	2		
OVERALL APPEARANCE		382	9.8
- Dirty buildings	76		
- Surroundings scruffy	52		
- Bad new buildings	41		
- Undistinguished buildings	34		
- Nasty shop windows	9		
- Bad old buildings	7		
- Good old buildings	52		
- Surroundings bright	34		
- Shop windows nice	27		
- Buildings clean	5		
- Good mix of old and new	1		
- Fine, unique buildings	10		
- Good new buildings	16		
- Street furniture ugly	16		
- Unpleasant look of traffic	1		
- Ugly adverts	1		
TOTAL FOR THE 8 SELECTED ATTRIBUTES		3,034	78.0

TABLE 1.1 COVERAGE OF THE ATTRIBUTES SELECTED BY PERCENTAGE OF IMPRESSIONS INCORPORATED WITHIN THEM

Attributes and Impressions	No. mentions per impression	No. mentions per attribute	% of all mentions
Rebuilding nuisance	171		4.4
Atmosphere grouping	(175)	not sufficiently specific for inclusion	4.5
- Bustle, lively	117		
- Atmosphere and people friendly	44		
- Cosmopolitan	7		
- Traffic lively	4		
- Social centre	2		
- Barrow traders	1		
Shops compact	191	Covered by SHOPS ACCESS and CROWDS	5.7
Shops too close together	12		
Shops scattered	17		
Dirt, dust and litter	46	APPEARANCE	1.5
Street clean	11		
Insufficient time to cross at lights	46	FREEDOM TO CROSS and SAFETY	1.2
Street confusing, distracting	42	EFFECT OF CROWDS	1.1
Lack of trees and flowers	21	APPEARANCE	1.0
Street closed in	6		
Broad street	13		
NUMBER AND PROPORTION OF IMPRESSIONS PARTIALLY COVERED	751		19.4

TABLE 1.2 IMPRESSIONS PARTIALLY COVERED BY THE 8 ATTRIBUTES

Attributes and Impressions	No. mentions per impression	% of all mentions
Air pollution group	76	1.9
People vulgar, destructive	15	
Lack of cover from weather	6	
Lack of WCs	4	
Lack of seating	3	
Lack of pubs	1	
NUMBER AND PROPORTION OF IMPRESSIONS LARGELY EXCLUDED	106	2.6

TABLE 1.3 IMPRESSIONS LARGELY EXCLUDED FROM THE 8 ATTRIBUTES

THE MEASUREMENT OF ATTITUDES

The structured interviews were carried out in October 1973 at the same locations in Church Street, at two other locations in Lord Street (also a shopping street) and in a "control" street. Although traffic was to be banned entirely in Church Street from March 1974, it was only to be restricted to buses and service vehicles in Lord Street. The control street chosen was in Warrington, a town about 20 miles (32 km) outside Liverpool, where no changes in the traffic system were proposed. The questionnaire, which was to be administered on-street, was in three distinct parts: (i) a fairly standard set of questions relating to person-type and trip-characteristics (age, sex, accompaniment by children, origin, mode, purpose and frequency of trip, etc); (ii) a simple example of the attitude measurement system we had devised, which the interviewers used for the purpose of explanation; and (iii) the full "matrix" of eight attributes which the interviewees then completed for themselves. This matrix and the instructions for completing it are shown as Figure 1. The same three-part quest-ionnaires were used in Lord Street and Bridge Street (Warrington).

The form of the matrix had evolved from previous pilot surveys (Rees et al., 1975). For several reasons we abandoned the semantic differential approach and chose instead the continuous subdivided 10 cm scale where the standard bi-polar descriptions of the attri-butes themselves were telescoped into the fewest possible words for their meaning to be conveyed. In addition, the interviewers were permitted to elaborate these descriptions to a limited extent,

HOW DO YOU RATE THE FOLLOWING FEATURES OF CHURCH STREET ?

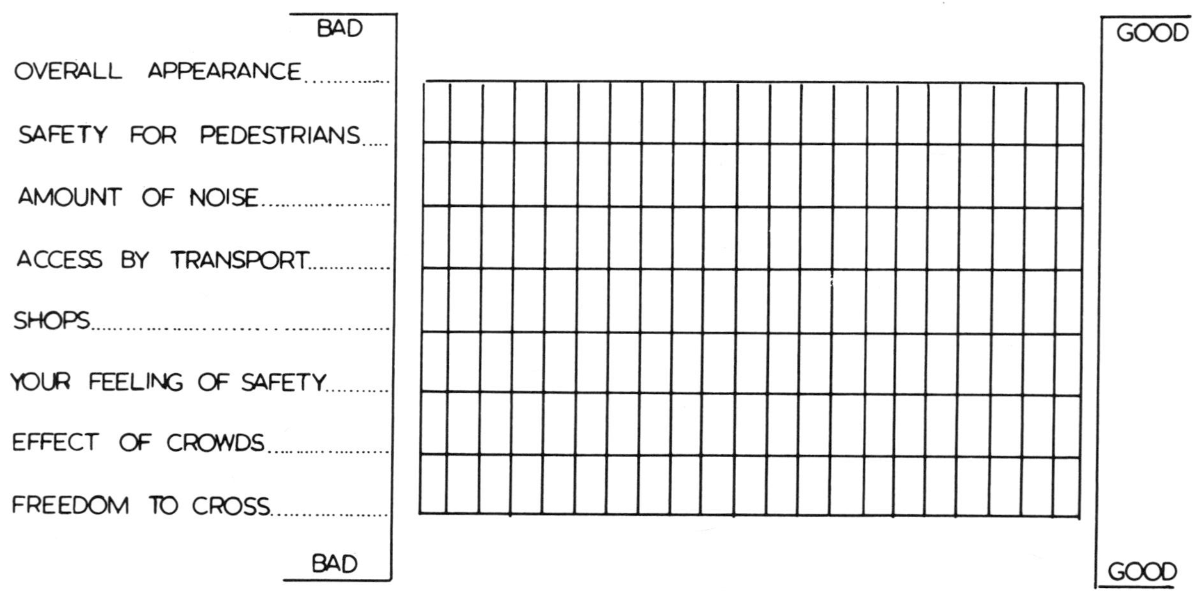

Remember, these are features of any street. Read through the list, and think of how good or bad each feature is in Church Street.

Mark down the feature that is BEST and the feature that is WORST in Church Street.

Now mark the other features in between – just put the crosses where you feel they should be.

FIGURE 1. MATRIX USED IN STRUCTURED INTERVIEWS FOR ATTITUDE-RATING

although interpretation was generally left to the interviewee. Initial fears that a self-administered technique of attitude scoring, as part of an on-street interview, would prove unwieldy were dispelled in the pilot surveys. In fact, the highest overall refusal rate in the three streets was about one-third, with males and younger age-groups responding more readily than females and older people.

The advantage of the matrix form adopted, reinforced by the sequence of instructions that respondents were given, was that the attitude expressed towards each attribute was not isolated, but was always within the context of the other seven attributes which together made up a "total" evaluation of the street. Thus, by having to appraise all eight and choose the one they felt was 'best' and then the one they felt was 'worst', they were in fact forming their own evaluative boundaries within which their attitude to the remaining attributes had to fall. Simultaneously, the relative strengths of affection for each, between the polar extremes of 'good' and 'bad', were determined. By this means, the data could be translated into "scores" ranging from 21 (good extreme) down to 1 (bad extreme) on an equal interval basis. The achievement of interval-scale data was to enable us to carry out fairly high levels of analysis on the data (regression and factor analyses).

Far from being irritated or bewildered, the majority of interviewees responded with interest and comprehended the task to the full. An important aspect of these particular attitude measurements concerned the immediacy of the responses. There was no recollection required nor were hypothetical situations put forward. The subjective response was merely to conditions as they prevailed at the time. A total of 3,700 questionnaires were completed (with less than 200 unfit for analysis) during the 'before' surveys. A similar number was achieved exactly one year later, in October 1974, after the changes to the traffic system in Liverpool had been accomplished. From these two sets of data relating to the 'before' and 'after' situations, we were able to deduce attitude shifts.

THE ANALYSIS OF ATTITUDE SCORES

The first task of analysis was to check whether there were any significant differences between the sample population in each street surveyed. Since the frequency distributions of attitude score were all other than normal a two-stage comparison was undertaken; firstly, a Kolmogorov-Smirnov test was performed which revealed significant differences (mostly at the 5% level) in about two-thirds of the samples tested; and secondly, for these samples where differences were revealed, straightforward comparisons of mean scores were made. Some of the differences were explicable on the grounds of observable differences in the streets themselves; others (e.g. attitudes towards Crowdedness) are more difficult to explain. Conceivably, the different mixes of person types and trip characteristics may have accounted for the differences between the samples. This was the next task of analysis.

The comparison of attitude scores 'before' and 'after' pedestrianisation would have been greatly eased if the frequency distributions had been at least para-normal. Unfortunately, despite an exhaustive breakdown of the categories, the only attribute to show strong signs of normality was Access by Transport, when disaggregated by origin; for the most part the distributions were distinctive and plausible (see Figure 2) but often rather far from normal. Thus there were no obvious reasons to suggest that the attitudes of any one sub-group of the population were dominant in influencing the overall profiles of attitude or that, towards the opposite poles, markedly different mixes of characteristic were to be found. To test this statistically, a multi-linear step-wise regression analysis was performed to find out what relationship (if any) existed between the peoples' characteristics and their attitudes.

In general, the coefficients of correlation obtained were very small (ranging from 0.134 to 0.381) and numerous variables (up to 10 or 11) tended to crop up in the equations, indicating that the effects of person-type and trip-characteristics were extremely weak and diffuse. Perversely, one of the more frequent variables to enter was that of Interviewer (which we thought should be investigated for its effect). Although there was no consistent pattern evident, it seems to underline the great importance in this kind of work of interviewer-selection and training. Of the other characteristics that entered, only a few similarities could be seen between the three equations for each attribute. The age groups 60+ and 46 - 60 did appear in several equations for the Liverpool streets, suggesting that age differences could contribute to attitude. On the other hand, in the regressions of Access by Transport the equations for all three streets revealed negative coefficients for car-drivers. We took this to reflect a real difference in accessibility for this group, rather than any psychological difference, as compared to bus-users and train-users.

THE EFFECTS OF PEDESTRIANISATION

Examples of the frequency distributions of attitude-scores for both the 'before' and 'after' surveys are plotted in Figure 2. In every case, the 'before' profiles of distribution are as one

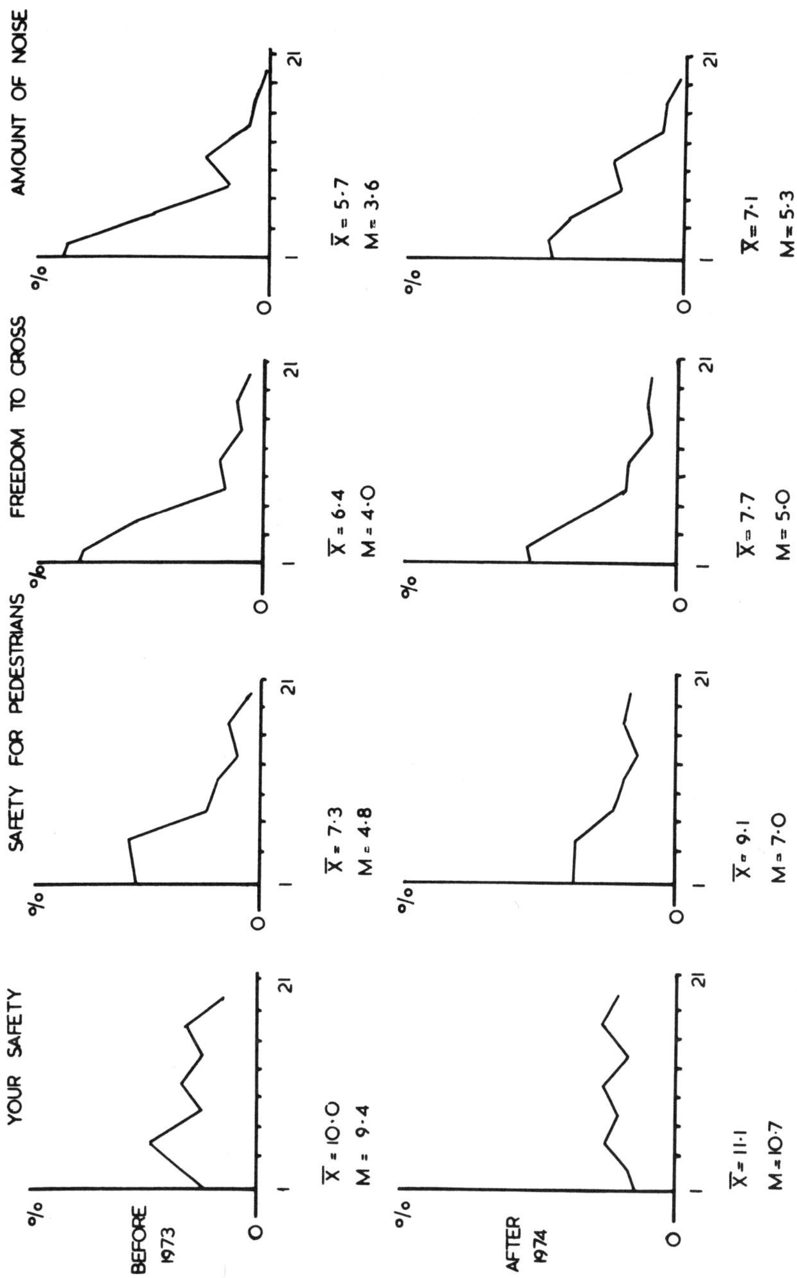

FIGURE 2.1. LORD STREET. RELATIVE FREQUENCY DISTRIBUTION OF ATTITUDE SCORES FOR TOTAL SAMPLE

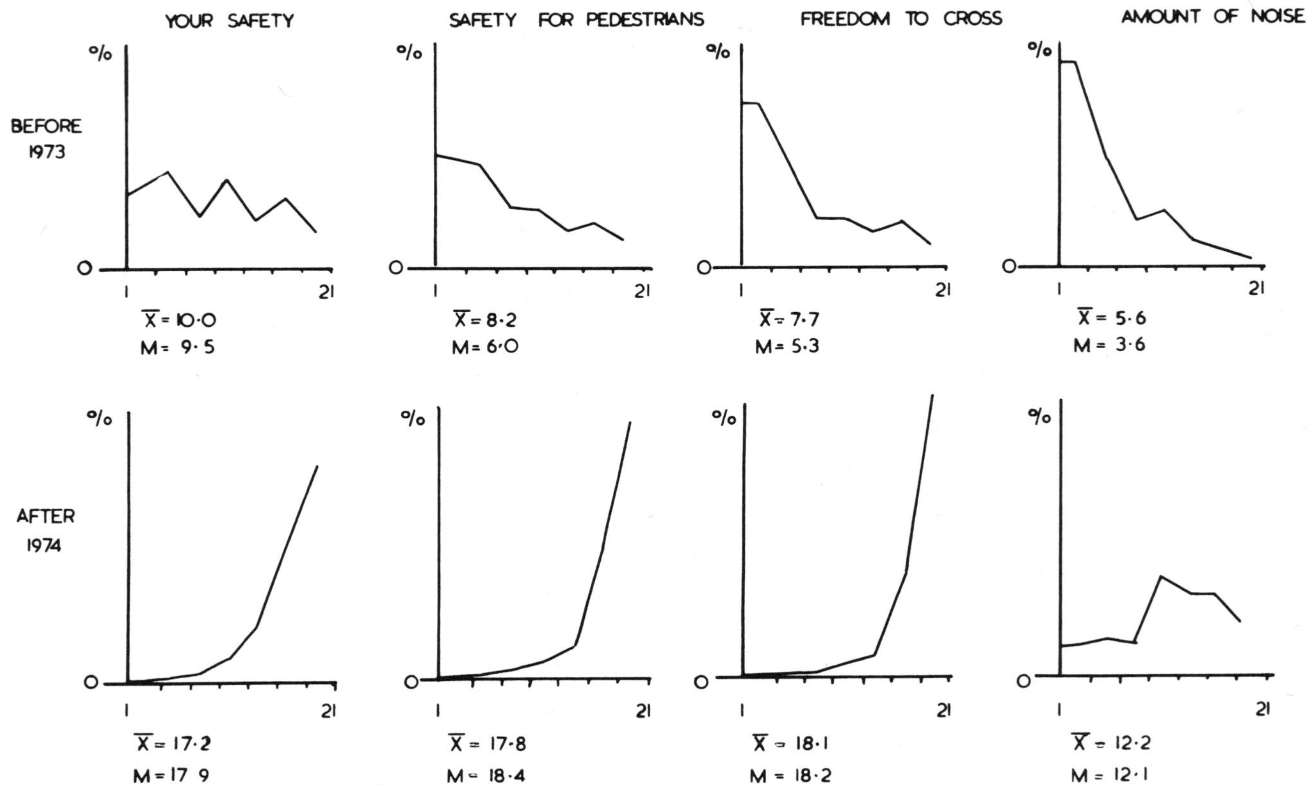

FIGURE 2.2. CHURCH STREET. RELATIVE FREQUENCY DISTRIBUTION OF ATTITUDE SCORES FOR TOTAL SAMPLE

would expect, with Noise, Freedom to Cross and Effect of Crowds strongly skewed towards the 'bad' end. Shops, by contrast, are highly regarded. Equally, the profiles between the two Liverpool streets for the same attribute are very similar in the 'before' conditions, which confirms the lack of significant differences within the sample populations.

Encouraging results are the very marked shifts in the profiles for Church Street when all the traffic had been removed and the very slight shifts in those for Lord Street when only cars had been banned. The median attitude-scores for Noise, for example, were identical in the two streets before the change (M = 3.6), but rose dramatically in Church Street to 12.1. While in Lord Street it rose to only 5.3 and the profile had scarcely altered. Because of the concentration of heavy vehicles remaining in Lord Street after the change, the L_{10} noise level had fallen only by 2 or 3 dB(A), whereas in Church Street free-of-traffic it had fallen 10 to 12 dB(A). The other dramatic shifts to occur in Church Street were on Freedom to Cross and the two safety attributes. In each case, the shifts in Lord Street were very slight and the profiles unaltered.

The interesting differences revealed between the two safety attributes supported the initial decision to carry them forward separately. This decision arose from the clear distinction made by many people in the tape-recorded interviews between concern for their own safety and an altruistic concern for the safety of others. The differences in profile, although maintained in Lord Street after the change, appear to collapse in Church Street once the danger from traffic is eliminated. The residual fears about safety may relate to the occasional (illegal) intrusion of service vehicles or to dangers unrelated to traffic such as a bomb threat (which was the subject of exaggerated publicity at the time of the survey).

So far as Church Street is concerned, the effects of pedestrianisation can be seen most clearly when the cumulative distributions of 'before' and 'after' attitude scores are plotted. Access by Transport is broken down by type of user in Figure 3. This attribute is of interest for two reasons: first, unlike the other seven which could be regarded as "environmental" attributes, Access by Transport is of a different genre; and secondly, it tends to be opposed in trade-off terms to the others since, generally-speaking, accessibility deteriorates as environmental conditions are improved. Given these substantial shifts in each of the attributes that we have taken to constitute the overall perceived environment, it would be tempting to assume that the value of each may be related in some way to the magnitude of the inscribed area (using some analogy with hysteresis-loss, for example) and that these values could then be summed to give an

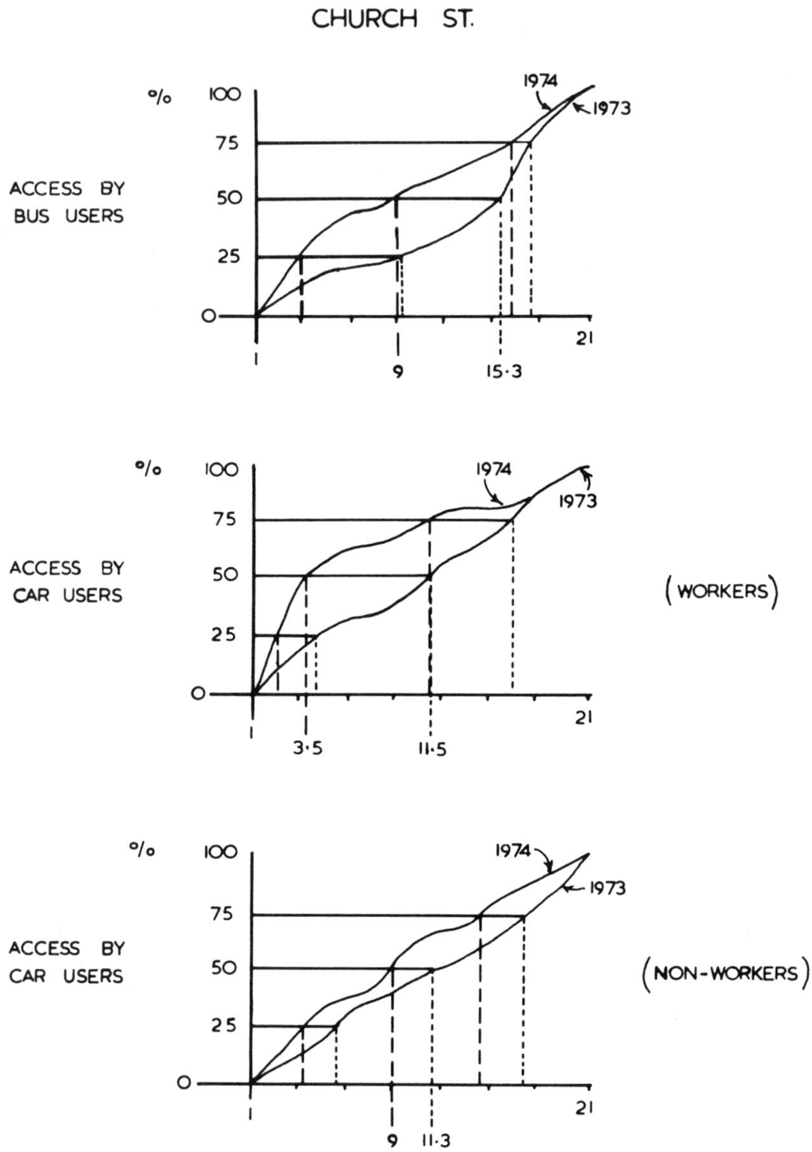

FIGURE 3. CUMULATIVE FREQUENCY DISTRIBUTIONS OF ATTITUDE SCORES IN RELATION TO ACCESS BY TRANSPORT BEFORE AND AFTER PEDESTRIANISATION.

overall evaluation of the change. Two problems impede a simple solution of this kind. First, although we have assumed the eight attributes to be of equal importance, they may well have to be weighted to allow for differences in their perceived importance to the environment as a whole. Secondly, values can only be summed if the attributes to which they relate are independent of one another. On the first problem, little can be said at this stage as its relevance was not appreciated at the outset of the project. On the second problem, we have carried out correlation tests and factor analysis on the attitude data to investigate the nature of interrelationships that may exist.

INTERDEPENDENCE AMONGST ENVIRONMENTAL ATTRIBUTES

From inspection one can see that the two aspects of safety are likely to be related to some degree, since few people expressing concern for the safety of others would be entirely indifferent to any risks to themselves. Similarly, one would expect there to be a safety implication in Freedom to Cross, since people are inclined to delay crossing or detour to a crossing place if random crossing of the traffic stream is too dangerous. As Figure 4 illustrates, this three-way interrelationship in fact emerged quite strongly from the correlation analysis. The highest correlation

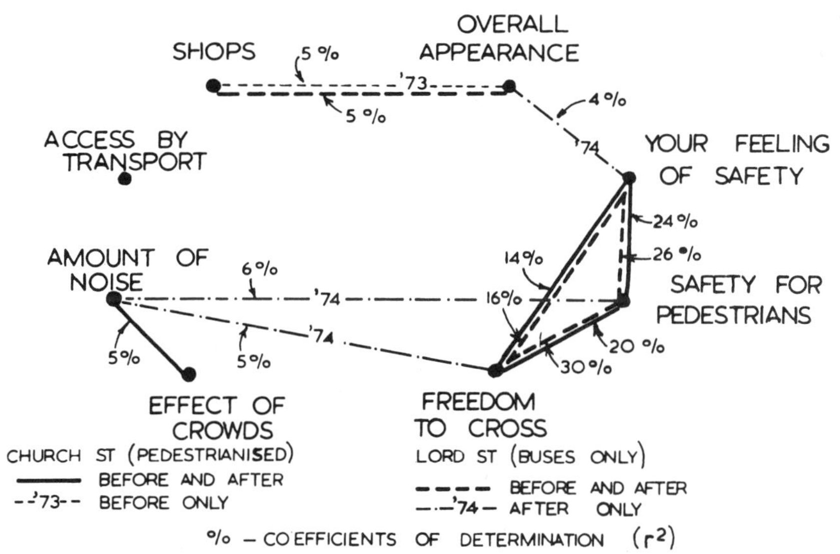

FIGURE 4. RELATIONSHIPS BETWEEN ATTRIBUTES FROM CORRELATION ANALYSIS.

coefficient obtained (just over 0.5) was between Freedom to Cross and Safety for Pedestrians, in Lord Street. In every case the same clustering occurred, which suggests a stable interrelationship. Unfortunately, the question as to the separation of personal safety from the safety of others is not resolved since the correlations obtained sustain the view that they should be distinguished, but only half-heartedly. More work will be needed on this question.

The link between attitudes to Shops and to Overall Appearance is consistent across all three streets, although also fairly weak; the maximum r value obtained was 0.33 in Bridge Street. Closer investigation of the tape-survey analysis confirms the link, however, as the "appearance" group of impressions were of three kinds: (i) two-thirds relating to the appearance of the buildings; (ii) one-quarter relating to the surroundings in general; and (iii) the remaining tenth specifically referring to the shop windows. The only other correlation of any significance was that between the attitudes to Noise and to Crowdedness in Church Street. Although clearly a fairly weak relationship, it did nevertheless recur in Bridge Street and in Church Street after the traffic was removed. What is interesting in this is the evidence revealed in the tape-recorded interviews. Many people referred to the streets as 'distracting' and 'confusing' - impressions which seem to be compounded of both noise and the effects of crowdedness.

The most encouraging result is the emergence of Access by Transport as a free-standing factor apparently unrelated to any of the seven "environmental" attributes. Its independence confirms its role in the overall trade-off between accessibility and the environment. Indeed, attempts to construct regression equations for each street using Access by Transport as the dependent variable produced R^2 values no higher than 5%. The only attribute consistently entered was Overall Appearance, for which no logical explanation can be offered.

Finally, the correlation matrix for the eight attributes was fed into a factor analysis program so as to identify any basic structure in the interdependencies that we had found. With a minimum acceptable eigenvalue set at 1.0, three independent factors emerged which were similar for all three sets of data and were composed of broadly the same clusters of attributes found previously (Figure 5). The almost complete independence of Access by Transport was confirmed (the highest factor loading obtained being less than 0.099). The only shared attribute was that of Safety for Pedestrians, which loaded slightly on the "bustle" factor comprising Noise and the Effects of Crowds; although this was only observed for Lord Street.

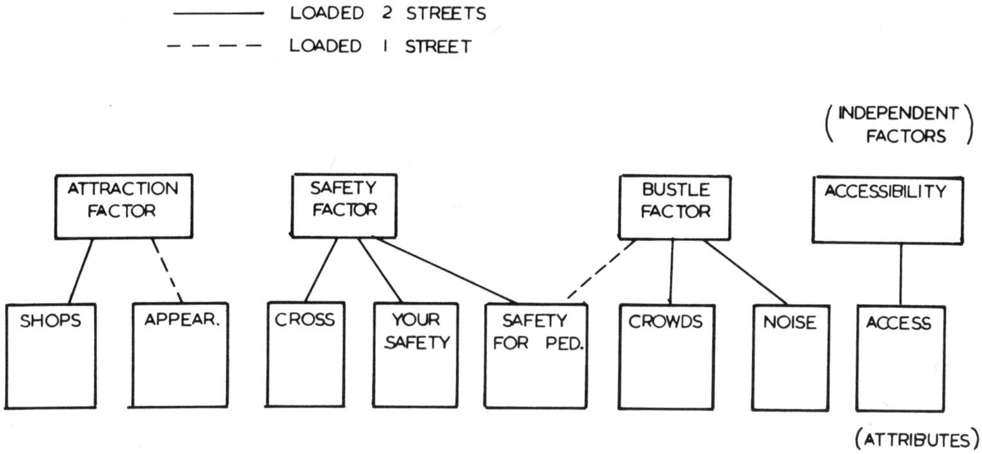

FIGURE 5. RELATIONSHIPS BETWEEN ATTRIBUTES FROM FACTOR ANALYSIS (FACTOR LOADINGS)

IMPLICATIONS FOR ENVIRONMENTAL EVALUATION

Even though some of the data on the after-surveys remain to be analysed and a further set of non-directive interviews have been undertaken in Church Street (in April 1975), some conclusions are already evident from the study. Most important, perhaps, is the conclusion that what are perceived as the most significant benefits from banning the traffic have not coincided with those factors that are conventionally measured. Of those that do (Noise, Safety and Access by Transport), there is no guarantee that the way in which physical measurements are formulated and expressed provide an adequate proxy for the conditions as perceived. A number of physical measurements were taken both before and after the street closure (e.g. sound levels, traffic flows, pedestrian delays and concentrations), so this can be investigated further. If the failure of conventional physical indices is confirmed, then new combinations of measures and new weighting systems will have to be sought.

The factor most likely to defy any ultimate objective measure is that of "attraction". The Shops component of that factor is arguably outside the planners' control but the overall appearance of the street is not. The effect on attitudes of a greatly enhanced appearance was not in fact tested in this case since, apart from removing the traffic and dumping some trees in upended sections of sewer pipe, nothing has been done to alter the townscape in Church Street. Preliminary analysis of the 'after' tape-recordings, however, indicates this as a major evaluative factor.

From a research standpoint, the most important outcome of the study is the development of a successful attitude-measurement technique for on-street interviewing. It is not yet possible to undertake a "total" evaluation using this technique but the relevant factors can be identified and assessed quantitatively. The irrelevance to evaluation of such aspects as air-pollution and vibration revealed in Liverpool has, of course, to be demonstrated elsewhere. But if it is, then a great deal of time and effort can be saved in elaborate, scrupulous and detailed physical measurements that have no bearing on the problem.

Lastly, the use of this technique to investigate other environments besides shopping centres may confirm the indication that stable and consistent attitudes are to be found towards perceived nuisances that are largely independent of the types or predispositions of the persons expressing them. In this respect, as in others, we may find that human beings are of a family in their responses to environmental disturbance and that the intangibility of many of the social costs and benefits of pedestrianisation schemes can ultimately be resolved.

REFERENCES

Buchanan Report (1963) Traffic in Towns. London: HMSO

Crompton,D.H., & Gilbert,D.A.M. (1971) Indices of environmental conditions in central area streets. Report for City of Edinburgh Planning and Transportation Study.

Dawson,R.F.F. (1974) Environmental Simulator: progress report. Report 659. Transport and Road Research Laboratory. Crowthorne.

Griffiths,I.D., & Langdon,F.J. (1968) Subjective responses to road traffic noise. Journal of Sound and Vibration, 8.

Hills,P.J. (1975) Research in transport studies at the University of Leeds. Traffic Engineering and Control, 16.

Joyce,F.E., & Williams,H.E. (1972) On assessing environmental impact of urban road traffic. International Journal of Environmental Studies, 3, 201-207.

Manheim,M.L., & Suhrbier,J.H. (1974) The automobile and the environment: implications for the planning process. Environment Directorate. O.E.C.D. Paris.

O.E.C.D. (1973) Effects of traffic and roads on the environment in urban areas. Road Research Group. Paris: O.E.C.D.

Rees,F.J., Clyde,C.A., & Dix,M.C. (1975) A structured attitude survey in two Liverpool shopping streets before the implementation of traffic schemes. WP 64 (unpublished). University of Leeds.

Wigan,M.R. (1975) Some environmental impacts as part of the transportation planning process. Supplementary Report 136 UC. Transport and Road Research Laboratory. Crowthorne.

SOCIO-ECONOMIC AND SITUATIONAL FACTORS IN URBAN TRAFFIC-NOISE ANNOYANCE LEVELS

Gerald R. Brown and Wm. Saur

Department of Civil Engineering

University of British Columbia

Traffic noise is one of the most prominent environmental quality problems in cities. In recent years considerable research effort has been devoted to urban traffic noise measurement and community response (Anon, 1963; Galloway et al., 1969; Griffiths & Langdon, 1968). Much of the research is related to measuring noise, developing scaling techniques to replicate the traffic noise stimulus as received by the human ear, and converting this into an effective tool to predict composite response patterns (Schultz, 1972; Johnston & Carothers, 1973; USEP, 1971). Much less research has been directed to the understanding of the variability of community response as a function of human factors (Galloway et al., 1969). This paper is an attempt to examine this variability using studies carried out recently in Vancouver, Canada by the authors.

The specific purpose of this research is to examine perceived noisiness and to determine if there are significant differences in annoyance response to traffic noise as a function of individual socioeconomic and situational factors. The literature of noise research implies that an individual's perception of and response to traffic noise depends upon at least two primary sources: (a) the nature and environment of the noise stimulus, and (b) the state of the receiver when he perceives the noise stimulus. As Shultz points out, the response of an individual to a given noise stimulus depends not only on the intensity of noise but also on the current and past state of the receiver (Schultz, 1972). Traffic noise intensity, frequency, and duration affect response patterns. These in turn are dependent on the composition and volume of traffic. The distance from the noise source and physical features such as the presence or absence of landscaping will also affect responses.

The personal characteristics and cultural history of the recipient will probably also dictate the intensity of response and annoyance. High status individuals (higher levels of income and career) may be more sensitive to traffic noise because of an increased expectation of a 'good' environment. Age, number of cars owned by the family, whether living in rented or owned accommodation and length of residency would appear to be factors in response patterns. The focus of this paper is the existing state of individual receivers of traffic noise stimuli and their corresponding annoyance levels.

The practical application of an understanding of this dimension of traffic noise response is in the area of traffic management and planning. The increased incidence of public participation in these processes dictates a need for better predictions of neighbourhood response to traffic improvements. It would appear to be useful in traffic management and planning decisions to have a better grasp of potential attitudes to traffic noise problems at the micro-urban level as a function of the participant's socio-economic and situational characteristics.

PRELIMINARY INVESTIGATION

Preliminary studies provided some insight into traffic noise perception and annoyance and resulted in some hypotheses to be tested. The first studies were directed toward determining the perception of environmental quality, including traffic noise, by pedestrian shoppers on portions of two suburban shopping streets (Pendakur & Brown, 1969; 1970). Sidewalk interviews were held with 165 randomly selected pedestrians who were asked to rate the quality of the environment with respect to traffic noise on a 5-point Likert scale from Very Good through Satisfactory to Very Poor. Almost 64% rated the area Satisfactory with mean noise-meter readings at curbside ranging between 60 and 70 dBA. There was sufficient evidence from this survey to draw the conclusion that shoppers in this type of environment had a high expectation and tolerance of traffic noise. It was also found that responses were to some degree correlated with the age of respondents.

It was decided to conduct a broad survey of resident's attitudes to traffic noise. A random sample of 3,500 persons was taken in conjunction with a travel pattern study of the metropolitan region of Vancouver (Roer, 1972). A single attitudinal statement was posed - "The reduction of traffic noise is the most important improvement needed in residential areas of Vancouver". Respondents were asked to indicate their level of agreement or disagreement on a 5-point scale. The median score for all residential districts was 3.52, on a scale from Strongly Disagree to Strongly Agree. Five of the thirteen districts analyzed had scores above this, and three districts had scores between 4.39 and 4.45.

Chi-square tests were made of the correlation between attitudinal responses and selected socio-economic and situational variables. The results indicated that significant correlations existed in many districts, with education, vehicle ownership, and household income showing the strongest relationship.

ATTITUDINAL RESPONSE SURVEY

On the mildly encouraging results of these previous studies it was decided to conduct a detailed attitudinal study in three districts of Vancouver: Kitsilano, West End and Fairview. These sites were selected because they represent those districts in which residents had expressed high levels of annoyance. They also appeared to represent relatively discrete socio-economic characteristics. Kitsilano is an older area, of mostly single family dwellings, currently experiencing an invasion from the 'counter culture' living in rooms and flats. The West End is a very dense, modern, high-rise apartment district adjacent to the downtown core. Residents are mostly single and live communally. Fairview is mixed, older, residential and commercial. Residents are predominately in the lower middle class.

The study included traffic and noise level measurement accompanied by face-to-face interviews of residents in about 20 blocks in each district. Study locations were selected to represent (a) light, moderate, and heavy traffic conditions, and (b) variability in socio-economic characteristics within the district. An average of 6 household interviews per block, selected at random, were conducted for a total of 330 respondents. Mapping of major landscape features and buildings at each study location was also done.

Noise readings were taken at curb-side in mid-block by a wind shielded Elektret condenser microphone connected to a Bruel and Kjaer A-weighted sound level meter, and were recorded on a Sony cassette recorder. Analysis was made through a Bruel and Kjaer Model 4420 statistical distribution analyzer and Model 2606 measuring amplifier, to obtain L_{10}, L_{50}, and L90 sound levels in dBA. A hard copy of noise levels was made using a Bruel and Kjaer Graphic Level Recorder. After calibration, readings were taken for one hour in each location. The TNI was calculated for each reading using the formula $TNI = 4(L_{10} - L_{90}) + L_{90} - 30$. This measuring system was selected because the dBA scale has consistently proved to be the most practical and reasonably representtive of subjective evaluations of traffic noise. (See Robinson

Noise Index	Mean dBA Sound Levels		
	Kitsilano	West End	Fairview
L_{10}	74.7	79.1	78.0
L_{50}	64.6	68.6	66.7
L_{90}	52.6	57.9	53.0
T.N.I.	111.0	112.7	123.0

TABLE 1. MEAN SOUND LEVELS BY DISTRICT

et al., 1961; or Galloway, 1969. The TNI is used after Griffiths & Langdon, 1968.) Results are shown in Table 1.

The character and speed of traffic were also recorded at each measurement station. It was assumed in the traffic and noise measurements that there was no variation in either flow or noise along the block.

A face-to-face questionnaire survey of attitudes and situational and socio-economic characteristics was conducted on each block. The questionnaire was designed to achieve several objectives: Part A, to obtain information on existing and previous residency and vehicle ownership rates (situational factors); Part B, to have the respondent rank the desirability of traffic noise reduction in relationship to other environmental quality improvements (this was done by asking respondent to allocate a hypothetical budget of $100 to a list of 10 possible improvements); Part C, to have the respondent locate on a 5-point scale his degree of agreement - disagreement with 15 attitudinal statements regarding traffic noise annoyance, and to have respondents rate the relative level of annoyance of various types of neighbourhood noise; and Part D, to obtain information on socio-economic profiles.

SURVEY FINDINGS

The survey data were examined to gain insight into the following research questions: (a) do residents of the 3 districts perceive traffic noise as a major environmental quality problem,

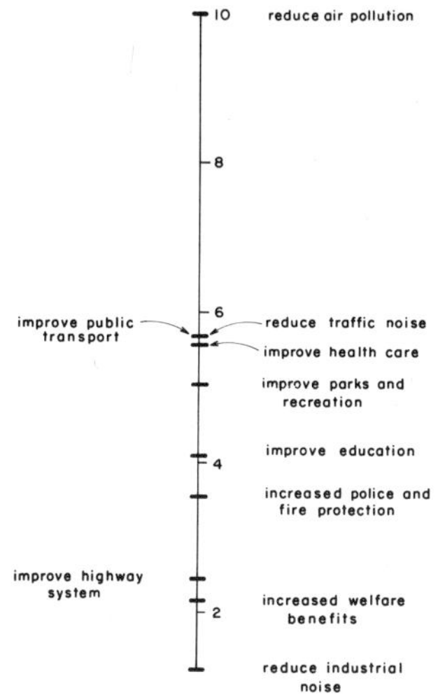

FIGURE 1. RELATIVE IMPORTANCE OF COMMUNITY IMPROVEMENTS TO RESIDENTS OF THREE STUDY DISTRICTS

and (b) is there any indication that perceived noisiness (or annoyance) of traffic varies with their socio-economic and situational characteristics?

Part B of the questionnaire, dealing with traffic noise ranking vis-a-vis other environmental quality problems was analysed to determine the relative importance of traffic noise reduction. Since most respondents allocated their hypothetical budget of $100 to only about 3 items and since many did not allocate the complete $100, the decision was made to simply rank 3 items in the order of importance given by the respondent. A score was obtained by using a reciprocal rating for each item ranked. That is, the rating for any item is $(n + 1 - x)$ where n is the total number of items ranked and x is the rank order. A rating scale 1-10 was developed in which air pollution, the highest priority item, was assigned the scale position of 10. The results of this are shown in Figure 1.

Figure 1 implies that traffic noise reduction in the three districts is an important environmental quality priority. With the exception of air pollution and public transit, the reduction of noise from traffic is seen to rank third in order of priority. The level of environmental quality is currently high in Vancouver. This scale does not imply that traffic noise levels are necessarily annoying, but that they are seen as a problem only in relationship to a selected set of other problems.

Attitudinal Response Category	Percentage in each Response Category, by District		
	Kitsilano	West End	Fairview
Disagree	24	22	25
Indifferent	6	7	13
Agree	70	71	62

TABLE 2. ATTITUDE TO NOISE BY DISTRICT

In response to the direct attitudinal statement "the noise from traffic in this neighbourhood is annoying" (Part C of the questionnaire) 66% of all respondents either mildly or strongly agreed. Only 8.2% were "indifferent" to the question indicating that by and large respondents were sensitive to the problems of traffic noise. Table 2 shows that for each district the responses were reasonably similar, except for Fairview which showed less annoyance than the other districts and had a high proportion of "indifferents". The results are consistent by district and show a reasonable concern for noise pollution in all districts.

The results of Table 2 also show that Fairview residents are less annoyed by noise while at the same time experiencing the highest levels. One explanation for this is that Fairview residents have a low level of expectation of 'good' environmental quality and are therefore less sensitive to traffic noise.

MULTIVARIATE STATISTICAL ANALYSIS

Analysis of the survey results showed apparent variations in perceptual levels and annoyance as a function of socio-economic and situational variables. The multivariate statistical technique of discriminant analysis was used to test which variable might contribute to possible variations. Discriminant analysis can be used to determine the statistical separation of different classes or groups of persons by means of the variables entered into the analysis. The hypothesis test is based on developing a discriminant function of the form $Z = \lambda_1 X_1 + \lambda_2 X_2 \ldots \lambda_n X_n$ in which X is the set of variables, λ is the weighting coefficient and Z is a classification index. Weighting coefficients are estimated from the data and a discriminant function valued for each group. The valued function is then used to re-assign original group members and the success in doing so is a measure of the degree to which each group is separate from all other groups. For a full description of the technique see Hope (1968).

The attitudinal survey respondents were separated into 5 groups depending on how each responded to the statement "the noise from traffic in this neighbourhood is annoying". Group classes and size are shown in Table 3. A set of 16 socio-economic and situational variables were identified and coded using combinations of ordinal and interval scales. The socio-economic variables included age, income, sex, education, number of vehicles in the household, residence ownership status, residence tenure, and occupation. The situational variables were T.N.I. L_{10}, L_{50}, shortest distance from residence to traffic stream, volume of traffic, presence or not of large trees, presence or not of large buildings, and presence or not of traffic flow interruption. The variables were then standardized.

Several step-wise discriminant tests were made in which each variable was entered separately and sequentially. Test evaluation was made by observing the degree to which each variable contributed to providing a statistically significant discriminant function and a reasonable increase in the success of the posterior classification.

Group Class	Attitudinal Scale Position	Number of Respondents in Group
Strongly Disagree	1	28
Mildly Disagree	2	58
Neither Agree nor Disagree	3	27
Mildly Agree	4	106
Strongly Agree	5	111

TABLE 3. ATTITUDE TO NOISE BY GROUP CLASS AND SIZE

The first series of tests were made on the assumption that each of the 5 categories on the attitudinal scale could be discretly defined and classified into a group characterized by its own discriminant function. Table 4 shows the results of this test.

It was thought that an improvement in results might be possible if some consolidation of respondent groups was undertaken and separation into only three groups attempted: those who were not annoyed, indifferent, and annoyed. The statistical results of this test are shown in Table 5. The indifferent responses were then removed from the sample ($<$ 10% of the sample) and a test made on only two categories: those not annoyed and those annoyed. The results of this test show the best re-classification accuracy and are displayed in Table 6. The 60-66% posterior classification,

Test	Significant Discriminating Variables	Variable Level of Significance	Percent Correct Posterior Classification
A	EDUC, DIST	$p < .05$	28.48%
B	EDUC, DIST, VEH	$p < .05$	30.30%
C	EDUC, DIST, VEH, TREES, CAROWN, L_{10}	$p < .05$	32.42%

Matrix of F_p's of group means separation for test A

Scale Position	1	2	3	4	5
1	x				
2	.22	x			
3	.06	.33	x		
4	.01	.10	.83	x	
5	.001	.004	.26	.13	x

TABLE 4. DISCRIMINANT ANALYSIS OF FIVE ATTITUDINAL GROUPS

Test	Significant Discriminating Variables	Statistical Level of Significance	Percent Correct Posterior Classification
D	EDUC	$p < .05$	47.58%
E	EDUC, DIST	$p < .05$	52.42%
F	EDUC, DIST, CAROWN	$p < .05$	56.06%

Matrix of F_p's of group means separation for test D

Scale Position	1-2	3	4-5
1-2	x		
3	.15	x	
4-5	.001	.49	x

TABLE 5. DISCRIMINANT ANALYSIS OF THREE ATTITUDINAL GROUPS

Test	Significant Discriminating Variable	Group Means Separation F_p	Variable Level of Significance	Canonical Correlation R_c	Percent Correct Posterior Classification
G	EDUC	$<.001$	$p < .05$	.21	60.07%
H	EDUC, DIST	$<.001$	$p < .05$	.28	65.68%
J	EDUC, DIST, OWN/RENT, VEH	$<.001$	$p < .05$	.33	63.04%

TABLE 6. DISCRIMINANT ANALYSIS OF TWO ATTITUDINAL GROUPS

while not showing a particularly strong group separation, does represent reasonable results when compared to other analyses of this type.

INTERPRETATION OF STATISTICAL TESTS

The results of the statistical tests are encouraging. They point to certain trends and directions which should be the subject of further research into the area of subjective noise assessment. Education is clearly a factor in attitudes towards noise, with those of higher level education expressing more annoyance than those of lower level. This finding is supported by Galloway et al. (1969).

As expected, the distance to the traffic noise source was a factor in differentiation between annoyed groups and non-annoyed groups.

The remaining statistically significant variables have some effect, but the most important role for these variables was in producing a more balanced re-classification. That is, the percentage correctly classified into each group was more accurate, while overall improvement in classification was not improved much. It was found that renters tend to be somewhat more sensitive than owners. This unexpected result may be due to the high incidence of young, environmentally conscious renters in the sample.

Respondents who have lived the shortest time in their residence are most indifferent to traffic noise. In time they become somewhat annoyed until finally after approximately 5 years they apparently become acclimated. Age, to the degree it is important, follows this same pattern.

As expected the sensitivity to noise decreases with car ownership. Lower job classification produces less sensitivity to traffic noise.

Finally, landscaping and traffic flows produce interesting results. It is not clear that well-landscaped areas produce less annoyance. It is also not clear what effect traffic volumes have on noise annoyance. Both the very annoyed and not at all annoyed are characterized by high levels of traffic in some tests. Notwithstanding, those in the 3 groups case who are annoyed experience higher volumes of traffic. These findings appear to corroborate the hypothesis that certain types of people like to live in heavy traffic areas, presumably because of increased accessibility.

CONCLUSIONS

The research reported upon here gives only tentative conclusions. However, it has become apparent that traffic and accompanying traffic noise, per se, may not be the only factors to consider in traffic management and noise control. There is some evidence here, and in Galloway (1969), that sufficient variation in subjective responses by socio-economic profiles may mask substantial variation in traffic noise levels as recorded and scaled. As a minimum requirement we must be aware of variations likely in community response to given noise levels before controls are imposed. Schultz has reached a similar conclusion when he writes that "no scale which predicts a fixed relation between stimulus and response can be justified" (Schultz, 1972).

This hypothesis needs further testing. Particularly, more attention should be paid to attitudinal measurement as related to socio-economic profiles and to the assessment of environmental quality expectations. Noise control measures applied indiscriminantly in isolation from a better understanding of the variation in noise attitude patterns will in the long run prove to be inefficient.

REFERENCES

Anon. (1963) Noise. Final Report of the Committee on the Problem of Noise. Command Paper 2056. London: HMSO

Galloway,W.J., et al. (1969) Highway noise: measurement, simulation, and mixed reactions. National Cooperative Highway Research Program Report No.78. Highway Research Board. Washington D.C.

Griffiths,I.D., & Langdon,F.J. (1968) Subjective response to road traffic noise. Current Paper 37/68. Building Research Station. Watford.

Hope,K. (1968) Methods of Multivariate Analysis. London: University of London Press.

Johnston,G.W., & Carothers,R. (1973) Annoyance measurements related to urban traffic noise exposure. Research Report No.13. Joint Program in Transportation. University of Toronto-York University.

Pendakur, V.S., & Brown, G.R. (1969) Accessibility and environmental quality. Journal of the Urban Planning and Development Division, American Society of Civil Engineers.

Pendakur, V.S., & Brown, G.R. (1970) Community values and street quality. Plan: Journal of the T.P.I.C., 10.

Robinson, D.W., Copeland, W.C., & Rennie, A.J. (1961) Motor vehicle noise measurement. The Engineer, 211.

Roer, P.O. (1972) Vancouver area travel study. School of Community and Regional Planning, University of British Columbia.

Schultz, T. (1972) Community Noise Ratings. London: Applied Science Publishers.

United States Environmental Protection Agency, Office of Noise Abatement and Control (1971) Community Noise. NTID 300.3. Washington D.C.

MOTOR TRAFFIC NOISE AND HOUSING

Conall O'Catháin

The Martin Centre for Architectural and Urban Studies

University of Cambridge

This paper describes an attempt to interpret the known behaviour of traffic noise and human response to it in terms of geometrical properties of housing layouts. It is based mainly on the results of research done in England by the Building Research Station (Scholes & Sargent, 1971), the National Physical Laboratory (Delany, 1972), the Greater London Council (1970) and the Transport and Road Research Laboratory (1970).

Their work on the prediction of noise-levels, attenuation by screens, and community response to traffic noise has been done from the point of view of increasing our understanding of these topics as physical phenomena. This work attempts to go one stage further towards practical application in environmental planning by relating traffic noise behaviour to housing configuration and density. This has, of course, implications for transportation planning strategies, inasmuch as it might be considered to be a contribution to a numerate description of such intangibles as 'environmental quality' and 'amenity', or even annoyance. This paper describes an architect's approach to the interpretation of published work on the subject. At the outset it was envisaged that this study would result in a theoretical or at least simplified description of the phenomena, which would make them amenable to manipulation early in the design process.

First, a word about noise characteristics. The model that is normally used is to consider the centre line of the road as a line-source with cylindrical spreading of noise. The sound-waves can travel over and around obstacles and this can result in attenuation of the sound if the obstacles are large. Motor traffic is considered to produce a randomly varying noise and it is usually measured in terms of L_{10}, that is, the sound pressure level which is exceeded for 10% of the time; although L_{50}, the mean level and L_{90}, the 'background noise' are also used. An acceptable correlation has been found between the average L_{10} over the eighteen hour period from 6.00 a.m. to midnight and the average population response, in experiments carried out in Britain. It is assumed that a predicted L_{10} will give a certain proportion of annoyance in the population. Actually, doubts have been cast on this assumption, since firstly it has been suggested that the population is not homogeneous in its response to noise (Bryan & Tempest, 1973), and secondly it is found by psychologists that the response of an organism depends not only on the stimulus, but also on the history of the organism.

A great deal of work has been and is being done on the prediction of L_{10}. This does not concern us - for our purposes we shall assume that an L_{10} at 30m has been calculated (or measured) and take it from there.

There are two main factors in attenuation: distance and barriers. The effect of distance from the road on L_{10} is that there is a reduction of about 3 dB(A) per double distance. Near the road the attenuation curve is steep, showing sensitivity at short distances, while it tends to level off at longer distances. This means that if a receiver is near the road quite small distances can have a big effect, whereas at long distances the same separation would produce virtually no discernible difference. The human ear cannot detect differences of less than about 3 dB(A).

Attenuation by barriers plays an important role in noise reduction. The estimated reduction is based on the difference between the length of the sound path over the top of the barrier and the direct path without a barrier, and also the angle of view of the barrier on plan.

In order to calculate the L_{10} at a point it would be necessary to calculate separately the contributions from both screened and unscreened parts of the road. For more complicated situations with more than one barrier or a varying noise source, so much calculation would be needed that it is hardly feasible without a computer.

Now let us look at barriers in more detail. Barriers can be conveniently, if arbitrarily, categorised as 'high' or 'low'. By 'high' I mean houses or buildings and by 'low', walls, usually built along the side of the road.

In general, the higher the barrier, the better. If the proposed barrier is rather low and the expected attenuation is of the

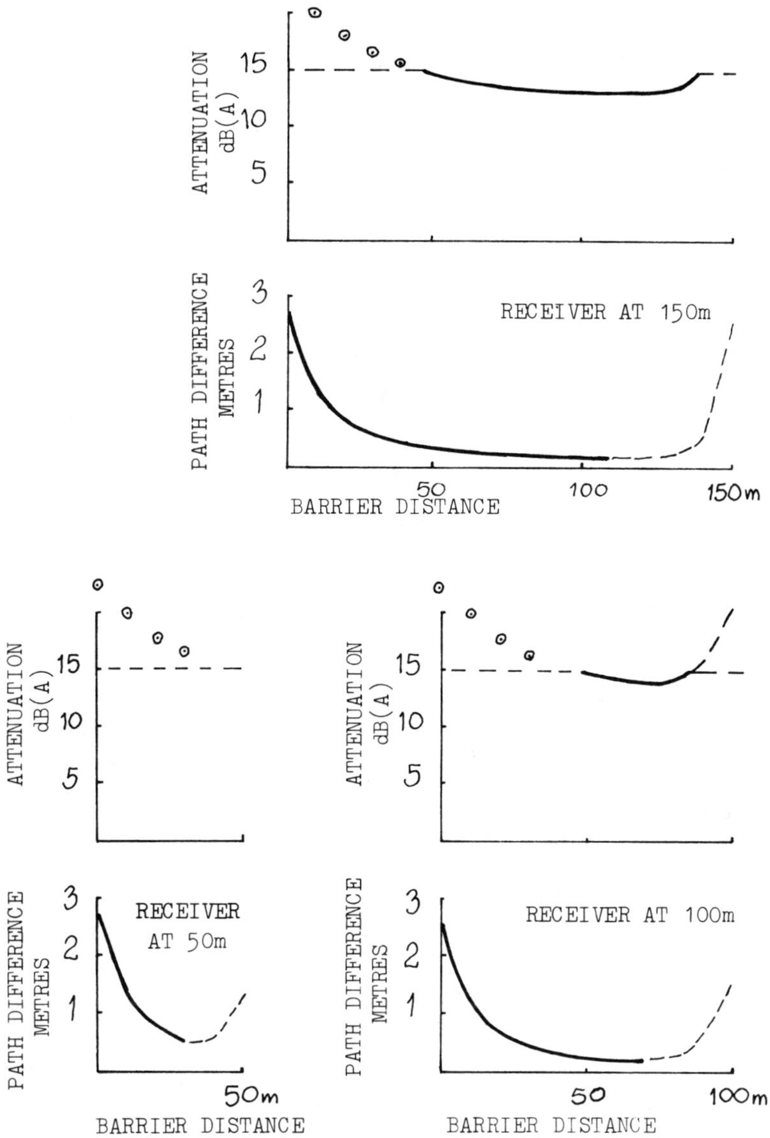

FIGURE 1. 7.5m BARRIER AT A RANGE OF DISTANCES. RECEIVER FIXED.

order of 5 dB(A) or less, then, where the intervening ground is soft, the ground can absorb about that much sound if the buildings are also low, i.e. one or two storeys; and so the barrier would be redundant. If barrier blocks are contemplated, where these have breaks in them it would be much better to replan them, closing or at least overlapping the breaks and thus creating a continuous, even though lower, barrier.

Figure 1 shows some examples of the behaviour of the path difference function for a two storey barrier block when the receiver is fixed and the barrier position varies - it is high at both ends of the range and fairly flat between. In practical terms this means that if the barrier can be relatively close to the receiver, or vice versa, then a larger attenuation results (though doubts have been cast on predicted attenuations from barriers in excess of about 15 dB(A)). But in the rest of the range the path difference is, to a good approximation, constant. If this value is not on the steeper part of the curve, then the resulting attenuation will be nearly constant.

Figure 2 shows the path difference for a series of receiver positions with the 'high' barrier fixed at 10m from the road. It shows the very small effect on path difference of distances above 20 or 30m. The dotted line in this and in the other figures at about 15 dB(A) reflects doubt about the ability of practical barriers to achieve attenuations greater than this. The barrier is assumed to be a row of two storey houses with a pitched roof, the ridge of which is 7.5m high, and the receiver is 5.0m above ground, i.e. the top of an upper floor window. Incidentally, the predicted value of L_{10} for a ground floor window, say 2m high, has not been found to vary by more than 1 dB(A) from that of the upper window. This means that only the upper one, which will, of course, be more exposed to the noise, need be calculated.

It appears that it is easier to make useful generalisations about 'high' barriers than about 'low' barriers. This is because, whereas the former give typical path differences of the order of .25m or more and thus fall on the flatter section of the attenuation curve, the latter usually give smaller values between 0.25m and zero and the attenuation curve is quite steep in most of this region. This is shown in Figure 3, which shows plots of path difference and consequent attenuation for the same range of receiver and barrier configurations with a 3.0m high barrier.

The foregoing illustrations have assumed an infinitely long barrier which is, of course, rarely the case. For short barriers or situations where groups of buildings or other objects act as multiple barriers, as an alternative to the computing technique described above, it is possible roughly to arrive at a 'net barrier length' and thus to compute and sum only two L_{10}s for each point

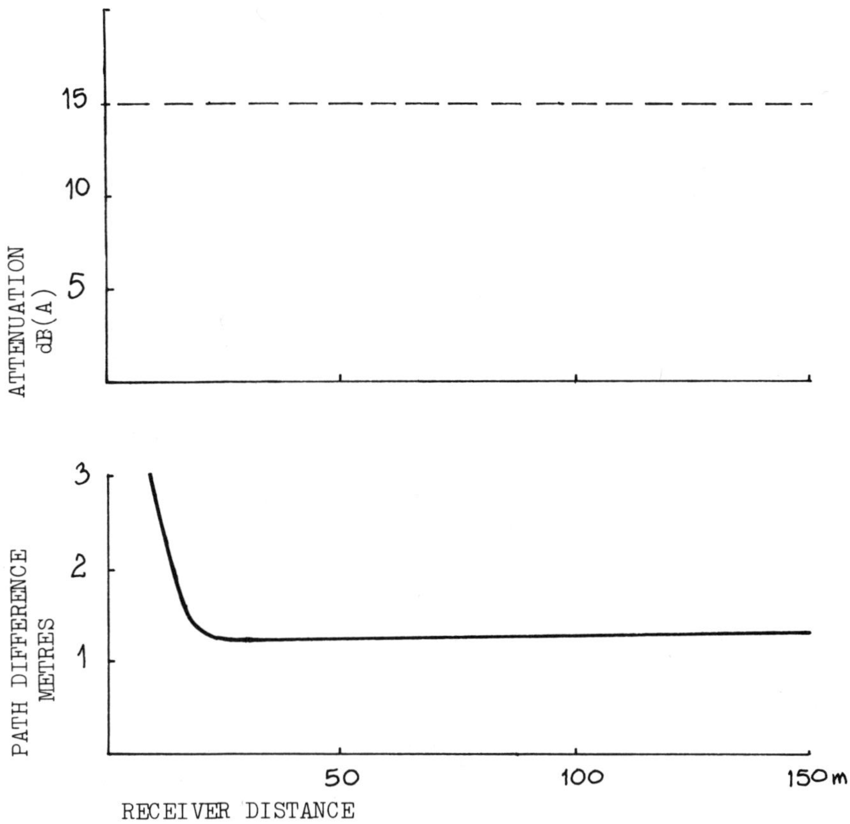

FIGURE 2. 7.5m BARRIER AT 10m WITH RANGE OF RECEIVER POSITIONS.

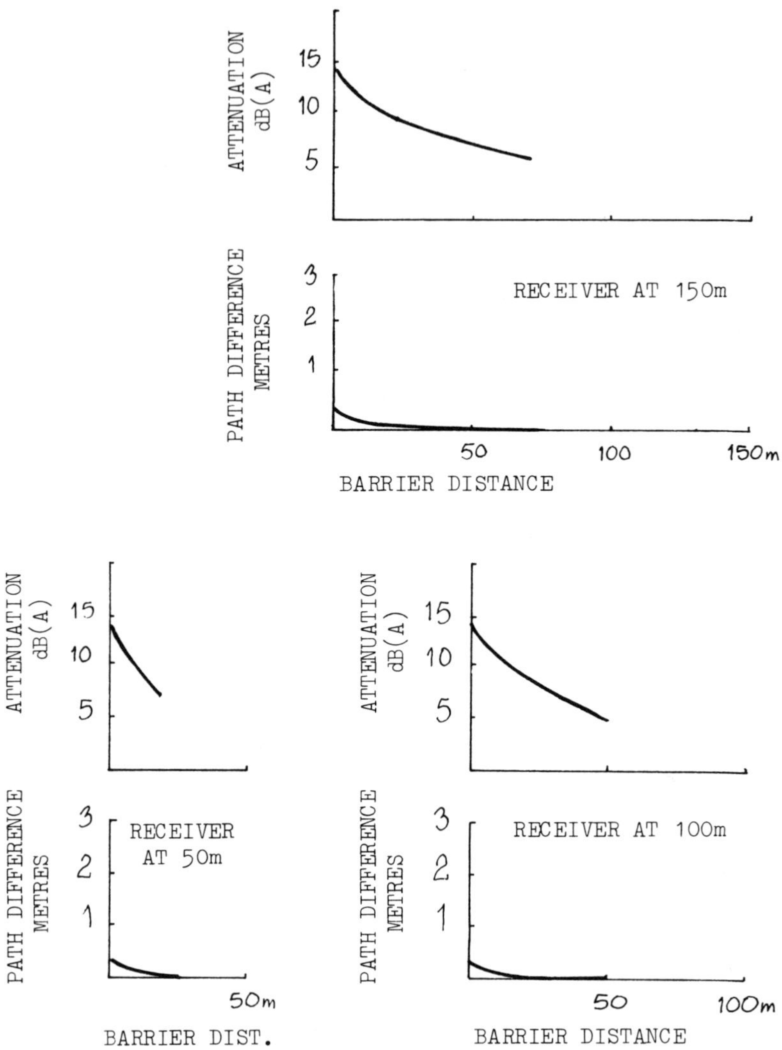

FIGURE 3. 3.0m BARRIER AT A RANGE OF DISTANCES. RECEIVER FIXED.

where it is desired to know the noise level. But this is inaccurate if the configuration is really complex (although computer prediction could be wrong, too, in complex situations, because so little is known about reflection). An important implication of the distinction between 'low' and 'high' barriers is this: while the former is simply a high wall shielding human habitation space, the latter wall is an inhabited wall, so it must be well insulated on the noisy side.

In the light of what has been said, it is possible to devise a strategy for layout planning which attempts to make the best use of our knowledge learned so far. In order to make definite statements it is necessary to define some noise level to be used as a standard of acceptability. There has been a good deal of discussion about this, but we will deal with the matter summarily by saying that we wish to place our standard at a point along a scale stretching from the unacceptable to the unattainable. In Britain, for all practical purposes, the Government defined the unacceptable when they provided for free insulation for houses where a new road resulted in a level of 70 dB(A) or higher. For the quiet end of the scale we could take the level of a quiet suburb, say 50 dB(A) or less. The ambient noise climate for different locations, or perhaps for different kinds of development, would also have to be taken into account.

Having said all this, let us look at a simple example to see what useful conclusions we can draw from a systematic study of the behaviour of noise, or at least of predicted noise levels. Figure 4 shows the output from a computer simulation of a 300m long barrier 3m high alongside a highway. The shape of the contours agrees well with measurements taken at a test site on the M1 motorway in England by the Building Research Establishment. They show clearly the reduced effectiveness of the barrier near both ends. They also show that the characteristic shape of the contour is largely independent of the height of the measuring point, except where this would be so high as to fall outside the 'noise shadow' cast by the barrier. The actual noise exposure increases, of course, with height in a fairly regular fashion.

I remarked earlier that noise levels seemed to be in general within 1 dB(A) of each other on ground and first floor levels. These contour plans have been calculated at the following heights: 1.5m, 3.0m, 6.0m, 9.0m, 12.0m. They seem to show a regular increase of about 1 dB(A) from each plan to the next. This corresponds very roughly to a 1 dB(A) difference per storey height, up to about four storeys.

The programme also computes the area enclosed by each contour, and thus allows us to make fairly precise calculations about density of development and site coverage.

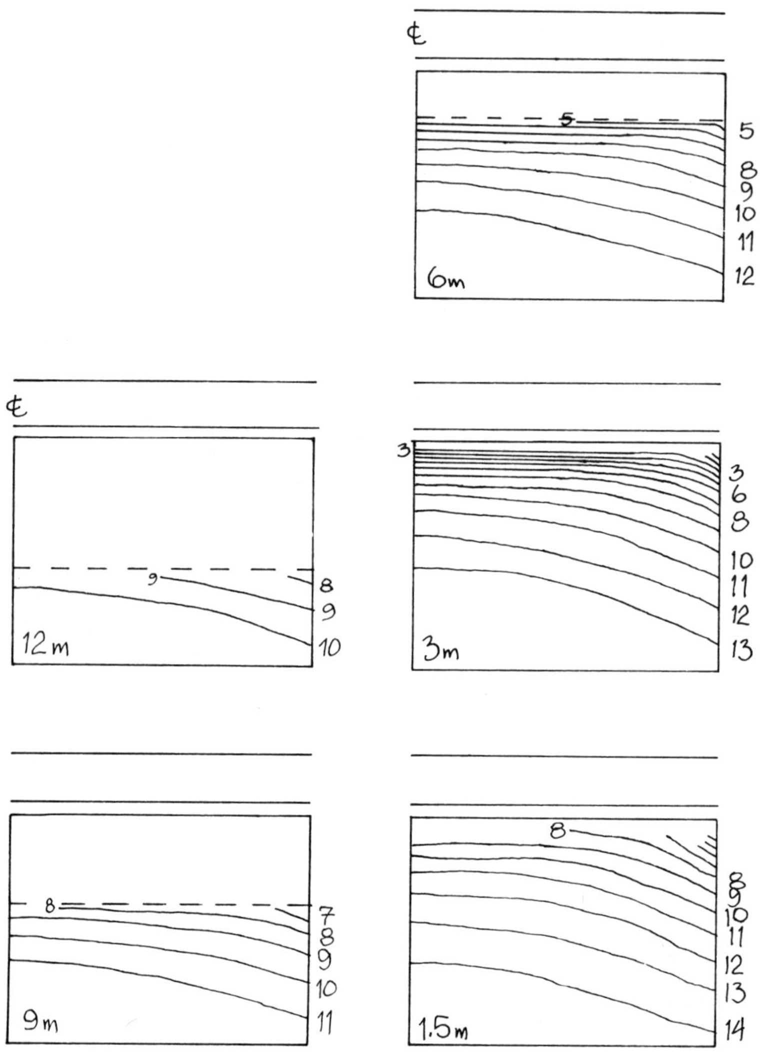

FIGURE 4. CONTOURS OF ATTENUATION FOR VARIOUS RECEIVER HEIGHTS.

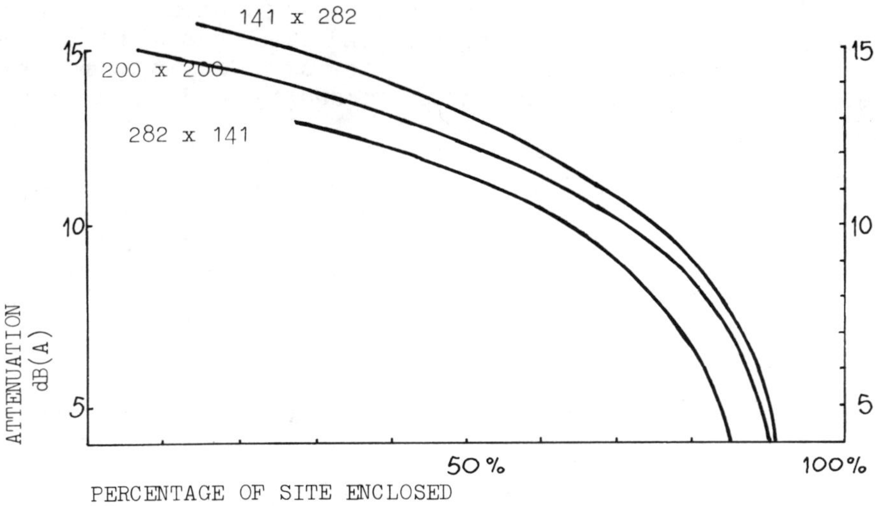

FIGURE 5. NOISE ATTENUATION FOR DIFFERENT SITE DIMENSIONS.

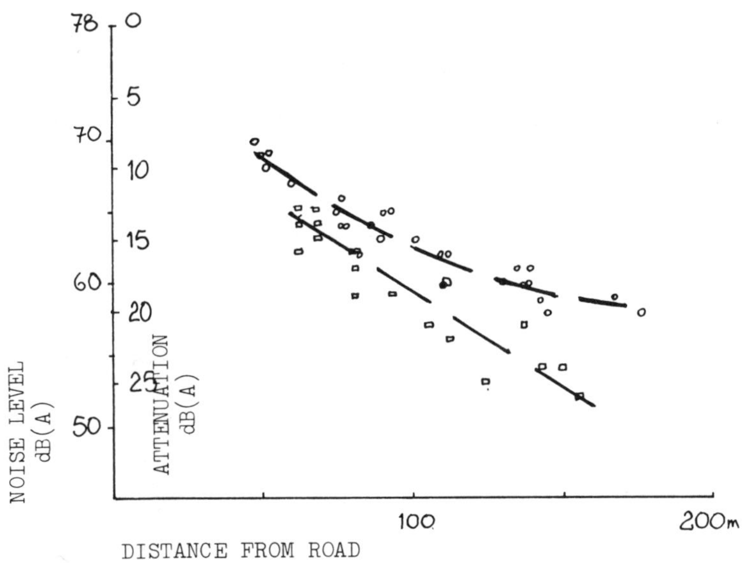

FIGURE 6. NOISE ATTENUATION FOR DIFFERENT BUILDING ORIENTATIONS.

In the following calculations I have assumed a 4 hectare site - 200m x 200m square adjacent to a highway, with a continuous 3m high barrier along the side of the road which has a 3.66m wide shoulder. Suppose, for example, we decide as policy that not more than 10% of the development shall be exposed to 60 dB(A) or more. A typical density might be 40 houses per hectare. We first of all compute the reference noise-level for that road - the L_{10} at 30m. Suppose this to be 66 dB(A). Therefore we need an attenuation of 6 dB(A). Figure 5 tells us that the 6 dB(A) contour encloses 87.7% of the site, about 35,000 m^2. This area is to contain 90% of the housing. Therefore the net density of that area of the site is 40 x 90/87.7 = 40 x 1.03 = 41 houses per hectare. In the remaining part of the site the net density is 40 x 10/12.3 = 40 x .81 = 32.5 houses per hectare.

The critical relationship is the percentage of development divided by the percentage of site. If this is near unity then the scheme is viable, assuming a homogenous development. If it is much greater than unity this usually indicates that a different built form will have to be selected (insofar as net density correlates with form). Of course, such a decision changes the rules of the game and we would need to backtrack and reconsider the changed situation. The point is that this fairly simple procedure gives us a good guide of the extent to which that part of a site closest to the road - the most noise-sensitive part - will be affected by early design decisions.

It is not possible to give a facile guide as to when the break point - if any - is reached when one has to resort to barrier development, but it can be found fairly easily for particular cases using this procedure. Figure 5 also shows how the curve changes if the site dimensions have different proportions, 2:1 and 1:2.

The values in the above example would have been:

	$>$60 dB(A)	$<$60 dB(A)
2:1	0.55(22)	1.1(44)
1:1	0.81(32.5)	1.03(41)
1:2	0.89(35.4)	1.01(40.6)

When considering buildings as noise barriers it is important to remember that people still live in them. Therefore they must produce an adequate climate for themselves as much as for other dwellings. For the purposes of this paper we shall classify housing into court, block and row forms. The most noise-resistant forms are the courtyard and the long, continuous block. From our discussions of 'high' barriers we know these give very considerable shielding, the courtyard being the better because its resistance is independent of orientation or access considerations. For a basic level of 78 dB(A) @ 30m, the noise level would be around 58-62 dB(A) inside a two storey courtyard, even when quite close to the road; which represents an attenuation of 16-20 dB(A).

The small block form is the worst performer, as expected. Not only does it generate a short effective barrier, but the noise is dispersed in all directions by reflection and consequently little protection is given to the open space around dwellings. But at high density, where the proportion of building to gap is high, this type begins to approach the performance of the third type - the row normal to the road. Here the noise travels between the rows, but is attenuated increasingly because they begin to act as barriers with increasing distance from the road.

The continuous barrier, parallel to the road, is theoretically very good, but its performance can be seriously diminished or its use precluded by considerations of access and orientation. Figure 6 shows a rough comparison of predicted levels for the two cases of long-block, normal and parallel-to-the-road, each slightly gapped, which is commonly the case. The L_{10} @ 30m in these cases is 78 dB(A). It is clear that for the range of distances where they are comparable (100m - 150m), the parallel case is about 5 dB(A) better; that is to say, the parallel row gives an L_{10} of around 55 dB(A) in this range of distance, while the normal row gives about 60 dB(A) or worse. The latter is not sensitive to orientation and neither is really affected by access considerations. The court form L_{10} would probably be about 50 dB(A) at that distance.

The inescapable conclusion is that buildings exposed to noise are becoming more pachyderm, or at least more introverted or turned away from the noise. This is the only possible response when those responsible for road-design work in a social vacuum. This is a lesson that has been learned and forgotten many times through history, from the Roman house with its atrium (and ancient Rome had traffic restrictions!), the Cambridge college with its square, the Italian Renaissance palazzo, and the nineteenth century grid city with its built-up streets and secluded back gardens. The Modern Movement swept away not only the architectural follies but also the accumulated architectural wisdom of the last century.

There are two possible approaches open to us now: either we consciously strive to produce highly protective buildings which have the minimum direct connection with the outside; or, better, we reorganise our design hierarchies so that this does not have to happen and design our highways with due consideration for their effects on the people they are supposed to serve.

REFERENCES

Bryan, M.E., & Tempest, J.W. (1973) Are our noise laws adequate? Applied Acoustics, 6, 219-231

Delany, M.E. (1972) A practical scheme for predicting noise levels. Report Ac57. National Physical Laboratory.

Greater London Council (1970) Traffic Noise. Urban Design Bulletin No.1.

Road Research Laboratory (1970) A review of road traffic noise. Report No. LE357. Crowthorne.

Scholes, W.E., & Sargent, J.W. (1971) Designing against noise from road traffic. Current Paper 20/71. Building Research Station. Watford.

THE CONCEPT OF SPATIAL MOBILITY AND AN ANALYTICAL APPROACH

Christopher Doubleday

Martin Centre for Architectural and Urban Studies

University of Cambridge

The urban transport problem has two aspects: one is the difficulty of movement in towns, the other is the environmental disruption caused by people moving around. The common source has been the incompatability between the relatively sudden change in urban travel patterns occasioned by the rapid diffusion of a particular technological innovation, the motor car, and the much slower changes in urban physical structure. It is suggested that the problem has assumed great importance in the public eye because the full potential of this innovation for increasing individual freedom, although having been realised in the early years of motoring, has now been frustrated. It is this, rather than difficulty of movement of itself, that has created the challenge.

Over the last two decades, the true complexity of the problem of movement in towns has been increasingly revealed. In the period up to the early 1960s, the problem was seen simply as one of vehicular traffic congestion arising from inadequate road capacity. An engineering approach of increasing capacity, either by more efficient traffic regulation or by new construction, was adopted in attempting solutions.

A wider view developed towards the end of the 1960s and the problem was set in an economic framework. A state of traffic congestion represented for society an economically inefficient balance between the demand for and supply of transport facilities. Planners might choose, in the light of the known elasticities of demand for transport and the cross-elasticities between modes, to modify demand patterns as an alternative to increasing the supply of transport facilities. The economists' approach also helped to make explicit the full range of costs and benefits associated with the development and operation of transport facilities. This allowed overall evaluations of transport proposals to be attempted.

However, severe theoretical and methodological problems have been encountered in making this form of policy evaluation, in particular the issue of equity. Economic evaluations have typically ignored this non-efficiency dimension, despite the significant redistributive effects which are often implicit in transport investments. This deficiency shows the need for a wider perspective of the urban transport problem, one which embraces a social dimension in addition to the conventional engineering and economic ones.

Several recent writers (in particular Hillman, 1973) have argued that as a result of social and economic trends, reinforced by certain aspects of current planning policy, there is growing inequity in the ease with which different groups in the population can move around in urban areas. In short, there is a growing problem of mobility deprivation. In this respect the availability of a private car is considered to be an important discriminating characteristic. It is thought that policies in both land-use planning and transport planning, operating in a situation of rapidly increasing private car ownership, have aggravated the problem. The costs and service spiral of public road transport in the context of increasing private car traffic and the absence of government subsidies is well-known (e.g. Plowden, 1972),but other urban trends are no less significant for mobility. These include the increasing separation of different land-uses, trends towards low density development and the decentralisation of activities, as well as policies aimed at providing higher quality public services in relatively fewer locations. All these effects have tended to emphasise the differences in mobility between those with cars and those without. They may be partly attributable to the increased overall car ownership in society.

This paper initially considers the concept of mobility in an attempt to determine how mobility deprivation might be assessed, proceeds to assess a practical approach to measuring its effects and concludes by discussing the relevance of transport modelling techniques to attempts to solve the problem.

THE CONCEPT OF MOBILITY

Although the term mobility is used widely in urban planning literature, it is typically left undefined or, at best, is defined with many implicit assumptions. Usually, it implies some idea of the speed with which a person can move about in the town or region in which he lives; in abstract terms, a particular space-time relationship. However, since, in general, movement is not undertaken for its own sake (i.e. it is a derived economic demand), it is appropriate to take a wider view and adopt an activity-space-time framework.

If it is assumed that the individual has a set of needs and desires, his pattern of use of space and time in the urban system can be seen as the result of his efforts to fulfil these by undertaking activities at various locations and times in the context of certain sets of restrictions and constraints. These are of two main types: external restrictions and conditions arising from the economic, institutional and legal arrangements in society, which apply to every individual; and personal ones.

External restrictions consist of such factors as working hours, shop-opening hours and the spatial distribution of shops. The level-of-service provided by the transport infrastructure is another important element. At the personal level there is a budget constraint, i.e. the individual's expenditure over some time period must equal his income. Additionally, there are usually conditions that certain activity-space-time elements are considered fixed (e.g. sleeping at home at night is fixed).

Increased mobility in the conventional sense serves to increase the range of activity-space-time elements which are feasible for an individual with a given set of constraints and restrictions. However, equivalent increases in choice can arise from changes in the activity-space restrictions (relocation of facilities in a more evenly dispersed pattern) or in the activity-time restrictions (changes in opening hours). Mobility, therefore, is more meaningfully compared on the basis of the freedom of choice of activities, locations and times which an individual enjoys than on the speed of available transport. It is then possible to include the effect of the various operative constraints in the evaluation.

In summarising this broad concept of mobility many issues necessarily have been omitted from the discussion. For example, the effect of distance on the relative importance of activity-space choices has not been considered. It is possible that distant choices are not perceived as adding to the range of choice when adequate nearby facilities are available. More fundamentally, variations in individuals' preferences have not been made explicit, although they are pertinent to the evaluation of the range of choice which an individual perceives. This assumes critical importance when measures of mobility deprivation are attempted.

While it is possible to assess deprivation in many areas of social and economic concern by reference to statistics for a single variable (e.g. wage-rate in the case of employment and pupil-teacher ratio in education) or to some minimum agreed standard (e.g. provision of basic sanitary facilities in the case of housing), the variation in individuals' preferences for different activities prevents any simple measure of mobility deprivation being defined.

It is clear that mobility is an extremely complex phenomenon and that insights from several disciplines are required in constructing an adequate concept. Nevertheless it is hoped that the framework presented here will make explicit the different sets of constraints which are implied in various uses of the term 'mobility' and that it will serve for the development of suitable measures. In particular, it enables the approach to mobility problems adopted in a current research project to be seen in a wider perspective.

ANALYTICAL APPROACH

It is evident that public policy in the fields of land-use and transport planning and, more generally, in the provision and operation of social facilities can significantly affect the degree of spatial mobility which individuals enjoy. It is appropriate therefore for research to be directed at determining the effects which different levels of mobility have on individual life-styles. In particular, it would be useful to find out whether and to what extent those with relatively high levels of mobility make use of the large range of opportunities open to them, either in pursuing a wider range of activities or in using a larger number of locations or times for undertaking particular activities. Such studies would require large amounts of detailed information both on the availability of opportunities in an urban area and on the activity patterns of individuals over extended periods.

It is possible, however, to take more circumscribed approaches which may be easier to implement. One such approach is to concentrate solely on travel behaviour and to infer how people react to changes in mobility levels from observations of their travel patterns. This method has already been employed in other studies and these are briefly reviewed elsewhere (Doubleday, 1976). The particular approach adopted in the present research project derived from the initial assumption that, in view of the substantial resources which have been devoted to the survey and modelling techniques employed in the conventional land-use-transportation planning process, it would be advantageous to study mobility effects within a similar framework.

The particular statistic of travel behaviour selected for

analysis was the mean trip-rate. The measurement and subsequent forecasting of this rate (within a trip generation sub-model) are critical steps in the conventional sequential transport demand model and it is important to determine whether and to what extent differences in mobility levels lead to variations in trip-rates. However, typical applications of the category analysis method of predicting trip generation, do not address this issue directly. They employ the household as the unit of trip generation, whereas it has been shown (Hillman, 1973) that, for a representative sample of individuals in each of five English geographical wards each with a distinctive density of development, there are substantial variations in mobility levels within households. In particular, household car ownership does not adequately represent the availability of cars to different members of the household. Furthermore, no direct account is taken of the ease with which a resident can reach relevant opportunities in the urban area. In the present analysis, therefore, particular attention was paid to these issues. Individual rather than household mean trip-rates were studied and three factors were selected to characterise mobility levels. These were car availability, the level-of-service provided by the transport infrastructure and the spatial distribution of relevant opportunities relative to the individual's home. The latter two factors were conveniently aggregated within an accessibility index. Several alternative forms of this index were constructed for the purposes of this analysis, including the well-known one due to Hansen (1959). His accessibility index for the zone of residence i, A_i is given by:

$$A_i = \sum_j X_j \exp(-b\, t_{ij})$$

where

X_j = measure of opportunities in zone j

b = distribution parameter

t_{ij} = travel-time between i and j

A variety of values of the factor X_j was used to reflect the different activities which are relevant to the various trip purposes studied. They were based on information which is acquired in typical transportation studies for the calibration of trip attraction sub-models. For example, shopping trips were studied with respect to the spatial distribution of floorspace devoted to retailing. In the other part of the accessibility index, two sets of values of the inter-zonal travel time t_{ij}, representing private car and bus travel respectively, were employed, thereby enabling realistic accessibility indices to be attributed to persons to whom cars were and were not available.

Two approaches were made to assessing the extent and seriousness of the effects of mobility deprivation. Firstly, a cross-sectional study of how car availability and accessibility affect individual trip-rates was made using information from the 1962 Reading Travel Survey and contemporary land-use and employment surveys. Within the limitations imposed by the data base, an attempt was made to control the effects of socio-economic group, employment status and stage in life-cycle (as represented by household composition), all of which were considered to be determinants of the particular pattern of restrictions and constraints which an individual experiences. Particular attention was given to the optional trips (i.e. those not made to or from work or school) of housewives and retired persons, since these groups have often been identified as being especially affected by relative deterioration in public transport services.

The second approach was a longitudinal one of examining changes in trip-rates for groups with different levels of mobility over a period of about a decade. This was made possible by a repeat travel survey of the Reading area in 1971 using procedures which were virtually identical to those of the earlier one. In this case no contemporary data on either the transport networks or on the land-use and employment patterns was available to this study. Comparative work had to be confined to variations attributable to the car availability factor.

It is apparent that substantial assumptions had to be made in implementing this approach. It is necessary to discuss these in order to assess its worth.

DISCUSSION

There are two important considerations in interpreting the results of the analysis. These are to assess to what extent the factors chosen to characterise mobility reflect the range of opportunities available to individuals; and secondly, to decide to what extent significant differences in mean trip-rates are good indicators of adverse effects on groups with low levels of mobility.

If the present approach is viewed within the conceptual framework previously outlined, it can be seen that several assumptions have been made about the operative restrictions and constraints. It is assumed that both the preferences of individuals of comparable socio-economic group, employment status and stage in life-cycle and the constraints under which they pursue their lives are identical, i.e. they consider the same activity-space-time elements fixed and they have identical budget constraints. In other words, within the defined social categories, there are no significant differences in preferences and in constraints between those with and without cars available and between those living at different points in the urban area. This carries the arguable implication that the acquisition of a car and the choice of residential location are neither responses to preferences for particular activities nor attempts to change the perceived day-to-day constraints.

If these assumptions are maintained, the level of mobility should be reflected in the factors used in the analysis. Several points of either theoretical or practical significance may be made about these factors. In the context of the data available to this study, only crude estimations of car availability could be made; in particular, no information was available on the holding of driving licences.

The accessibility indices may be criticised on several grounds. No account is taken of the availability of opportunities by time of day. Consequently an important element in the activity-space-time framework is ignored. Secondly, the computation of the indices requires some parameter to represent the ease of travel in the urban area. In the case of the Hansen index this is a decay function parameter which may be approximated from the mean trip-time. The accessibility measure, therefore, is not wholly independent of the trip behaviour patterns to which one is trying to relate it. Finally, the level of spatial aggregation at which the accessibility indices were computed (traffic zones of approximately 1000 households) and the motorised networks used in evaluating inter-zonal times may render them insensitive to certain mobility effects which occur at the local level. In particular, it has been shown (Hillman, 1973) that for certain purposes, notably shopping, a substantial proportion of trips are made over short distances and on foot.

The second important consideration in interpreting the results is the significance to be attributed to differences in trip-rates. The fundamental issue is whether or not choices to pursue a greater diversity of out-of-home activities are reflected in higher mean trip-rates. Although this would appear to be a reasonable hypothesis, there are several factors which complicate the argument. Perhaps the most important of these is the ill-defined description of travel behaviour which the term 'trip' provides. Although well-suited for use in conventional modelling procedures, it is difficult to use in detailed examination of mobility effects on account of its vague relationship with travel decisions. A further difficulty is the large variance found in individual trip-rates when they are computed on a single day's travel pattern.

This discussion has highlighted the difficulties associated with studying aspects of mobility deprivation in the framework of conventional large scale transportation analysis and suggests the need to develop alternative approaches. One such approach is to take the activity pattern view of human behaviour, i.e. describing an individual's life over a given time-period as a series of activities between which travel may form a transition. Thus, travel is viewed as one link in the pattern of an individual's life, rather than as a trip which is measured in isolation from this pattern. This approach would be congruent with the activity-space-time view which has been presented and would allow mobility seen as the range of opportunities available within a particular set of constraints to be more directly assessed.

IMPLICATIONS FOR URBAN TRANSPORT PLANNING

The perspective on mobility adopted in this paper has implications for certain aspects of the urban transport planning process. At the outset it should be noted that this concept of mobility suggests that transport planning should not be considered in isolation from wider urban issues, nor indeed from certain fields of social policy, in attempts to ameliorate problems of mobility deprivation. Policies on zoning and development control and on the degree of spatial concentration of social facilities have implications for the available range of choice. While social policies on such matters as the provision of children's nurseries and retired persons' day centres can alter the constraints which are experienced by individuals.

Several deficiencies in the urban transport modelling procedure have frequently been described. These have been summarised by Jones (1974). Some of them are such that the conventional model is insensitive to certain mobility effects. The present research was partially stimulated by an awareness of these inadequacies. Two aspects seem especially important: the implied inelasticity of travel demand with respect to levels of effective mobility, and the need for a greater degree of disaggregation. Existing models fail to distinguish, except at a very general level, between those persons with significantly different levels of mobility and, as has been noted, are not sensitive to local effects.

In view of these difficulties, Jones (1974) has suggested that optional travel should be modelled within the activity-space-time framework which has been described, and has outlined a prototype activity-travel simulation model. It explicitly considers the choice between travelling to new activities and pursuing alternative ones which do not involve travel.

CONCLUSION

This paper has attempted to provide an adequate conceptual framework within which various uses of the term 'mobility' can be understood and has described a simple approach to measuring the effects of different levels of mobility in the community. The substantial assumptions which have been made in performing this empirical analysis have pointed to the need for a radically different approach to the description and modelling of travel if solutions to mobility problems are to be given priority as objectives of social policy.

ACKNOWLEDGEMENT

The research is supported by the Social Science Research Council.

REFERENCES

Doubleday,C. (1976) The effects of differing levels of spatial mobility: a discussion. Transactions of the Martin Centre, 1.

Hansen,W.G. (1959) How accessibility shapes land use. Journal of the American Institute of Planners, 25, 73-76.

Hillman,M. (1973) Personal mobility and transport policy. Broadsheet 542. London: Political & Economic Planning.

Jones,P.M. (1974) An alternative approach to person trip modelling. P.T.R.C. Summer Annual Meeting. London: P.T.R.C.

Plowden,S.P.C. (1972) Towns Against Traffic. London: Deutsch.

SECTION 7. TRAFFIC ACCIDENTS

The next four papers might appear to represent something of a break with the remainder of the contributions. The attention to road accidents as a topic is more specific and concentrated than the cover given in other sections of this volume. It is not treated, however, simply as a posterior analysis and search for the determinants of undesirable events in the urban environment. Previous papers by Navin and by Ashford and Kirk have already shown that preventive design is a part of the current approach to accidents. The same point is made in the papers which follow.

Shaoul is particularly critical of the more traditional studies which focus on such factors as driver characteristics, information presentation and acquisition, skills, stress, driver education, and so on. She argues that they are unlikely to help planners and engineers in their goal of designing a system to cater for human capabilities. Instead of equating man with a limited-capacity, information-processing channel and treating driving as a negative feedback control system, researchers should assume man to be an open, adaptive system and driving to be only one of a system of social activities in which safety may even be relatively insignificant.

The social concomitants of driving are also emphasised by Knapper and Cropley in their studies of attitudes towards seat-belts and accidents. Their findings suggest that interpersonal factors, such as the presence or absence of other people, or the judgements drivers make about others whom they meet in the driving context, are important situational components of driver attitudes and behaviour.

The papers by Shaoul and by Knapper and Cropley examine attitudes as a part of a much wider system. The statistical research of Hodgkinson and Raymond, by including a range of physical and social variables to describe the urban structures associated with different frequencies and types of accident, is built on the same assumption. Apart from its methodological contribution their study is intended to pin-point certain variables as accounting for the variation in accidents between different areas, and to suggest appropriate counter-measures to reduce the number and seriousness of accidents.

The current increase of interest in the bicycle as a means of urban transportation is probably under-represented in this volume. But it is true that the interest is generally tempered by the concern over bicycle safety which is the subject of Drury's paper. His human factors approach involves the collection of trip data in terms of cyclist functions, and the analysis of these functions to determine the human/machine/environment mismatches which cause the errors responsible for bicycle accidents. Survey and laboratory data suggest how the cyclist can be integrated into the urban transportation system and how the urban environment and legislation should develop to allow the advantages of cycling to be obtained without the present level of risk.

THE SYSTEM CONCEPT AND THE STUDY OF ACCIDENTS

J.E. Shaoul

Department of Civil Engineering

University of Salford

It is an indication of the prevalence of systems theory, which transcends established intellectual disciplines and demands a re-orientation of our knowledge, that papers on accident research should be included in a scientific conference on transportation and urban life. It implies a recognition that accidents are not only part of a transport and urban system, but a determination of their causes is to be found within that system rather than some external system, such as a belief in mysticism or the malfunctioning of human beings and their personality, which exists independently of other systems.

Despite their name, accidents being part, albeit an unwanted part, of the driver, vehicle, road and traffic system, which is in turn a subsystem within the transport system as a whole, they are the result of natural laws. They are necessities such that a determination of their causes should result in prediction and control, or the freedom from the necessity of being involved in them. Thus, the ultimate object of research is to produce countermeasures that will increase safety. It is a measure of our lack of understanding of the phenomena we call 'accidents' that the same name can be given to all these events within all the different types of transport systems (road, rail, air, sea) and other systems. They are clearly very different with respect to the elements of technology involved, level of human interaction required and their causes. This paper will very briefly discuss the implications of the systems concept and recent research, for the development of our understanding of accident causation.

THE SYSTEMS CONCEPT

Modern science is characterised by its ever-increasing specialisation, necessitated by the enormous amount of data and the complexity of techniques and of theoretical structures in every field. However, this is opposed by the fact that independently of each other, similar problems and conceptions have evolved in widely different fields.

Scientific investigations towards an understanding of accident causation, reflecting this trend, have been conducted in such diverse fields as highway and traffic engineering, ergonomics, psychology and driver education, in an attempt to determine the objective laws governing the occurrence of accidents. Perhaps because of this, such research is not always accessible to the planners and engineers who are charged with the task of designing a safe transport system for human use. But in part it was also due to the fact that accidents were seen as human errors, not in the system sense of the word, but in the legal and moral sense. Research on drivers, therefore, tended to focus on the driver as an individual rather than as a driver who interacts with a complex system. This concept of accidents led research in the direction of accident proneness, faulty attitudes, personality and so on, and pointed to countermeasures that took the form of punishment and deterrence. It never produced any findings that were conclusive.

In recent years, the direction of research has altered as the system concept in the social sciences has gained increased attention. New sciences have arisen, the core of whose notion is 'system'. Until fairly recently, the world was seen as chaos, in which atoms appeared to represent ultimate reality, with life as an accidental product of physical processes, the outcome of random mutations and survival through natural selection. Now the world is seen as organisation and this trend is marked by the emergence of new disciplines such as cybernetics, information theory, general system theory, etc. They are different in their basic assumptions and techniques; but they agree in being concerned with 'systems', 'wholes' or organisation. They are predominantly developments in the engineering sciences, necessitated by the complexity of systems in modern technology, man-machine relations and programming.

These techniques have been applied in the behavioural sciences because of increasing dissatisfaction with contemporary psychological theory, which was reductionist in approach. Hence the emphasis, noted earlier, on such concepts as 'personality'. Despite the claim that this kind of research in the behavioural sciences belongs to the 'systems' sciences, little attention has been paid to the 'systems' literature, as such, written over the last 30 years, in particular to "General System Theory". (For an introduction, clarifying its scope, see von Bertalanffy, 1973).

General System Theory postulates that the formulation and derivation of those principles which are valid for 'systems' in general is possible. Just as physics is concerned with systems of different levels of generality, e.g. special systems for engineering, special laws of physical disciplines such as mechanics or optics, the laws of great generality, e.g. principles of thermodynamics which apply to systems of intrinsically different nature, mechanic, caloric, mechanical, etc., so principles applying to systems in general are required, irrespective of whether they are of physical, biological or sociological nature. A general theory of system would reduce duplication of effort by providing models that could be used in and transferred to different fields and by avoiding the vague analogies which have often marred progress. Much of the work carried out in the name of systems research in the behavioural sciences has confused techniques with the reorientation required by the systems concept. Inappropriate models for the social sciences have been used and attention has not been paid to the contributions to science made by other disciplines.

THE LIMITATIONS OF THE HUMAN FACTORS RESEARCH

Human factors is one of the newer disciplines to attempt to assess design implications so that an improved physical environment might offer fewer inducements to unsafe behaviour and protect the road-user from the most severe consequences of such behaviour. This approach is gradually gaining acceptance in traffic safety in preference to punitive measures. Likewise, increased attention is being paid to accident counter-measures which seek to influence the driver directly, e.g. propaganda and driver education. These measures focus on changing individual behaviour so as to reduce exposure to risk or to inhibit actions that are believed to be related to the precipitation of crashes.

The work covered in the name of human factors (e.g. Forbes, 1972) includes research on biographical and physiological characteristics of drivers, factors in sign legibility, methods of presenting information to drivers, skills, judgement and information acquisition in controlling vehicles in traffic, impairment of driver performance due to such factors as stress, and driver education and improvement.

In addition, the research journals contain many articles which purport to show a relationship between a variety of variables and accidents. It is difficult to make a judgement about most of the studies. They are characterised by a lack of explicit statements on the assumptions used, hypotheses to be tested and methods employed. The superabundance of research at this primitive level suggests that the overall research strategy errs in presenting the mere existence of a statistical relationship (its substantive meaning is never examined). Ideally the construction of scientific theories to cover broader fields is based on the synthesis of the separate research results in this field. A coherent synthesis cannot be forged from a collection of relationships between ill-defined variables of unknown strengths and magnitudes. The necessary conditions for a synthesis include an evaluation of the results available in the field, a coherent interrelating of the magnitudes found in those results and in the construction of models based on those magnitudes, which permit effective countermeasures to be conceived and introduced. However, existing investigations do not permit such a synthesis.

There are several areas of concern about the research on driver behaviour. The first is that the variables being studied are not always behaviour, but some summary variables such as age, sex, socio-economic status, etc., which have little explanatory power. Information collected at such levels is unlikely by itself to provide a close or accurate description of how the observed relationship with accidents came about. Accidents are known to vary in terms of age or sex. Yet if detailed explanations are required of how such relationships arise, it is necessary to resort to other data or theories.

The main area of concern, however, is the extent to which such approaches can achieve the stated objective of designing the man/machine/environment task and/or system to fit human capabilities wherever possible. In principle it appears to be a very interesting approach. But most of the analysis is carried out in a vacuum quite distinct from actual accident occurrence. No attempt is made to show how this type of formal model, especially with regard to car following or overtaking, has any predictive validity. Indeed, the conclusion is often reached that drivers do a remarkably good job, even in a system where breakdown appears inevitable.

The cybernetic (i.e. closed system) model, proposed in most research to describe the driver/car/road/environment system, equated man with a limited-capacity, information-processing channel and driving as a negative-feedback control-system. It seems unlikely that a theory of driving which makes such simplistic assumptions about the driver will be very useful. Such a theory denies the possibility that he may initiate any action and ignores contributions to our understanding of man of the developmental psychologists such as Piaget, who showed that man, with respect to the cognitive factor, was not a passive receiver of stimuli coming in from the outside but in a very concrete sense creates his universe; and the view of man as an open, adaptive system, as propounded by the originator of the General System Theory, von Bertalanffy. Driving is only one of a range of social activities which is influenced by the prevailing social mores which may be antipathetic to safety. Given this uncertainty about the validity of the model, the ability of this type of approach to help planners and engineers design a system to cater for human capabilities may therefore be very limited.

THE RELATIONSHIP BETWEEN SAFETY AND ACCIDENTS

A systems study of safety implies primarily an investigation of principles governing the behaviour of elements within the transport system and the transformation of their activities into accidents, i.e. a study of change. Yet most of the studies which claim to be taking a systems approach tend to study the fixed elements within the system, e.g. total number of miles of the road system or density of population. Furthermore most studies focus their attention on safety rather than the behaviour of the system, and assume that this is measured by accident records. The word 'assumed' is used here since accidents appear to constitute an indirect measure of safety - safety in other words is measured by the results of its absence.

An accident is usually defined as certain types of outcome of a collision between two objects in the road transport system, which will result in damage to property and/or injury to a road-user. Thus accidents are defined by their outcomes rather than their antecedent behaviour. They belong to the wider class of collisions. Various factors such as energy-absorbing bumpers, seat-belts, efficient ambulance services, or an unoccupied vehicle may in fact prevent a collision from being classified as an accident, because they reduce the consequences of such a collision.

Since the assumption is made that unsafe practices on the part of the road-user may lead to an accident, these considerations lead to the use of collisions rather than just accidents as criteria for safe practices. Even this criterion cannot be viewed as a representative sample of safe practices in a study concerned with the behaviour and activities of a driver on the road, but rather as an intermediate criterion for the ultimate criterion - namely safe, risk-free driving practices.

An interesting study which showed a fairly constant ratio of serious conflicts to injury accidents at intersections with fast moving traffic was carried out by the Transport and Road Research Laboratory (Spicer, 1971; Russam & Sabey, 1972), although the method of assessing the severity of the conflict was not defined. It is worth noting that despite the fact that accidents belong to the wider class of conflicts, the justification for such studies is that it may permit research workers and engineers to use the frequency of occurrence of serious conflicts as an 'intermediate' criterion of safety in place of the less frequently occurring accidents. The apparent paradox is resolved by viewing conflicts and accidents as a measure of the same series, namely unsafe driving practices.

The use of accident records as a criterion of safety involves several assumptions: that the accident is in fact reported and there are no systematic biases in the accident reporting procedure; that it is known whose behaviour is represented by the effects observed; and that these results could only have arisen from behaviour as such. In the case of the latter, it is usually very difficult to accurately reconstruct the events leading up to an accident and so determine the degree to which its cause was behavioural or due to some factor in the driver/car/road/traffic system. Accident records as they relate to individual drivers are notoriously unstable from year to year. Indeed, the evidence as it relates to accident reliability (Haakinen, 1958; Burg, 1970; Campbell, 1971) suggests that they cannot be used for measuring individual performance but only the system's performance, e.g. along a particular stretch of road. Even so, accidents cannot be equated with faulty performance of the system, but the results of some faults in the system's performance.

The assumption was made that the driver's behaviour will lead to an accident. Our investigation into accidents (Shaoul, 1975c), including those where the driver in our study could not, according to legal criteria as defined by the Highway Code, be held to blame, indicated the deficient nature of many of the procedures which are laid down for negotiating hazards and/or inadequate emphasis on certain aspects of the procedures. Accidents are not just a result of an 'error' in procedure for carrying out a manoeuvre. Conversely, an analysis of the driving data collected during a drive, some time after passing the licensing test, showed that incorrectly carrying out the procedures for a manoeuvre did not lead automatically to an accident.

Thus it can be seen that the procedures themselves may be deficient. The procedures may be carried out correctly, but the manoeuvre should never have been carried out at all. For example, with regard to driving in the fog, no specific procedures will guarantee safety. Therefore, the general rules for the driver are to minimise driving under such conditions and to endeavour to compensate for the effect of fog by the increased use of lighting and warning devices. The only other procedure which can be laid down is the general one of reducing speed.

The results of our research (Shaoul, 1975b) indicate that the relationship between the correct performance of procedures and the absence of accidents cannot be substantiated. There is no evidence in the research literature for the relationship between improper driving and accidents based on explicitly stated, objective criteria. Improper driving can only be implicated if it can be demonstrated that it occurs more frequently as a causal factor among accident-involved drivers than among those not involved.

Thus, one of the chief problems in the study of the principles governing safe behaviour is the selection of the relevant behaviour to be observed. This is often termed the intermediate behaviour or "criterion", since in the field of accident causation some of our most useful knowledge will be acquired by experimental studies of an accident countermeasure evaluated by means of intermediate criteria. This may sound paradoxical since scientific knowledge is supposed to precede the design of a countermeasure. However, by altering our environment by means of a closely monitored experiment, thereby producing some observable effect, we see and understand the causal chain. The experiment forms the test of causality. This implies that much can be learned by evaluating many new features in our cities, such as pedestrianized shopping-precincts, off-street car parks, or one-way streets, in terms of the behavioural changes they were designed to promote and the effects on accidents.

RECENT RESEARCH ON THE WAY THE ROAD STRUCTURES THE DRIVER'S ACTIVITY

An interesting study of the use of transverse yellow bar patterns on the roads to influence the driver's perception of speed (Irving, 1973) will permit such an investigation to be made between the accident countermeasure, the specific behavioural objective, and the effect on accidents. Rumble strips, alternate rough and smooth road surfaces which produce a loud rumble and some vibrations when driven over, are a possible means of alerting and slowing traffic on roads carrying fast-moving traffic. Preliminary results showed that on one installation 50m from a junction the proportion of cars travelling at over 64 km/hr was reduced from 21 per cent to 7 per cent (Watts, 1973). By means of in-depth accident studies, attempts are being made to assess which aspect of the road-configuration providing information for the drivers was misleading the drivers and thereby resulting in accidents. Most of this work is characterised by little explicit theoretical discussion. But the focus of interest on the modification of driver behaviour by manipulating the information received from the road environment is paramount and does permit a study to be made of their effectiveness, both in producing behavioural changes and promoting safety. Although the major interest is in safety, as reflected negatively in accident rates, it is assumed that safety will be affected through these mediating practices. Although the value of these practices in reducing the frequency and severity of accidents has still to be demonstrated.

The conclusions reached in our own experimental evaluation (Raymond et al., 1973; Shaoul, 1975a,b,c,) and others suggest that where the task is seen as one of affecting the attitudes, skills and knowledge of the driver directly, by either propaganda or education, and where the relationship between them and accidents is not known to be a causal one, the likelihood of achieving, or even being able to measure, a substantial change in driving performance and safety is limited. The major determinant of aspects of driving performance was found to be the road configuration. In particular, the activities carried out by the driver when negotiating a specific manoeuvre, such as a right turn at a traffic-light-controlled junction, were found to vary with the design of the road. A variety of other manoeuvres was found to vary in this way. Thus it can be seen that the notion of driver stability or "lapses" has little meaning, since the driver adapts in a very concrete way to his environment.

Therefore, the only type of changes whose outcomes we can be sure of measuring are changes in our physical environment. This implies studying the underlying structure of the road/driver system to see how it affects the manoeuvres a driver has to make. This represents a departure from previous research which has attempted a formal analysis of the driving task. Most analyses have been specifically concerned with but one facet of driving.

Such a procedure implies the formulation of a model in the sense of a set of interrelated variables that include the driver, the vehicle and the environment. It could take the form of comparing the structure of two different road systems, to see how the road system, economic system and other systems affect the imperatives to alter speed and direction and consequently affect accidents. It soon becomes clear by comparing different road systems that it is necessary to look at the road system within other transport systems, or economic and social systems. Constraining the task of the human to that of guiding and controlling the vehicle may give greater feasibility to the technical job of understanding the capabilities required to operate a vehicle efficiently and safely. But it overlooks a number of important determinants of the way the car is used in our society.

Another approach in which we are currently engaged at the University of Salford, involves, as part of a study of the relationship between training and young people's accidents, assessing by various means the driving performance of 300 young drivers several years after qualifying as a driver. One of these was an observation check-list. This method involved noting whether or not a driver carried out the prescribed sequence of activities for negotiating a particular hazard at a particular point along the route. Some work has already been carried out to assess its internal and external validity (Shaoul, 1975b).

This method of observing driver behaviour consists of noting what happens for each road configuration along the 25-mile route, rather than of judging the safety with which it was carried out. A complete record is available of 300 drives. So far, the data have only been analysed for the groups as a whole. It is proposed to look at each individual drive and note each person's departure from the average. This examination of behaviour with respect to the advised procedures permits us to find out the use made by drivers of the advised procedures as laid down by the Highway Code and Driving Manual.

On the one hand, widespread failure to carry out particular procedures e.g. looking in the mirror before slowing down at a 'stop' or 'give way' sign, would tell us a great deal about the way the road design structures the driver's behaviour. For example, drivers are required to look right/left/right before driving through traffic lights on green. From our observations, at most traffic lights, less than one third of the drivers do so. However, for a few traffic-light-controlled junctions, this figure is very much higher. This usually occurs when the junction is of a type other than a cross-road. Individual failures to carry out a particular sequence which is carried out by the overwhelming majority of drivers would yield useful information about individual drivers. On the other hand, if certain types of road configuration regularly produce activities on the part of the road-user over and above those required by the advised procedures, this would go some way towards indicating where and in what circumstances the procedures are deficient.

The aim is to build up a compound profile for each driver which could then be related to all the available data regarding his experience, exposure to risk and accident record, in an attempt to validate the usefulness of the advised procedure. In addition, an attempt will be made to relate these observations with recorded accidents of the general population along this route. It is rare to have such a complete record as it relates to a driver's actions with respect to a particular route (as opposed to a series of judgements which are not fixed to particular points on the road network). Such an analysis could be of great use to those whose concern is the training of drivers and those whose concern it is to improve the physical environment so that if offers fewer inducements to unsafe behaviour. This should provide us with information based upon the activity of drivers as currently practised and enable us to see how the road design structures individual behaviour.

Although traffic engineers have recognised this implicitly in their work on improving traffic flow by altering the road, changing its layout, controlling the drivers by means of traffic lights, or the pedestrians' activities by means of barriers at corners and pedestrian crossings, this has not always been explicitly or even implicitly realised by those concerned with traffic safety.

REFERENCES

Bertalanffy,L.Von. (1973) General System Theory. Harmondsworth: Penguin.

Burg,A. (1970) The stability of driving record over time. Accident Analysis & Prevention, 2, 57-65

Campbell,B.J. (1971) Who really causes the accidents? Traffic Safety, 71, 22-24.

Forbes,T.W.(ed.) (1972) Human Factors in Highway Traffic Safety Research. New York: Wiley.

Haakinen,S. (1958) Traffic accidents and driver characteristics. Scientific Research No.13. Finland's Institute of Technology, Helsinki.

Irving,A. (1973) The perceptual problems of the driver. First International Conference on Driver Behaviour. Zurich.

Raymond,S., Jolly,K.W., Risk,A.W., & Shaoul,J.E. (1973) An evaluation of the effectiveness of driver education in reducing accidents to young people. Mimeograph. University of Salford.

Russam,K., & Sabey,B.E. (1972) Accidents and traffic conflicts at junctions. Report No. LR514. Transport and Road Research Laboratory, Crowthorne.

Shaoul,J.E. (1975a) The use of a test of driving knowledge and driving practices as criteria for assessing the effectiveness of driver education. Interim Report. University of Salford.

Shaoul,J.E. (1975b) The use of driving tests as an alternative criterion to accidents for evaluating driver education. Interim Report. University of Salford.

Shaoul,J.E. (1975c) The use of accidents and traffic offences as criteria for assessing the effectiveness of driver education. Mimeograph. University of Salford.

Spicer,B.R. (1971) A study of traffic conflicts at a rural dual carriageway intersection. Report LR410. Road Research Laboratory, Crowthorne.

Watts,G.R. (1973) Road ramps for the control of vehicle speeds. First International Conference on Driver Behaviour. Zurich.

A QUASI-CLINICAL STRATEGY FOR INVESTIGATING ATTITUDES IN THE TRANSPORTATION DOMAIN

C.K. Knapper and A.J. Cropley

Department of Psychology

University of Regina

In a context where transportation decisions increasingly involve political, rather than merely scientific or technical considerations, the study of social pressures and psychological motivations is especially pertinent. Indeed, a number of contributors to the present volume have commented on the need for a more thorough understanding of social and psychological factors in the field of transportation (e.g. Krampe, 1976; Leibbrand, 1976; Michon, 1976). The present writers have attempted to develop a strategy for investigating social psychological aspects of transportation, and have applied it in two studies, concentrating particularly on the behaviour of automobile drivers.

In studying any type of behaviour the psychologist usually has the aim of making predictions about some future eventuality. Two basic approaches to this predictive task have been traditionally adopted by behavioural scientists. The first of these, which stems from "behaviouristic" psychology and has as its paradigm the experiment, operates by manipulating independent variables to determine what effects this has upon dependent variables. Unfortunately, this type of manipulation cannot easily be carried out on the large scale frequently necessary in transportation studies, because of human and financial costs and because the social consequences of such experimental manipulation are often politically unacceptable. This is not altogether unreasonable since, if large scale changes are introduced (say the replacement of a tramway system with buses), then any assessment of the psychological impact of such a change, once the change has actually been made, may well be of only academic interest. One alternative which is frequently employed in the transportation domain is to simulate the change on a much smaller scale. However, it is by no means certain that effects demonstrated in simulations of this sort will be replicated in real life situations.

An alternative, indirect approach is to try to develop knowledge of the inter-relationships among the human and situational aspects of the domain under investigation, and subsequently to predict the effects of changes in one on the other. In this approach, the human beings and the situational factors are treated as a system. Typically, in investigations of driving behaviour, this approach has involved measuring certain characteristics of people, and attempting to relate these to subsequent behaviour when behind the wheel. Such characteristics might include aspects of personality, or intelligence, or even a straightforward expressed opinion about what a person believes he would do in a particular situation.

Studies of this indirect type in the area of driver behaviour are extremely common, but unfortunately they are plagued by just as many shortcomings as those involved in direct manipulations. One approach, according to Linares-Maza (1971), has attempted to identify variables that seem likely to be highly correlated with skilful driving such as the psychophysical traits of reaction time, visual acuity, or ability to cope with glare. Another tactic has been to try and construct a picture of the characteristics of the good driver in terms of his responses to aptitude tests, personality test batteries, or projective tests like the TAT and the Rosenzweig Picture Frustration Test.

In contrast, other investigators have attempted to identify the potentially bad driver through the use of psychoanalytic techniques. Linares-Maza (1971) has pointed out that such studies suggest the existence of three basic types of dangerous driver: the irresponsible driver, the arrogant driver and the extra-punitive driver. However, although psychoanalytic theories purport to explain accidents in terms of the psychological make up of the driver, a full understanding of the nature of the relevant personality syndrome can only be discovered by costly and complex individual psychoanalysis. Hence the psychoanalytic approach has not lent itself to widespread application in the area of driver behaviour.

Similar problems have arisen with other indirect psychological approaches. For example, although Egan (1967) was able to show that drivers with very low IQs were involved in more accidents and committed more driving offences than those with average and higher intelligence, the same author was unable to demonstrate any direct, linear relationship between IQ and good driving. In the area of personality, Undeutsch (in Wagner and Wagner, 1968) was able to identify various personality patterns which were related to poor driving: the tendency to be impulsive, to be over-routinized, and lacking in reflection, to be self-centered and rigid, or to be anxious, and indecisive. Parry (1968) also stressed the importance of aggressiveness and over-anxiety in bad driving. However, in spite of these apparently promising findings, Linares-Maza (1971) pointed out that the psychological deficiencies which may lead to dangerous driving generally go undetected in brief interviews conducted by psychiatrists and psychologists. Indeed, Hoyos et al. (1965), on the basis of their comprehensive review of personality studies, concluded that the best predictors of good and bad driving behaviour were relatively superficial characteristics, such as the number of years of driving experience, history of court appearances, and so on.

Consequently it appears that the practical usefulness of much early psychological work in the area of driving behaviour is of questionable utility. Bocher (in Wagner and Wagner, 1968) demonstrated that the correlations between predicted behaviour and actual driving performance are usually quite low and rarely more than .50 (the higher correlations being obtained when batteries of psychological predictors are combined). Thus it would be fair to say that most indirect psychological studies of driver behaviour lack predictive validity.

Another indirect approach involves use of the concept of "attitude". However, the apparently straightforward strategy of trying to predict behaviour on the basis of expressed attitudes and opinions (what drivers say they do and feel) is fraught with methodological problems - so much so that this approach has been virtually abandoned by many social scientists in recent years, except for predictions in highly circumscribed situations, such as voting behaviour or simple consumer choice-patterns. Among the chief difficulties to cope with are the discrepancies between what Argyris (1975) calls "espoused theories of action" and "theories in use". Expressed in more colloquial terms, the problem here is to distinguish between what people say they believe, what they really believe, and the way in which beliefs will translate into action.

Compounding the problem that attitudes and behaviour are often not isomorphically related is the additional difficulty that there are great variations in the sophistication and complexity with which different people are able to articulate particular attitudes. Furthermore the subtlety with which measuring instuments are capable of probing the attitudinal domain also varies considerably. These two aspects of attitude measurement may be conceptualized in terms of two types of general measurement error. The first involves a situation where the attitude being measured is simple and the measurement technique too complex. This might be represented by an attitudinal statement such as: "Wearing a seat belt gives me the feeling of being trapped and confined in my car", with which a respondent is asked to express a degree of agreement or disagreement. A statement of this sort may well have the effect of "reading into" the response a quite inappropriate level of complexity and subtlety.

The second type of error is where a complex attitude is measured by a technique which is inappropriately simple. An example here might be the attitudinal statement: "I prefer my own car to public transport", where the level of expressed agreement says nothing about any qualifications the respondent may wish to make concerning his answer, and hence is likely to predict behaviour extremely poorly.

The present writers believe that the constructs of attitude and opinion are useful ones, and that they should be capable of promoting understanding of both feelings and actions. However, if this is to be done with any sensitivity and accuracy it is essential that the principles articulated by the early students of attitude measurement be recalled (e.g. Likert, 1932; Thurstone & Clave, 1929). The strategy developed by the present authors attempts to go back to the basic principles of attitude measurement and to avoid the two basic types of error described above. In the first place, the technique places stress on probing the underlying, relatively subjective aspects of attitudes, and achieves this by using an eclectic approach derived from many different areas of psychology, both novel and traditional, clinical and quantitative. A second important feature of the strategy is to allow respondents to specify the attitudinal domain for themselves, by responding spontaneously in a freely structured situation and in their own words, rather than having put into their mouths ideas which reflect merely the prejudices and preconceptions of the investigators.

This quasi-clinical strategy of attitude measurement has been applied in a number of settings. In the field of transportation, two basic types of study have been carried out. The first involved the investigation of probable public reactions to a specific contingency aimed at inducing safety-related behaviour - the introduction of legislation making the use of car seat-belts compulsory. The second type of study was of a more general nature, and was in fact an attempt to formulate a social psychology of driving behaviour, focusing on the relationship of the automobile driver to other people in the driving situation, such as car passengers, pedestrians or drivers of other vehicles.

Both sorts of study involve considerations which have vital consequences for policy and planning decisions in the area of traffic safety. Motor-vehicle accidents are now among the leading causes of death in the highly industrialized nations. In the United States, over two million people have died in automobile accidents in the last fifty years, while another 110 million have been maimed or seriously injured. In 1973 alone, about 57,000 people in the U.S.A. were killed in motor-vehicle accidents, and four million injured (American Safety Belt Council, 1974). In Canada, motor-vehicle accidents killed 6,237 in 1972, and are the leading cause of death between the ages of 15 and 35 years (Hauser, 1974). Although various countermeasures exist, some of them known to be highly effective in ameliorating the consequences of accidents (seat belts, for example), the public has been notoriously resistant both to the urgings of public-safety officials to make use of such countermeasures, and also to public education campaigns aimed at promoting safer driver behaviour (see Cohen, 1975, for a summary of some of the available data). Consequently, the specific question arises of why people do not make use of available countermeasures, along with the more general one of why they resist efforts to promote safer driving. It was these two questions to which the studies reported here were addressed.

THE METHOD

The strategy for attitude measurement (described more fully in Knapper et al., 1973) involves a sequential, funnel-like set of steps, and, as stated above, attempts to combine elements of both quantitative and clinical psychology. In the initial stages, small samples of respondents are interviewed intensively in both small-group and individual, quasi-clinical settings. The intention here is to provide a variety of channels through which respondents can express not only the commonplace truisms so frequently elicited, but also the unanticipated, more subtle fears and prejudices that form a vital component in the realm of attitude and opinion. In later stages a larger respondent pool is used, selected by random sampling procedures, and here the full weight of quantitative techniques of investigation and analysis is used to maximum effect, including such sophisticated procedures as factor analysis. A major feature of the approach is that precise hypotheses are not postulated in advance, nor are specific forms of measurement for the final sample decided at the outset.

Although it follows from the above description that the precise contents of each stage of the strategy will differ from study to study, the following characteristics are common elements. The first step involves interviews with experts in the area (for example police, safety officials, government personnel, and so on) as well as the usual comprehensive literature-review: the idea at this point is to gain some familiarity with "official" opinions. In the second step, group sessions are carried out with members of the public; the sessions comprise from six to twelve respondents, usually selected from a panel maintained by the University of

Regina, and are as unstructured as possible to allow for free and spontaneous response from the participants. A variety of devices is used by the groups leaders to stimulate discussion and explore and probe opinions in depth. These include techniques derived from encounter groups, "brainstorming" exercises and adaptations of projective tests, where groups members are asked to imagine themselves in a particular situation (e.g., in an accident) and then "act out" their response. A record of the session is kept and subjected to content analysis. These two steps (interviews with experts/literature review and the group sessions) together comprise the Pre-Pilot Stage.

On the basis of the analysis of material generated in the Pre-Pilot Stage a schedule is prepared for use in the Pilot Stage and administered to a group of respondents selected by random sampling procedures, interviewed individually in their homes. The schedule is not handed to respondents directly, but instead serves as a guide for a team of specially trained interviewers. The instrument comprises a number of "probes", some in the form of questions, some simple "projective" tasks in which respondents are asked to put themselves in particular situations, describe certain types of driver, etc. Even word association and sentence completion tasks are not ignored and, in addition, respondents may be asked to complete certain pertinent personality measures (selected on the basis of the literature review) in their own time. Throughout the Pilot Stage, interviewers always first ask questions in the most general possible way to allow respondents to specify for themselves what is relevant and salient. (Similarly, in initial contacts with respondents for all stages no mention is made of the precise focus of the investigation, although careful efforts are made to explain clearly the purpose of the study to respondents - verbally and by letter - after the interview has been completed.) Once the topic has been broached in this general way, interviewers then proceed to ensure that a number of basic areas are covered, using the schedule probes as guides. Responses to this stage are subjected to content and statistical analysis.

In the final Survey Stage a much larger number of respondents is used, once again randomly selected from the relevant general population. Interviews are shorter than in the Pilot Stage and based upon a questionnaire, which is highly structured on the basis of statements suggested by analysis of Pilot Stage responses. What is required here is a relatively objective instrument which, while not being totally self-administering, is within the capacity of a wide range of members of the public to complete for themselves, with only minimum help from an interviewer. The items used in the Survey Stage questionnaire all preserve the actual wording of responses given in previous stages, so that the contents of the instrument are phrased in the everyday English of the ordinary person.

The format for presenting these statements needs to combine an easily comprehensible form of presentation and the capacity to yield comparable data from subject to subject, yet requires a simple response that will yield data amenable to expression in numerical form for subsequent statistical treatment. This is generally achieved by using one of two techniques. The first is an adaptation of the semantic differential in which statements are converted to adjective pairs of opposite meaning (e.g., "The person who always wears a seat belt is: 'manly' .... 'effeminate'"). The second is to present items in the form of assertive statements (e.g., "I feel anxious in heavy traffic") and ask people to indicate how strongly they agree on a seven- or nine-point scale. Analysis of results of the Survey Stage is done solely statistically, usually by correlation and factor analytic procedures. A summary of the procedure is presented in schematic form in Table 1.

Stage	Respondents	Complexity of Materials	Kind of Analysis
Pre-Pilot	Experts and opinion leaders	Unstructured interviews	Subjective "content analysis"
	Members of the public	Group sessions - semi-structured interviews	Content analysis
Pilot	Members of the public randomly selected	Structured interviews using a schedule[1]	Content analysis and simple statistical analysis
Survey	Members of the public randomly selected	Highly-structured interviews using a questionnaire[1]	Complex statistical analysis including multivariate procedures

[1] A schedule differs from a questionnaire in the present context in that a schedule is a set of instructions to interviewers, on the basis of which they conduct an interview recording subjects' responses either in summary or verbatim form. A questionnaire is a set of questions handed to subjects for them to fill out for themselves by selecting alternatives, marking boxes and so on.

TABLE 1. A SCHEMATIC ORGANIZATION FOR ATTITUDE MEASUREMENT

As stated above, the strategy has been employed in a wide variety of settings. So far public reaction to this interviewing approach has been gratifyingly favourable. Not only are response rates high (in excess of 80%), but informal comments from respondents appear to indicate a liking for the opportunity to give their views freely (Pre-Pilot and Pilot Stage), and (in the Survey Stage) a positive reaction to the wording of statements which were often believed to express an idea "perfectly".

RESULTS

Seat Belt Study

A questionnaire concerning attitudes to seat belts was administered to 465 residents of the City of Regina selected by a random sampling procedure (Kish, 1965, p. 398). A preliminary finding involved the levels of seat-belt usage reported by respondents. Only 39 out of 465 (8.3%) claimed that they always used their belts. A little over half of the respondents (54.2%) claimed to use their seat belts at least some of the time. As might be expected, reported level of use in the city correlated significantly with reported use on the highway, the correlation being .69 ($df = 463$; $p < .01$). Attitudes expressed towards seat belts were highly favourable, despite the relatively low levels of actual use claimed by respondents. In general, three groups could be discerned, comprising the 8.3% who claimed that they always used their belts and who were very enthusiastic about them, another group of about 10% who were intractably opposed to seat belts under any circumstances, and the remaining group of people who rated seat belts highly and wanted to see them worn more (by other people), but who seldom or never fastened their own belts. This finding resembles that of Marzoni (1971) who reported favourable attitudes to seat belts in 90% of his subjects.

The information collected in the steps just described was used to examine three issues: people's opinions about seat belts, their opinions of habitual users of seat belts, and their reactions to other people who reminded them to fasten their belts. The findings may be summarized in the following way. There is a very high level of acceptance among the public that seat belts are effective. The major reasons why they are not worn involve lack of concern for safety, absent-mindedness, and the failure of any authoritative pressure to fasten up. These findings suggest that the best method for increasing levels of seat-belt use would be to take the decision to fasten up out of the hands of the individual driver. This could, perhaps, be achieved through the development of some kind of automatic seat belt such as those being devised by some European motor-vehicle manufacturers. However, changes in car design would take several years to become universal, whereas there is a need for measures that could come into effect almost immediately. Consequently, laws requiring the use of seat belts might well be a better short-term solution. Such a law appears to have good prospects of being successful, at least among the residents of Saskatchewan, Canada. The survey's findings suggest that people would recognize that the intention of such a law would be to protect them, and they would agree that the measure being mandated was one which would be effective. Furthermore, it seems likely that a high proportion of the public would accept that the government is a legitimate authority which has a "right" to issue directives concerning motor-vehicle behaviour.

Social Psychology of Driving

A second study was based on the responses of a separate group of respondents who were also selected in the way outlined above. This study investigated interpersonal factors in drivers' attitudes. The data were not obtained by observing behaviour in real situations, but were elicited by the use of the technique already described. Respondents were asked to imagine that they had been involved in an accident involving another vehicle. Each respondent was then asked to try to conjure up the situation as vividly as possible in his mind's eye and to describe his reactions. At the risk of oversimplification, the findings can be summarized by saying that, according to the survey, psychological reactions to an accident are conditioned by the behaviour of the other driver, on the one hand, and by the presence or absence of persons other than the two drivers, on the other. For example, the presence of a passenger will tend to inhibit extremely aggressive behaviour on the part of the innocent party; similarly, apologetic or embarrassed behaviour on the part of the offending driver mediates embarrassment, sympathy or tolerance on the part of the injured party, and the behaviour of onlookers can further effect the driver's reaction.

A second interesting finding is that drivers make inferences about the attitudes and motivations of other drivers, and react in terms of those inferences. For instance, the young or the old tend to be perceived by other drivers as incompetent, selfish, careless or overly-cautious. Young male drivers who have girls with them are viewed with a jaundiced eye by other drivers, are assumed to be at fault if an accident occurs, and elicit aggressive responses. Long-haired male drivers, particularly if they are informally dressed or driving a brightly-painted vehicle, predispose other drivers to believe that they have bad driving habits. In an accident situation, such a driver is likely to be blamed, and to provoke angry or aggressive reactions. Out-of-town licence plates and large or small cars have similar effects. (There is evidence that this tendency can even extend to more experienced observers of driver behaviour, such as traffic police. A study by F.K. Heussenstamm, conducted in 1970 in California, found that drivers of cars bearing prominent 'Black Panther' bumper stickers suddenly attracted great numbers of traffic citations, even though none of the drivers had had previous citations - see Arons, 1975.)

A general conclusion from this study is that drivers continually impute motives, opinions and values to other drivers, on the basis of a number of visible characteristics that are, however, not directly related to the driving process at all. Examples of such characteristics have already been given. Just as everyday social interactions are mediated by inferences based on observable key attributes (the process of "stereotyping"), interaction among drivers, especially in accident settings, are subject to a similar process. Practical consequences of this state of affairs include the fact that driving manoeuvres such as speed-setting, positioning of the vehicle, reactions to other vehicles, and so on, are not based solely on the practical, physical attributes of the situation. Driver behaviour also involves a psychological process, in which reactions are based on motives and attitudes inferred to exist in other drivers. These findings strongly support the conclusion by Michon (1976) that one important contribution of psychology to understanding human behaviour in the transportation domain lies in understanding that the road-user is, among other things, a "social being" engaging in a series of social interactions with other users.

Discussion

The set of steps proposed here for investigating psychological factors in the transportation domain does not contain necessarily novel individual elements. However, it is the authors' belief that there is need for a systematic re-statement in a transportation context of the possible role of attitude surveys, and demonstration of how they may strategically be employed. Furthermore, by spelling out a 'modus operandi' which it is claimed will avoid some of the problems inherent in the area, it is hoped to provide an illustration of one pattern of good practice, and some guidelines for continued application of the concept of attitude. Nonetheless, the practical experience of the present authors suggests that increased interest in analysis of the transportation domain from the point of view of social science will make one point increasingly clear: input from the general public, whether by means of the present strategy or other techniques, will remain only one of the elements in a process which will continue to be heavily dependent upon political factors. Indeed, the social scientist will need to be alert to the possibility that his work will be used merely to give an air of legitimacy to political decisions, or indeed for merely manipulative purposes.

ACKNOWLEDGEMENT

The findings summarized here were derived from two empirical studies supported by contracts with the Ministry of Transport of the Federal Government of Canada. The authors are grateful for this support. However, the views expressed here are entirely those of the authors, and in no way reflect official opinion or policy.

REFERENCES

American Safety Belt Council. (1974) Solution for Highway Carnage: Safety Use Legislation. Washington,D.C.: American Safety Belt Council.

Argyris,C. (1975) Dangers in applying results from experimental social psychology. American Psychologist, 30, 469-485.

Arons,S. (1975) The rise of police logic. Saturday Review, April 5, 12-13, 51.

Cohen,R. (1975) The case for wearing your seat belt: Why you should. Canadian Consumer, 5, 24-26.

Egan,R. (1967) Should the educable mentally retarded receive driver education? Exceptional Children, 33, 323.

Hauser,D.J. (1974) Seat belts: Is freedom of choice worth 800 deaths a year? Canadian Medical Association Journal, 110, 1418-1423.

Hoyos,C.G., et al. (1965) Psychologie des Strassenverkehrs. Bern: Huber.

Kish,L. (1965) Survey sampling. Wiley: New York.

Knapper,C.K., Cropley,A.J., & Moore,R.J. (1973) A quasi-clinical strategy for safety research: a case-study of attitudes to seat belts in the city of Regina, Saskatchewan, Canada. Report submitted to the Federal Department of Transport, Canada. Car Safety Research Project, University of Regina.

Krampe, G. (1976) Cluster analysis of West German towns with special reference to personal rapid transit systems. In P. Stringer and H. Wenzel (eds.) Transportation Planning for a Better Environment. New York: Plenum.

Leibbrand, K. (1976) Recent changes in transportation and urban planning. In P. Stringer and H. Wenzel (eds.) Transportation Planning for a Better Environment. New York: Plenum.

Likert, R. (1932) A technique for the measurement of attitudes. Archives of Psychology, Whole No. 140.

Linares-Maza, A. (1971) Psicologia clinica, psiquiatria y conduccion de automoviles. Revista de Psicologia General y Aplicada, 26, 30-55.

Marzoni, P. (1971) Motivating factors in the use of restraint systems. National Analysts Inc. Available as report DOT HS-800 585, National Technical Information Service, Springfield, Virginia 22151.

Michon, J.A. (1976) The mutual impacts of transportation and human behaviour. In P. Stringer and H. Wenzel (eds.) Transportation Planning for a Better Environment New York: Plenum.

Parry, M.H. (1968) Aggression on the Road. London: Tavistock.

Thurstone, L.L. & Clave, E.J. (1929) The Measurement of Attitudes. Chicago: University of Chicago Press.

Wagner, D., & Wagner, J.J. (eds.) (1968) Handbuch der Verkerhrsmedizin. Berlin: Springer.

THE RELATIONSHIP BETWEEN ROAD ACCIDENTS AND URBAN STRUCTURE

Stanley Raymond and Michael D. Hodgkinson

Department of Civil Engineering

University of Salford Hampshire County Council

SPATIAL VARIATION ANALYSIS

The study of the variation in road accident statistics between different nations has become a well-accepted part of road accident analysis over the last few years. Little work has been completed, however, on the internal variations within one country, between different urban areas. The research outlined in this paper attempted to look at this variation within England, Scotland and Wales, using a multi-stage statistical approach. The variation was studied both spatially and temporally using data from the years 1960-1970. Such studies perform a two-fold function. Besides being valuable for use in deriving basic properties, and much needed system models, they are also useful in investigating the relative effectiveness of countermeasure treatments.

As Katz states, such studies are "... well adapted to reveal the effects on accident generation of differences in environment, geography, and transport arrangements, that exist between areas. Once these structural factors are identified - that is a model exists - then they can be put under statistical control in order to compare the changes in accident and injury from year to year, in the same geopolitical area" (Katz, 1971).

In order to study these variations a sample of 100 urban areas was taken from the 412 local authority areas in England, Scotland and Wales (prior to Local Government changes in April 1974) with a population of over 20,000 (1969), giving a sample size of approximately 25 per cent. A random stratified sampling technique was used taking into account variations in the population level, by using the population levels of 100,000 and 50,000

No.	Variable	No.	Variable
1	Population level	10	Town classification index
2	Area of urban district	11	Compactness index
3	Density of population	12	Degree of open space
4	Peak traffic flow	13	Standard of housing index
5	Estimate of through traffic	14	Age vulnerability index
		15	Driver vulnerability index
6	Total traffic flow	16	Total dwelling index
7	Car ownership	17	Housing tenure index
8	Mean social index	18	Change of housing stock index
9	Sex ratio index		

Note In addition, for 51 larger areas it was possible to include a Road density index, Police prosecution index, Police severity index and Rainfall index.

TABLE 1. LIST OF INDEPENDENT VARIABLES (1970)

as strata limits. For each area sampled a total of 18 compound variables were derived using published data from various sources, together with other data obtained from Local Authorities relating to the physical and social structure of those areas (Table 1). These variables were taken as the original data matrix describing the urban structure system. The response of this system was taken to be equal to the number of personal injury road accidents over the period 1960-1970. This information was obtained from the Transport and Road Research Laboratory's computerisation of the 'Police Statistics 19' forms and was disaggregated according to the type of accident, location of accident, etc., giving a total of 45 dependent variables. A sample of these is given in Table 2. Use of this data source gave spatial uniformity of data, though unfortunately restricting the temporal analysis because of the major police reorganisations of 1967-1968.

The basic aim of the multi-stage analysis was to study the spatial variation in road accident statistics between different urban areas by observing the difference between a calculated, "expected" number of road accidents or 'norm', and the observed number of road accidents for that same area. The technique used to attain this 'normal' level was that of multiple regression analysis. Unfortunately such a technique makes certain assumptions concerning the structure and composition of the original data matrix, and necessitated the transformation of the data matrix in the first stage of the analysis. It is therefore more convenient

1.	TC70:	Total Casualties
2.	TA70:	Total Accidents
3.	TFS70:	Total Fatal and Serious Casualties
4.	TPED70:	Total Pedestrian Casualties
5.	TCPED70:	Total Child Pedestrian Casualties
6.	TJUN70:	Total Junction Accidents
7.	TCRD70:	Total Accidents on Classified Roads
8.	TCPP70:	Total Casualties per 10^3 Population
9.	TAPP70:	Total Accidents per 10^3 Population
10.	TCVM70:	Total Casualties per 10^6 Vehicle Miles
11.	TAVM70:	Total Accidents per 10^6 Vehicle Miles
12.	SR70:	Fatal and Serious Casualties/Total Casualties
13.	YDRCR70:	Young driver and rider casualties/Total Driver and Rider Casualties
14.	TAMC70:	Total Accidents on Classified Roads per Mile of Classified Road
15.	TPEDPP70:	Total Pedestrian Casualties per 10^3 Population

TABLE 2. SELECTED DEPENDENT VARIABLES (1970)

to divide the explanation of the methodological approach used into the various stages of analysis (Figure 1). Limitation of space prevents full treatment in this paper and no reference is made here to the trend surface analyses.

METHODOLOGY

Stage 1. Transformation of Data

One of the major assumptions of the regression model is that the variables included in the analysis are independent and pairwise un-correlated. Since the 18 variables in the original data matrix were highly correlated some transformation was necessary. This was completed by using a principal components analysis, which besides orthogonalising the original data also identified important components which could be idealised as sub-systems within the total road accident system. In order to facilitate easy identification of the various derived components, a Promax rotation was included, which, whilst losing the absolute orthogonality of the derived data matrix, maintained an acceptable level of inter-correlation between the components. The 'scores' obtained from this Promax rotation were then used in the second stage of the analysis, as the input to the multiple regression analysis.

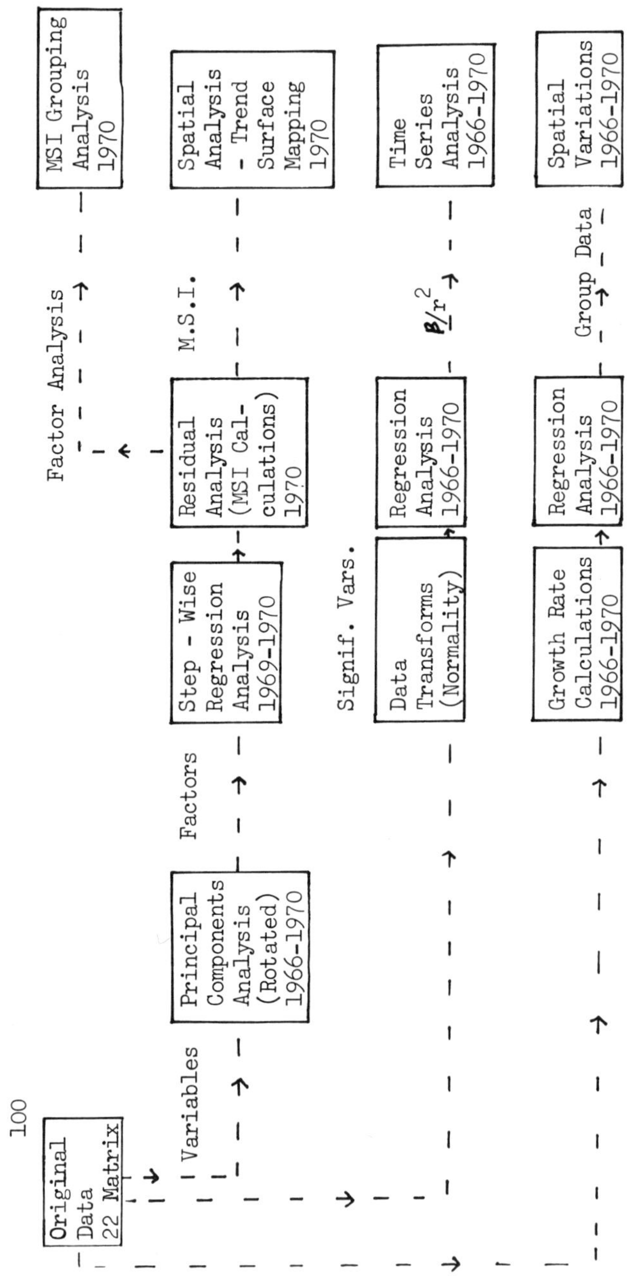

FIGURE 1. FLOW DIAGRAM OF METHODOLOGICAL APPROACH

Stage II. Multiple Regression Analysis

In the multiple regression model (Cooley & Lohnes, 1962; Poole & O'Farrell, 1970) Y_o is the dependent variable, X1 ... n the independent variables and α and β the parameters of the regression. For each dependent variable used a linear regression plane was derived, along with the respective deviations from this plane, for each individual observation (urban area). This allowed the calculation of a "mean safety index" for each year, each urban area, and each dependent variable, as defined in (1).

$$\frac{Y_{oi}}{Y_{oc}} \quad x \quad 100 \text{ (per cent)} \quad (1)$$

where Y_{oi} is the observed value of the dependent variable at location i and

Y_{oc} is the calculated value of the dependent variable location i

This index gives a variation about 100, being greater than 100 when the observed value is greater than the computed value, and less than 100 when the situation is reversed. Through this index it was possible to study the spatial variation in road accident statistics.

Within the regression model the β parameters are also important since they are partial regression coefficients. Each one gives the rate of change in the dependent variable, for a unit change in that particular independent variable, while the remaining independent variables are held statistically constant. Thus it is possible using these β coefficients to study the varying influence of any particular variable over time, by comparing the standardised β coefficient of any independent variable for any two different years.

In this research a modified regression model was used, that of 'step-wise' regression, which is an algorithmic procedure to obtain an optimum collection of statistically significant independent variables. The significance level used was 10 per cent ($p = 0.1$) which allowed the inclusion of variables which are only marginally significant and which might otherwise have been ignored.

Stage III. Temporal Analysis

Since the β coefficients derived in Stage II of the analysis refer to different components for each year, in order to get a true indication of the temporal change in the relative importance of the independent variables, it was necessary to return to the original data matrix. A new regression model was used which incorporated each dependent variable, and the relevant original, statistically significant independent variables as ascertained in the step-wise regression model. For this new regression model the standardised coefficients were compared for any two years and the variation noted. However, because the road accident data were very difficult to obtain prior to 1966 and the disaggregation of the data was very limited, it was only possible to conduct this time analysis for the dependent variables: Total Casualties, TC/000 & TC/Vm. 1966-1970.

The first method provided results as to the variation in the importance of certain independent variables over time, whilst the second approach indicated what changes in any one independent variable lead to similar changes in the dependent variable. That is, the second method had more implicit causal connotations within its approach, even though no such implication can be proved statistically.

THE NATURE OF THE DEPENDENT VARIABLES

Before a selection of results from this analysis are presented it is necessary to look at the nature of the dependent variables. These variables fall into two simple categories:

a) Absolute Variables - which use an absolute figure such as the Total number of Casualties.

b) Compound Variables - which express an absolute figure as a rate or ratio with some other factor, e.g. Total Casualties per 1000 population. These rates include both static group indicators as well as some exposure group indicators. These latter indicators are to be preferred, but due to lack of exposure data they are overshadowed by the static rates.

SAMPLE OF RESULTS

The results of the principal components analysis are that, of the original variables subjected to the analysis, six components were derived with an eigenvalue greater than one. They accounted for over 80 per cent of the total variance of the original data matrix (see Table 3). Put another way, the principal components analysis indicated that the original data matrix contained only six independent dimensions, each of which following identification

Component No. (see Table 2)	Component Eigen value	Cumulative Explanation of Variance (%)
1	5.863	32.570
2	2.733	47.751
3	2.019	58.968
4	1.777	68.838
5	1.298	76.048
6	1.034	81.795
7-11 (incl)	2.514	95.767
12-18(incl)	0.762	100.000

TABLE 3. EIGENVALUES AND VARIANCE OF PRINCIPAL COMPONENTS

could be related to various sub-sections of the road accident system. The first component in fact could be related to a 'size' or 'interaction' component, and accounted for over 32 per cent of the total variation of the data matrix on its own. Other important components identified related to social structure, socio-economic factors and shape variables. These components also seemed fairly static over time, implying that the urban structure as measured by these variables has altered little since 1966, the same significant "dimensions" being present at each time-period. This constancy within the data matrix is of great importance with respect to the temporal analysis. It means that the change in importance of any one component cannot be due to a change in the overall urban structure, but solely to that variable itself. The importance of these results is two-fold. They help in the identification of problems in the use of the regression model; and they support some of the 'a priori' hypotheses concerning the structure of the road accident system made at the outset of this research.

In the regression stage of the analysis the 45 dependent variables were introduced with widely differing results, indicating that the structural sub-system of the road accident system affects different parts of the system in different ways and to differing degrees (Table 4). It is interesting to note that where absolute figures were used as the dependent variable, an explanation level of over 80 per cent ($r^2 \geqslant 0.80$) was usually obtained, with most of the explanation being accounted for by the 'interaction' component. This can perhaps be explained in two ways. These results could simply suggest that as the size of the urban area increases, there is a corresponding increase in the volume of road accidents.

Yo Coding	R^2	$\bar{Y}$	S.D.	S.E.E.	$\bar{Y}$% of S.E.E.
TC70	0.986	804.43	1015.21	123.32	15.3
TA70	0.987	636.72	808.17	98.06	15.4
TPED70	0.988	257.41	369.55	41.69	16.2
TCPP70	0.292	5.74	1.102	0.95	16.6
TAPP70	0.285	4.49	0.879	0.78	17.4
TCVM70	0.784	18.48	13.140	6.38	34.5
TAVM70	0.739	14.47	10.450	5.67	39.2
TPEDPP70	0.640	1.67	0.444	0.28	16.8

TABLE 4. RESULTS OF SOME OF THE BASIC REGRESSION EQUATIONS 1970

Alternatively, they might indicate that since the volume of road accidents increases proportionally to the size of the urban area, and not to its structure, the road accident problem is not entirely an engineering problem, but more a 'psychological' problem due to the collection and interaction of individuals within a complex system. Following this line of argument, some further information can be obtained by observing the response of certain urban areas, specifically designed structurally to accommodate the traffic/pedestrian conflict. The obvious areas to study here are the new towns included within this analysis. Such a comparison for nine dependent variables is given in Table 5 for the year 1970. It can be seen in this Table that while these areas tend to be below the norm in most instances they appear to behave less favourably than would be expected relative to other areas, especially in the later new towns; although this could be due to the fact that these areas have not completed their development and are reacting correspondingly.

When compound variables are introduced into the regression analysis, the explanation level is drastically reduced, usually being around the level of 50 per cent (r^2 = 0.50). This is especially true when the compound variable used is the population factor, that is Total Accidents per thousand population, etc.. It can be argued that once the size component has been held constant these explanation levels are truer representations of the impact of the urban structural variables. The response of the new towns to these compound variables is shown in Table 5.

Two other important factors in the regression analysis deserving more detailed attention are that the impact of through traffic

NEW TOWN				MEAN SAFETY INDICES					
AREA	TC	TA	TFS	TPED	TCPED	TAPP	TCPP	TCVM	TAVM
Cumbernauld	139.33	127.62	34.99	67.59	300.00	103.11	108.54	123.92	144.66
East Kilbride	72.77	69.12	66.33	75.78	110.40	73.82	98.14	67.81	66.27
Glenrothes	60.43	63.20	184.63	58.39	66.56	88.65	106.24	160.78	144.40
Harlow	92.95	85.32	127.13	93.10	115.29	91.38	105.63	86.82	82.56
Hemel Hempstead	92.18	88.98	84.13	88.75	74.46	81.42	78.19	66.03	65.97
Stevenage	100.06	105.47	86.49	99.93	85.42	101.02	84.11	88.15	76.50

TABLE 5. THE MEAN SAFETY INDICES FOR 9 INDEPENDENT VARIABLES RELATING TO THE NEW TOWN AREAS (1970)

on the road accident level is perhaps more important than often realised; and that the impact of police surveillance and control can help in many road accident situations. Where police action as regards motoring offences is very stringent the response of the system is usually positive, even if only marginally.

Although the results from the temporal analysis have not yet been fully analysed, it would seem at present that the level of explanation due to the urban structural variables has been declining over the period studied. This tentatively suggests that the increased attention given to road accident problems in urban areas from an engineering point of view is being marginally successful, although no such definitive statement can be proved at present. Alternatively, it could be that increased propaganda relating to accident prevention or training has taken account, especially among the younger sections of the population. In this context it is worthwhile noting that in 1969 the structural variables accounted for more than 80 per cent ($r^2 = 0.85$) of the variation in pedal cycle accidents, while in 1970 this level was reduced to 64 per cent ($r^2 = 0.64$).

CONCLUSIONS

While this research has helped to prove the complexity of the road accident system, and its varying response from area to area, it can justifiably claim to have also presented two constructive results. It has suggested a methodological framework for the study of road accidents, and also the interaction within the road accident system. It has also helped to pin-point certain variables which account for some of the variation between different areas.

Both of these implications are also the foundation for further research. To understand the road accident system totally, the other non-structural sub-systems must be studied and placed within the context of those sub-systems studied in the present research. Similarly, there must be more exhaustive follow-up studies of individual areas to account for the anomalies indicated by this research. Such studies should involve the more detailed modelling of individual areas. Until such additional research is completed and until the road accident problem is accepted as an integral part of the urban environmental problem, little progress can be hoped for in road accident prevention.

REFERENCES

Cooley,W.W., & Lohnes,P.R. (1962) Multivariate Procedures for the Behavioral Sciences. New York: Wiley.

Katz,A. (1971) Towards a methodology of traffic safety measurement and program evaluation. Accident Analysis and Prevention, 3, 15-43.

Poole,M.A. & O'Farrell,P.N. (1970) The assumptions of the linear regression model. Transactions of the Institute of British Geographers, 52.

A SYSTEMS APPROACH TO URBAN BICYCLE SAFETY

Colin G. Drury

Department of Industrial Engineering

State University of New York at Buffalo

Over the last five years the U.S.A. has experienced a rapid growth in the sale and use of bicycles. Current estimates are that there are 15 million bicycles per year sold (more than automobile sales for 1974) and that one American in three uses a bicycle.

These increased sales and ownership statistics are the result of a number of factors. The market for costly and impressive ways of enjoying leisure has spread from boats, pools and motorcycles to ten-speed bicycles. Even a cursory observation of the roads of urban U.S.A. will reveal many European and domestic racing bicycles. There is the rediscovery of the bicycle as an aid to health. Riding for health is largely self-paced with a wide choice of routes if the rider has no fixed destination. The energy crisis has helped the bicycle boom as many commuters look for alternatives to 10 m.p.g. automobiles for urban use. The bicycle has traditionally been the transportation of youth (under automobile driving age). With racing and touring bicycles of high performance now available in large numbers, this segment of the population is bicycling more often and becoming involved in greater numbers with bicycle clubs, touring, camping, and racing.

The greatly increased use has been reflected in the accident, injury and death statistics. Estimates for 1974 are 371,000 injuries requiring hospital emergency room treatment and 1150 deaths from bicycle-related accidents (National Safety Council, 1974). This represents a 35% increase over 1971. The accident pattern is following the usage pattern in terms of age. There are now a greater proportion of accidents to people of fifteen years and older than before, reflecting the increased use by this age category.

The problem facing planners and legislators is to promote bicycling as an a priori healthy, economic and ecologically sound transportation mode without incurring the proportionate rise in accidents, injuries and damage. This paper is a summary of ways being used at the State University of New York at Buffalo to develop effective intervention strategies.

THE EPIDEMIOLOGY OF BICYCLE ACCIDENTS

Many studies (e.g. Brezine & Kramer, 1971; National Safety Council, 1958; Pascarella, 1971; Williams, 1974) have collected statistics on bicycle accidents. The caveats on accident statistics in the review by Hale & Hale (1972) are particularly appropriate here.

Most non-fatal accidents occur in residential areas, on straight roads, at driveways, intersections, etc., in clear daylight and on dry roads. By comparison, more of the fatal accidents occur outside residential districts, on straight level roads, away from driveways, intersections, in clear dry conditions but with a greater emphasis on night accidents in unlit streets. Typically fatal accidents occur to older riders and the cause of death is head injuries (Consumer Product Safety Commision, 1973). Interestingly, only 5-8% of all bicycle accidents involve collisions with another moving vehicle (Chlapecka et al., 1975).

Despite the considerable data on accident numbers, only recently have accident rates been available. These change the crude picture of accident numbers greatly. Whereas males have about nine times as many accidents as females, their accident rates are very similar (Chlapecka et al., 1975). Pascarella (1971) found rates of 1.49 accidents per 1,000 miles for males compared with 1.79 for females. Similar results for bicycle type show very little difference in accident rates between standard, lightweight and high-rise types, although there is considerable difference in accident frequencies.

Overall, the pattern of accidents seems to follow closely the pattern of use of bicycles, with some differentiation into recreational bicycle use (younger children, off road or by home, less severe accidents) and transportation use (older children and adults, on roads further from home, more severe accidents). Bearing this in mind, different strategies are required to improve different aspects of bicycle safety. Reduction of death and serious injuries will usually mean intervention into traffic aspects of bicycling.

Reducing accidents or milder injuries will mean more emphasis on bicycles as recreational vehicles.

FUNCTION AND TASK ANALYSIS

Whilst analysis of accident data can point out what Swain (1974) calls "accident prone situations", it cannot provide a logical framework for choosing strategies to eliminate or mitigate injuries and accidents. Accidents analyzed are those reported, either to police or hospital emergency rooms, and the data is collected from the memory of respondents after the fact. If accidents have any one characteristic in common, it is that they are all different. Another way to approach the problem is by the human factors route of Task Analysis.

A current project was based on this methodology. Task Analysis (e.g. Swain, 1962; Van Cott & Kinkade, 1972) is a detailed analysis of human overt and covert actions in reaching some stated goal. For this reason the current study is being restricted to traffic-type bicycling where goals and sub-goals are more obvious than in recreation. The first level of analysis was to categorize the major segments of a journey - a Function Analysis. This was accomplished by breaking the trip into Start, Stop, Intersections and Roadway between Intersections and then further subdividing these by type of intersection, the bicyclist's route through the intersections, the type of traffic control devices in use at the intersection, etc. This produced a total of 851 functions (see Table 1).

From the functions it has been possible to characterize any bicycle trip into a sequence of unique functions, one following the other. An analysis of many trips will privide a Function/Frequency Analysis. Two methods have been developed so far for collecting Function/Frequency data.

A roadside census form has been used on over 60 subjects, stopping them on their bicycle on particular routes and collecting information about them, their current trip and their previous trip. From the date and length of the previous trip it is possible to estimate annual use and mileage. In addition to the questionnaire (developed by E.B. Daniels and G. Kobas) photographs of bicycle and rider in standard positions are taken so that anthropometric data on the fit of bikes to riders can be surveyed.

The second method for obtaining Function/Frequency data is a mailed questionnaire developed by M.C. Drury. This collects considerable information on the time, route and conditions of the respondent's last trip. It has so far been tested on 35 suburban families. A comparison of the frequency of some of the major

GENERAL FUNCTIONS

1. Start of Trip
2. End of Trip

FUNCTIONS BETWEEN INTERSECTIONS: Drive 1 block on any of:

1.	Single lane,	One way	With Traffic
2.			Against Traffic
3.	Two lane,	One way	With Traffic
4.			Against Traffic
5.		Two way	With Traffic
6.			Against Traffic
7.	Multi lane,	Two way	With Traffic
8.			Against Traffic
9.	On sidewalk		
10.	Other		

FUNCTIONS AT INTERSECTIONS: Turn onto each of 10 above from any of:

1.	4-way intersection	
2.	Traffic circle	
3.	'T' intersection	'T' road on right
4.		'T' road on left
5.		From 'T' road
6.	'Y' intersection	Major road on right
7.		Major road on left
8.		From major road
9.	Merging lane on opposite side of road	
10.	Merging lane on same side of road	

EACH OF THE ABOVE CAN BE RIGHT, LEFT OR AHEAD AND WITH EITHER SIGNAL, STOP SIGN, NO STOP SIGN OR ALL WAY STOP SIGNS EXCEPT WHERE LOGICALLY IMPOSSIBLE.

TABLE 1. COMPOSITION OF 851 FUNCTIONS IN BICYCLE TRIPS

TYPE OF STREET		COLLEGE STUDENTS	SUBURBAN FAMILIES
		BLOCKS PER TRIP	
Single lane, one way, with traffic		0.0	0.0
Single lane, one way, against traffic		0.0	0.1
Two lanes, one way, with traffic		1.5	0.0
Two lanes, one way, against traffic		0.2	0.0
Two lanes, two way, with traffic		5.4	17.9
Two lanes, two way, against traffic		0.2	0.0
Multi lane, two way, with traffic		5.7	4.7
TYPE OF INTERSECTION		INTERSECTIONS PER TRIP	
4-way intersection		4.1	5.7
'T' intersection:	'T' road on right	2.1	5.9
	'T' road on left	0.4	3.6
	From 'T' road	1.7	6.5
'Y' intersection		0.3	0.0
Merging traffic		0.7	0.0

TABLE 2. COMPARISON OF TRIPS BY TWO GROUPS OF BICYCLISTS

functions between this sample and a sample of urban riders from the roadside survey is given in Table 2.

While these two data collection methods are of great interest to urban planners for their data on bicycle use patterns, their prime purpose is to concentrate the Task Analysis onto the more frequent Functions. The next stage of the Task Analysis is the analysis of 16 mm. movie films taken of experienced bicyclists performing each of a selection of the most frequent functions. Skills analysis from a frame-by-frame count of these films (Salvendy & Seymour, 1973) will be used to write down what behaviour should take place. Standard methods of Task Analysis with detailed accident reports from such data banks as National Electronic Injury Surveillance System can then be used to find the ways (and probabilities) that errors can occur, e.g. where task demands exceed human capabilities.

In practice, the relatively short bicycling season in the

northern U.S.A. precludes waiting for results of these analyses before proceeding with experiments and surveys which are known a priori to be important to bicycle safety. These investigations involve the three human-related aspects of safe bicycling: the basic motor skills, knowledge and practice of safe bicycling in an urban environment and the perception of bicyclists by other road users.

MOTOR SKILLS OF BICYCLING

Considering the enormous research effort devoted to control of aircraft and motor vehicles (e.g. Sheridan & Ferrell, 1974) there has been surprisingly little work of a similar nature concerning bicycles. A recent computer study of bicycle stability (Rice & Roland, 1970) used a crude model of the human operator's dynamics. Other, more detailed modelling, derived from the work of Van Lunteren & Stassen (1969) is reported by Sheridan & Ferrell (1974). While this shows to some extent the dynamics required it gives no information on the abilities related to the skill of bicycling (e.g. Fleishman & Rich, 1963). A series of projects is under way to determine the basic correlates of bicycle skill so that training programmes and testing instruments can be devised to augment motor skill. Two preliminary experiments have measured the effects of riding fatigue on skill measurements and the effect of rearward vision on control of a bicycle.

The first study (E.B. Daniels & G. Kobas) involved negotiating a course, having both straight and slalom sections, without error in four conditions: fresh, 1 mile, 2 miles, 3 miles. The "miles of bicycling" were performed on a bicycle ergometer set at 150 watts load - a representative level for fast commuting. Measures of speeds, errors and stopping accuracy showed no change with distance ($p > 0.05$, F test). Bicycle control, like automobile control is quite resistant to fatigue. This result greatly simplifies data on abilities and control skills.

The second skills-related study had riders negotiate a straight course and, at the sound of a car horn, glance over their left shoulder at the car. Measurements of the track followed by the wheels were taken for five subjects. Results showed a mean maximum deviation of 11 inches from the path when the subject looked to the rear, compared with 4 inches in a control condition. Calculations by the authors (Agate, Apte, Giovasis & Llinas) showed that this represented a considerable collision hazard if a bicyclist reacts by looking behind at the warning sound of a car horn. Skill in bicycle control can thus have a major affect on accident rates. (Supporting evidence comes from Chlapecka et al. (1975) who show that "loss of control" accounted for 68% of all bicycle accidents.)

The main programme of work on the analysis of bicycling skills will involve correlation of performance on twelve non-riding tasks (control precision, multi-limb coordination, reaction time and extent flexibility) with riding performance and will include a self-estimation of riding skill.

SAFE TRAFFIC PRACTICES

Skill in bicycle riding is a necessary but not sufficient condition for safe riding. The operator must also know the law and safe practices and actually observe them on the road. A simple knowledge test (for New York State only) has been devised (E.B. Daniels & M.M. Zajkowski) and pre-tested on college students and school children. It reliably differentiates between these two groups and also between U.S. and foreign students.

This is being coupled with a survey of the law observance behaviour of bicyclists at a variety of intersections. Table 3 shows details of fifteen intersections and their violation pattern. The most striking fact about this table is its almost complete absence of a recognizable pattern. The percentage of non-violating riders appears to vary randomly between intersection types. One reason for this was the detailed differences in opportunity of violation. For example, if bicyclists were approaching the junction on the major of two roads they will have longer green phases on the signals and less chance to go through the light at red. The presence of sidewalks, car parks or filling stations at the corners, encourages bicyclists to by-pass the traffic control at the intersection illegally.

The large number of violations, more than one per rider per intersection on average, was pointed out by the investigators (Chermak, Brazee & Odejimi) as being of prime importance for bicycle safety. Williams (1974) showed that over 75% of bicycle/motor vehicle collisions had the bicyclist responsible by breaking a law or a safety rule. These two facts together suggest strongly that accidents will be reduced by reducing the number of law violations. The purpose of the law knowledge questionnaire is to determine whether it is lack of knowledge of the law or deliberate disobedience which is a main problem, so that the optimum mix of education and enforcement can be chosen.

	Intersection					% Bicyclists Breaking				
	Area	Type	Control	Traffic Density	No. of Bicycles	0 Laws	1 Laws	2 Laws	3 Laws	4 Laws
1	BUSINESS	5-way	Signal	Heavy	35	40	34	6	14	6
2	RESIDENTIAL	5-way	3-way stop	Medium	35	14	40	17	23	6
3	BUSINESS	4-way	Signal	Medium	20	5	35	35	25	0
4	BUSINESS	4-way	Signal	Heavy	113	66	24	8	1	1
5	RESIDENTIAL	4-way	Signal	Light	90	41	27	29	3	0
6	BUSINESS	4-way	Signal	Heavy	36	45	36	19	0	0
7	BUSINESS	4-way	Signal	Medium	40	5	30	48	15	2
8	BUSINESS	4-way	Signal	Heavy	71	40	47	9	4	0
9	RESIDENTIAL	4-way	4-way stop	Light	57	5	37	40	16	2
10	RESIDENTIAL	4-way	2-way stop	Light	30	40	33	20	3	3
11	RESIDENTIAL	3-way	1-way stop	Medium	27	11	37	19	30	3
12	RESIDENTIAL	3-way	1-way stop	Light	14	14	58	14	14	0
13	BUSINESS	3-way	1-way stop	Medium	76	25	45	18	7	5
14	RESIDENTIAL	3-way	1-way stop	Heavy	30	17	50	30	3	0
15	BUSINESS	3-way	1-way stop	Medium	13	8	8	40	30	8

TABLE 3. DETAILS OF LAW VIOLATIONS AT 15 INTERSECTIONS

PERCEPTION OF BICYCLISTS BY OTHER ROAD-USERS

The road-users able to inflict the major damage on bicyclists are motor vehicle drivers. Their experience over many years of few bicyclists has reinforced attitudes to the bicycle as a toy, for use by children near home. For any traffic system to work, each user must be able to predict to a reasonable accuracy the future actions of other users. If motorists perceive any unpredictable feature (e.g. dog, child, bicyclist) their lack of predictive ability causes them to slow down and/or sound a warning.

Whilst much of our work is concentrated on making the bicyclist a more predictable, and hence safer, road-user this is only useful if the motor vehicle driver perceives the bicyclist. Most of the work on improving perception has involved legislation for reflectors, flags, lights, etc., to improve the visibility of bicyclists at night. But night-riding typically involves older riders, accounts for very few accidents (less than 10% even including dawn and dusk periods, Pascarella, 1971) although they tend to be more severe. The problem of daylight visibility is much larger.

An experiment (by P. Pietraszewski) compared colour slides of the same cyclists embedded in a traffic scene, but with different bicycles, clothing assemblies and flags. Motorists were shown the slides in a brief exposure and asked whether or not a bicyclist was present. Distances at which each bicyclist was seen 50% of the time ranged from 200 to 1100 feet. There was no systematic variation of mean visibility distance with clothing reflectivities from 3% to 68% nor with the addition of an international orange pennant.

The same study was extended to study the effect of expectancy of bicyclists on their perception by motorists. A selection of the above slides was exposed to motorists under two additional conditions. Instead of the original condition of 0.6 seconds exposure with the subject being alerted to report the presence of a bicyclist, in the two other conditions subjects were merely asked to list all traffic events on the slide. Experiment 2 used a 30-second exposure and Experiment 3 the original 0.6 second exposure. Table 4 shows the probabilities of detecting bicyclists in each experiment. It is obvious that bicyclists can be seen fairly well if either the motorist is alerted or he/she is under no time-pressure. But for an unalerted motorist under the time-pressure equivalent to about two-fixations the probability of detection falls to a third of its former value.

Expectancy is obviously a larger factor than visibility aids in daytime perception of bicyclists. In time, as more bicyclists

Slide Number	0.6 Sec. Exposure Alerted	30.0 Sec. Exposure Unalerted	0.6 Sec. Exposure Unalerted
1	0.71	0.67	0.50
2	0.64	0.63	0.00
3	0.71	0.56	0.00
4	0.36	0.38	0.00
5	0.93	1.00	0.17
6	0.64	0.89	0.64
7	0.36	0.56	0.00

TABLE 4. PROBABILITIES OF PERCEPTION OF BICYCLISTS UNDER DIFFERENT VIEWING CONDITIONS

use the roads, expectancy will increase. But if accidents are to be reduced in the interim, road signs and advertising campaigns will be needed to increase expectancy among motor vehicle drivers.

INTERVENTION STRATEGIES

There are basically three ways in which planners can intervene to reduce accidents: change the humans, change the machines, of change the environment.

Changes in the human operator of either the bicycle or the motor vehicle look attractive but need constant application. They are of doubtful effectiveness in the safety field (Swain, 1974). Traditionally, selection and job training have been more successful than exhortation in industrial and military contexts. With a youthful population the same effects are by no means certain, as witness the enormous cost of drug abuse education programmes. However, driver education appears to result in measurable improvements in motor vehicle safety, so that improvement is possible. The mention of selection brings with it the prospect of licensing bicyclists after some form of test - a move resisted by many legislators. It may, however, be wise to recognize the dual roles of bicycles as transportation and play vehicles and restrict use on the public roads to licensed operators as is done with, for example, trail motorcycles. The effectiveness of bicycle training schemes was measured by Aaron (1971) who found little effect of the mode of presentation. There have been no large scale studies of the effect of training schemes on accident reduction.

Enforcement of the law is another way to change the human operator. One problem here is the constant negative reinforcement of breaking laws with impunity and of freedom from accidents, which are after all rare events. Any programme of enforcement would have to be massive and sustained, with its own effects on an already overworked judicial system, to counter such reinforcement effectively over a long period. This solution only takes money and the willingness of parents to have their children charged with the driving offenses they commit in the same way as other, less vulnerable road-users.

Changing the machine involves changing the bicycle. Since its present designs hardly differ in accident rate, this is perhaps a less important factor than most ergonomists would like to think. Obvious improvements are in the strength and reliability of components to remove the 18% of accidents due to entanglement or mechanical failure. These are eliminable accidents by redesign of details. More thought should be given to elimination of the 8% of accidents due to carrying packages, stunting or riding double. Designed-in carrying platforms are essential for commuting bicyclists. Modification of the equipment by protective clothing would lead to a major enforcement problem, due to continuous negative reinforcement. However, it can be done (e.g. motorcycle helmets) and could effectively reduce the 76% of fatalities due to head injuries (Consumer Product Safety Committee, 1973).

Changing the environment means changing the streets, roads, vehicles, signals and laws which affect the bicyclists. One obvious method is bicycle paths which have been shown to reduce accidents (Jorgensen, 1969), although mainly between intersections and with even slight increases at intersections. A major use of bicycle paths is for scenic routes to keep recreational bicyclists from adding to the congestion on the roads. However, there is no possibility of a complete separation of motor vehicles, pedestrians and bicycles in central cities or dense residential areas. The ideal condition for bicycle paths is where there is a very small density of intersections. In other places, pedestrians, motor vehicles and bicycles will have to share the same facilities. This means increasing the bicycle expectancy of other road-users by highway signs, road markings, etc.

Changing other road vehicles to fit the bicyclist is being done, but for other reasons. Decreases in the size and weight of automobiles from fuel consumption considerations means marshalling less energy in collisions with bicyclists. Improved aerodynamics, for similar reasons and yielding bumpers for crash protection, also help the bicyclist when an accident has occurred. Much more could be done to deliberately design the front of automobiles for pedestrian and bicyclist protection. The current trend towards an increased glazed area in automobiles, particularly in the rear quarters, will remove one reason for bicyclists not being seen by a driver opening a car door.

Some of the legal aspects affecting the bicycle also need amendment. New York State laws require bicyclists to obey traffic laws as other vehicles and to ride as far to the right as possible. There is an obvious conflict between these requirements at left turns and on one-way streets. One law which appears to cause problems for bicyclists is the law about coming to a complete stop at stop-signs. The subjective cost of stopping is much higher for a bicycle than for the automatic-transmission automobile typical in the U.S.A. (Perhaps this accounts for the very large number of stop signs in U.S. residential areas.) Public debate on this law as it applies to quite slow vehicles such as bicycles may be appropriate.

ACKNOWLEDGEMENTS

The author wishes to thank the Department of Industrial Engineering (Chairman, Warren H. Thomas) for facilities to carry out this research and the Institute for Public Policy Alternatives, Albany, New York for their financial support of the project.

REFERENCES

Aaron, J.E. (1971) Selected bicycle safety instructional methods and content with applications to the highway transportation system. Contract SEE 35-092-71. Safety Center, Southern Illinois University.

Brezina, E., & Kramer, M. (1971) An investigation of rider, bicycle and environmental variables in urban bicycle collisions. <u>Proc. 14th Annual Conference Am. Assoc. Auto. Med.</u> Ann Arbor, Michigan.

Chlapecka, T.W., et al. (1975) Bicycle accidents and usage among elementary school students in the United States. N.S.C. Research Report. Chicago.

Consumer Product Safety Commission (1973) Hazard analysis related to bicycles. NIIC-1202-73-H001. CPSC Bureau of Epidemiology.

Fleishman,E.A., & Rich,S. (1963) Role of kinaesthetic and spatial-visual abilities in perceptual-motor learning. Journal of Experimental Psychology, 66, 6-11.

Hale,A.R., & Hale,M. (1972) A Review of the Industrial Accident Research Literature. London: HMSO

Jorgensen,A. (1969) The Effects of Bicycle Tracks on Rider Safety. Radat Trafekukkevhedscorskning. Denmark.

National Safety Council (1958) Circumstances of Bicycle Accidents. Chicago.

National Safety Council (1974) Accidents Facts. Chicago.

Pascarella,E.A. (1971) A study of youthful bicycle riders in an urban community. Highway Safety Research Center. University of North Carolina.

Rice,R.S., & Rowland,R.D. (1970) An evaluation of the performance and handling qualities of bicycles. Technical Report. Cornell Aeronautical Laboratories.

Salvendy,G., & Seymour,W.D. (1973) Prediction and Development of Industrial Work Performance. New York: Wiley.

Sheridan,T.B., & Ferrell,W.R. (1974) Man-Machine Systems. Cambridge, Mass.: MIT Press.

Swain,A.D. (1962) System and task analysis: a major tool for designing the personnel sub-system. Report SCR-457. Sandia Corporation. New Mexico.

Swain,A.D. (1974) The Human Element in Systems Safety. London: IN-COM-TECH.

Van Cott,H.P., & Kinkade,R.G. (1972) Human Engineering Guide to Equipment Design. Los Angeles: TAM's Books.

Van Lunteren,A., & Stassen,H.G. (1969) On-line parameter estimation of the human transfer function in a man-bicycle system. Proc.4th Intl. Fed. Autom. Control Congress. Warsaw.

Williams,A.F. (1974) Factors in the Initiation of Bicycle-Motor Vehicle Collisions. Insurance Institute for Highway Safety. Washington, D.C.

SECTION 8. CITIZEN PARTICIPATION AND OPEN PLANNING

Mobility and mobility deprivation, concepts introduced in an earlier section, are at the root of urban transportation problems, Uzzell argues, and only a participatory approach can solve them. Inner city residents often have less choice of facilities, poorer accessibility to them and yet have to bear higher costs than the suburban population. At the same time the latter's mobility has weakened their spatial involvement and commitment. The separation of local government from the locus of implementation widens further the gap between inner and outer urban concerns. Participation may be a means to narrowing the space between the values of inner city residents and planners. It is an active recognition that inner city communities are not closed systems.

The demand of the inner city poor for more open planning is also exemplified in Gakenheimer's analysis of transportation planning in Boston. The incursive and negative effects of new roads upon this group have introduced concerns which were neglected by classical transportation studies. Participation focusses on the immediate issues of land-use design. At the same time it tends to ignore planning at the scale of metropolitan form or activities allocation, which have been the traditional concerns in transportation planning. Gakenheimer emphasises the need to integrate these three scales of land development and to introduce intensive participation at levels beyond the local.

The viewpoint of the practising planner on participation is given in the papers by Keefer and by Fielding. These viewpoints have their origin in a considerable body of recent experience of citizen participation in transportation planning. Without minimising the problems involved, they are far from pessimistic. Keefer sees the major practical purposes of participation as being the production of plans that can be implemented and the achievement of credibility in the planning process. The means used in Pennsylvania, whence his experience is drawn, are citizen advisory committees and groups. A particularly interesting aspect of the programme is the development of citizen participation in statewide plans for road, rail and air systems.

Group involvement is seen also by Fielding as the primary means of promoting participation in a pluralist society. And again project implementation is the goal. A model of citizen participation is proposed and its use for a master plan in Orange County, California described. The central process of the model is the determination of community goals and the evaluation of alternative plans. The transportation planner takes on a role as 'co-ordinator - catalyst'. Despite the frequent complaint that participation slows down the planning process, Fielding would claim that "without the citizen panels, it is unlikely that the master plan could have been completed, and partly implemented, within three years".

POPPER AND PARTICIPATION. A SOLUTION TO URBAN SEGREGATION

D. L. Uzzell

Department of Psychology

University of Surrey

As urban settlement increases in size, there is a tendency for average journey lengths to increase, and consequently the necessity for faster means of transportation. Slowly we have been caught within a technological spiral. The solutions to urban problems have always been couched in terms of technological solutions, despite the fact that our present urban problems are not necessarity technological ones. As Schaeffer and Sclar write, "It is not surprising that new technology has no effective answer to the urban transportation problem, for the old technology did not create it" (Schaeffer & Sclar, 1975). Our present equation is 'congestion equals too many cars, not travelling fast enough'. If this were true, we could say that we were getting closer to the solution of our urban transportation problems every time we increase the speed of travel modes and decrease the travelling time. But this is not the problem. The whole history of improvement in transportation technology has shown that, by increasing speed and decreasing travelling time, we have invariably reduced the number for whom this benefits. Speed allows people to travel farther not faster, thus compounding the problems of increasing sprawl, segregation and environmental deterioration.

A recent report on traffic and its effect on the urban environment in London illustrates the point. The conclusions of this study were that seemingly conservation-orientated policies of abandoning the motorway ringroads around London may in the long run be more detrimental to the environment than the impact of the proposed motorway would have been. If car numbers continually rise and there are no main roads for them to use, they will be forced to use residential roads. Thus a dilemma is faced between knocking houses down to build faster and more efficient roads into the city centres or leaving the situation to find its own level, causing congestion and residential decay. Either way environmental damage would seem to be the outcome, even before planning decisions have been implemented (PICC, 1975).

But the creation of technological solutions to our problems has also created a differential effect. Improved transportation has widened the locational choice for some, but for others it has been narrowed. Stegman argues that the decentralization of offices and shops and the construction of fast roads is making urban services and activities more accessible to suburban dwellers than to central city households (Stegman, 1969). A simple trade-off model of household location based upon accessibility/rent is in many ways no longer applicable. The suburban and rural fringe dweller can have both accessibility and low rent, while the inner city resident incurs the cost of their provision. That is, in the traditional model, low suburban land-costs are traded off for longer journeys to work and reduced accessibility. Now, because of improved transportation and fast roads, we have a situation whereby suburban land-costs are comparatively low but urban accessibility has been greatly improved. Investment has been geared towards the private sector, at the expense of the public sector, and thus public transport has deteriorated especially in the inner areas. But inner city rent has not declined. We have a paradoxical situation whereby rent declines and accessibility increases with increasing distance from the centre (Garrison et al., 1959; Garner, 1967; Richardson, 1971).

Over the last two decades car ownership has risen from two million in 1950 to fourteen million in 1972. Like all new innovations, personal transport in the form of the car has slowly filtered down from the most affluent to the less so, and now is a feature common to most socio-economic groups. But in many ways we have created the myth of the mobile society - a society where location and ties to community are no longer a constraint to movement. Mobility, fast and individually convenient, is not available to everyone. In England and Wales only half the households own a car. When it is remembered that owner occupiers who constitute 50% of the households own two-thirds of the cars, it becomes apparent that not only is the distribution of non-car-owners selective, but amongst other things it becomes an indicator of social class (Census, 1974). But there are also other groups in society who do not have access to personal transport through no fault of their own - the young, the infirm, the disabled, those who cannot afford it. These are the groups which must rely upon public transport. From 1950-1972 bus passengers declined in number from seventeen million to six million. As more people possess cars, the service provision of public transport declines, because of higher operating costs and higher fares. Three important points emerge from this.

Those who use public transport exclusively are likely to be those with the least effective and persuasive voice in altering government public transport policy. Secondly, this group is also the least likely to make public transport a viable financial proposition, as low income groups have less surplus income to spend on transportation. Finally, private transport is becoming a ubiquitous necessity rather than a scarce luxury.

In a society which prizes mobility as a necessary condition of life, rather than a conspicuous luxury, by implication the person without mobility is a poor man. Eyles (1974) suggests that mobility and lack of mobility is a minor social indicator.

If a map of the modern city is examined in terms of isobars of accessibility, it is apparent that the city shape takes on a completely different form to conventional maps of the built-up area. The disparity becomes even greater when considering private as against public transport (for example, see Shepherd et al., 1974).

What are the implications of increased mobility? We are creating a highly transient society where involvement and commitment to people and places becomes less each year. Socially and spatially our ties with institutions are severed. For example, one of the traditional foci of community life has been the church, drawing support from its contingent parish. However, social and spatial mobility has resulted in the parish system breaking down, so that it is no longer a viable unit from which to draw a congregation. An interesting anecdote to come from one inner city parish in Liverpool relates how the Bishop, because of the worsening economic situation, told a local parish priest that he must reduce the number of curates he has in his parish. To this the priest replied: "You can take away my curates, but don't take away my bus stop".

It is not only cheaper and more convenient to travel from the suburbs to the city centre, but it is positively more difficult to travel from the centre to the suburbs. Public transport is largely based upon the idea of a monocentric settlement pattern. The opportunities for movement and the choice of alternative locational destinations is that much greater if one is travelling into the town centre, than if one is travelling in an outward direction. The further one moves away from the C.B.D. the more individualized forms of transport have to be. Yet according to Myers, in many large American cities in 1969 "less than 10% of the total person trips begin or end in the C.B.D." (Myers, 1969). Although transportation systems operate on a basically monocentric routing pattern, over the last decade or so there has been much emphasis on decentralization - in terms of both out-of-town working and shopping facilities. How well are these served by the public transport system compared with private transport opportunities? The implications are that not only is the mobility gap widening, but this in turn affects employment opportunities for inner city compared to outer city residents. From this it could perhaps be argued that we should increase people's mobility potential so that a wider variety of employment opportunities are available. This may seem to be a perfectly laudable objective, but in reality is a mistaken assumption. One is trying by these means to graft a physical solution onto a socio-economic problem. Social and economic problems can only be solved by social and economic solutions. The provision of improved transportation, while in itself a positive step, is not the complete answer. It must be integrated into a wide variety of complementary measures. Improved transportation by itself only shifts the problem, making it more mobile and less discernible.

What should our objective be? If improved transportation is not available to all sections of the population, what is the solution? The first reaction is to suggest a policy of improved transportation for as many people as possible. This would seem to be an ideal objective - a utilitarian sentiment expressing the greatest happiness for the greatest number. But is this socially just and justified? Surely this will lead to an even greater differential between groups? Minority groups in particular would suffer; not just the numerically small, but minorities in terms of power and influence. This would result from a public policy which supposedly has democracy and equality as its core.

How therefore does this seemingly paradoxical situation come about: altruistic goals ending in social injustice? It is because our over-concern for rationality and scientism has allowed us to ignore moral criteria as factors of equal validity.

If we adopt a Popperian standpoint, it becomes increasingly apparent that any maximization approach has inequality built into it (Popper, 1968). Maximization immediately leads us into conflictual situations, with moral decisions concerning how resources should best be allocated. From this position it is far too easy to make decisions in favour of those who have the most already. In a society which has planned obsolescence, conspicuous consumption and status as its objectives, emphasis will always be on maximization. Far more worthwhile is a policy of setting a minimum standard at which planning resources should be aimed. A philosophy of minimization has equality built into it and points us straightway in the direction of the socially most just solutions. To restate the original problem: to improve transportation for most of the people is idealistic in theory and difficult to implement in practice. The objective should be to reduce immobility rather than increase mobility. Transportation resources should be directed towards those who need them, to those who have no alternative to public provision. As present resources are geared towards providing more and more alternatives for those who have less and less need of them, rather than providing those with fewest alternatives with the ability to travel to the greatest number of destinations. This would result in a more comprehensive city transportation system, lessening the gap between the actual needs and requirements of residents and their needs as perceived by planners designing within a middle-class ethos.

There is much emphasis in the literature on building for the future. Is it not time we called a halt and took stock? We do not even have the city we want now, so how can we possibly start contemplating the city of the future? Bubbles under the Atlantic and scaffolding in space are futuristic designs which may or may not have relevance for future habitats. Obviously one has to plan for the future. But this in some ways presumes that the future is scientifically predictable. As Popper argues, this is logically incoherent because "it cannot give a consistent answer to the question, Who plans the planners?" (Magee, 1973). Furthermore, this presupposes that we can stipulate the basic values and objectives of society. The conflict that exists within contemporary urban society illustrates only too vividly that this is not the case. Temporary organizations, associations and pressure groups are increasing in response to the perceived assaults on the freedom of individuals to make decisions concerning their own life and environment. Pope Paul IV writes: "The reorganization of towns requires courageous decisions. We cannot flinch from our duty to re-examine the so-called traditional values, which in reality are nothing but an arbitrary preservation of individual interests" (Bonczek, 1968).

The transportation problems we now face are not so much technological problems as moral problems. We have the means to cure congestion and pollution, but not the economic and political will.

Problems come in many shapes and sizes. It is not always easy to distinguish between a symptom and a cause, because of the complex nature of the urban environment. Knowledge requires a deep understanding. Despite our data banks and the whole plethora of information and communication technologies, the identification of urban problems remains one of the greatest hurdles we face. As T.S. Eliot put it:

> "Where is the wisdom we have lost in knowledge?
> Where is the knowledge we have lost in information?"

How can we best learn the problems of inner city areas? One method used increasingly now within local government departments in general, and planning departments in particular, is that of small area analysis. These studies are usually carried out with the aid of indicators - that is, a whole battery of social, economic, political and cultural indices, which supposedly formulate an objective picture of the state of the environment and of the people who live there. But the decisions we come to, even the indicators we choose, are in part determined by our own evaluation as to what is a social and economic problem, and indeed what is worth examining. We always approach any subject weighed down by our own cultural baggage. It is for this reason we should be extremely careful when we as planners try to solve the problems of city groups. For an extreme example of this one only has to turn to the Third World. There, the imposition of a First World planning ethos onto a Third World settlement and transportation tradition has wrought social and economic havoc.

Our lack of knowledge and understanding is one of the most, if not the most, important factor in our failure to solve urban problems. Herbert Thela once said that, "knowledge is the currency of interaction" (Schaeffer & Sclar, 1975). I would like to reverse that and suggest that interaction is the currency of knowledge. If we are to understand the problems of our cities, we must be prepared to communicate with those for whom the city in its present form creates most problems. Knowledge can only be obtained by interaction. Interaction can only occur within the wider framework of a participatory system of democracy.

To plan within any culture, one must understand that culture. One must understand the values, norms, interests and preferences of the people for whom one is planning. How many environmental decision-makers and legislators actually live in, or even know, the area for which they are responsible - planners, councillors, politicians, social workers, police. This has tremendous implications for public participation. It is one of the canons of public participation that we recognize the plurality of values of urban residents. Indeed the 'raison d'etre' of increasing the participation of the public in planning matters is that recognition of the plurality of values has been ignored until now.

The essential point is that environmental decision-makers do not have a monopoly on values. It is more important than ever that, as they do not live in the culture for which they are planning, they at least recognize the existence of differing value systems. The space between the value systems of inner city residents and outer city residents is great enough. Public participation is one means, arguably the only means, by which that distance can be narrowed.

With increasing mobility, locational constraints are no longer a prime consideration. The locus of power, influence and control has been removed from those areas with which it is concerned. The space between the locus of power and the locus of its implementation, is widening. Extra-community systems now control intra-community affairs (for example, see Bernard, 1973). Indeed, it is ironic that we have increased public participation at a time when local government reorganization has accentuated the spatial and administrative separation of public and decision-makers. Decisions taken outside a community can affect the total social and economic life of that community. Investment decisions, such as relate to public transport provision, can have repercussions ranging from employment opportunities to a rise or fall in the birth-rate. When environmental decisions are taken, what criteria are considered? Are the social and psychological effects upon individuals and groups given equal weight alongside technological and economic factors in the final analysis? In questions of decision-making, objectivity of means is often confused with objectivity of ends.

It would seem that the philosophy behind systems analysis offers a great potential in providing an insight into urban transportation problems. Schofer's point that "Systems analysis is not a tool for answering questions, but a viewpoint from which to ask questions" (Schofer, 1970) would appear to be a good point of departure. The only means by which we can ask relevant questions is to have the entire system under review. Anything less is inefficient and could lead to socially unjust policies and outputs. The neighbourhood unit concept is a good example of this. Inner city communities are not closed systems, although planners often treat them as such. Neighbourhood relationships should be seen within the wider framework of the urban system, as local policy decisions are intimately tied up with central policy and decision-making.

Looking at the city and communities within it, from a system analysis point of view, community groups were once essentially closed systems in terms of power and social structure. But these closed systems have been slowly replaced over time by open systems in which extra-community systems play an increasingly dominant role in decision-making. This has resulted in the parochial power structure and social structure being broken down, with anomie and apathy setting in. Inner city communities cannot contract out of the urban system. However, an attempt has been made to try and participate in the decision-making. By means of increasing participation (ranging from squatting to pressure groups and neighbourhood councils), an open system situation of impotence has been turned, if not to their advantage, at least into a more balanced position.

Communities must now be looked at and understood in this light. They are no longer small, self-contained closed system pressure groups petitioning solely parochial affairs. Rather, they are assessors of society counterbalancing the traditional forms of urban government and challenging the previously unquestioned goals of city decision-making. The implications are great. They now play a different role in the urban system, having an effect on both environmental decision-making and, perhaps even more important, on themselves as individuals. It is when there is participation within an open system framework that a truly participatory democracy is realized.

In order to achieve a community-based participatory system, several important conditions have to be realised. Having discussed these conditions, I will suggest how these conditions can be fulfilled and bring about the type of benefits a participatory democracy offers. Firstly, it is necessary to improve the flow of information between planners and public, and public and planners. There must be a two-way interaction. The channels of communication must be available and attractive to all interested groups. Individuals and groups cannot be expected to contribute constructively to a planning procedure if they are the possessors of incomplete information. But education of this sort should not be occasional, spasmodic attempts to inform the public of what is going on. Rather it should be within the framework of a total environmental educational programme. We should be trying to create an aware citizenry, aware of urban problems, their possible solutions and the implications of those solutions. Perhaps even more important we should try to instil a sense of belonging and commitment to the urban environment, to make people feel they have a role to play in its future, and to share the responsibility of the decisions made. Secondly, probably nothing has done more to create apathy and disinterest in public affairs than the feeling of powerless in the face of professional expertise. Professionalism has excluded the qualities of spontaneity, enthusiasm and willingness from those people who potentially have a great amount to offer. Professionalism is useful in that it produces certain skills in small numbers of people. But skills are purely technical tools, and there is a great danger of slipping into the paralogism of assuming that expertise in technique is commensurate with expertise in values. It is very easy to confuse the two, and the demand for participation now is that this distinction has not always been drawn. Concepts of professionalism have no place in any discussion on the goals of urban life and the value system determining those goals.

It was suggested above that community groups are alienated from the decisions which affect them. If this is the case, our efforts should be towards restoring the balance, towards giving communities the initiative to identify and solve their own problems and make decisions with regards to their own future. If participation is to have an educational base, participation means helping people to help themselves. Participatory growth cannot be imposed from above under conditions of obligation. One cannot legislate for participation, because statutory arrangements tend to become the high-water mark of efforts rather than the springboard. That is why participation under the present legislative obligations will never achieve the type of goals I have suggested. If legislation, then it must be enabling rather than definitive, flexible rather than rigid. For this reason participation must come from below upwards, created within conditions of potential willingness. By these means community groups can and should be the core and focus of public involvement in planning. For example, schools have had a long history of importance in urban design, of being key concepts around which plans have been formulated. The central positioning of schools in the creation of neighbourhood units has surely more than locational advantages. It should represent the focal role schools can play in any community. Children live in practically every road in an area - they can carry out surveys, ascertain the needs and problems of the community and even contribute to their solution. By these means one has the community working for the community. The planners, the community and the children would all benefit (Ball, 1973).

The community approach is an interactional approach and fulfills the conditions I outlined earlier. Planners and planned would learn of each other's problems, needs and values, and an interactional information system would develop whereby the initiative for the giving and receiving of information is held by all. Furthermore, the community would learn that it still has an important role to play, and can contribute effectively to its own future. Secondly, community involvement would exist not in competition with professional planning efforts, but alongside. Finally, if decision-making is allowed to be more open - open to criticism and amenable to change, then there will be more faith in public institutions and a greater responsiveness from the public, maybe even ridding us of the apathy which plagues present participatory efforts. Participation is not solely a question of what plans people approve or disapprove, or even what sort of environment we want. It asks what sort of society we want.

CONCLUSION

Planning is essentially a problem-solving exercise. Planners are presented with the problem of restraining growth from developing in a haphazard fashion, and trying to channel that energy along the policy lines of the city's or planning department's philosophy. Planners are in a position to distribute and redistribute resources, and thus they are faced with political, social, economic and moral decisions. Problem-solving involves the putting forward of trial solutions and subsequent error elimination by means of criticism. On the basis that neither individuals nor society in general is predictable, it must be recognized that many human actions have unintentional consequences. For example, did Henry Ford think when the first Model T rolled off the assembly line, that the descendents of his innovation would be responsible for 25 million deaths, more deaths than caused by war this century. Planning should not be so rigid as not to allow change - change in the light of criticism. And criticism should come from those communities which planning decisions affect most. Councillors and politicians are elected, at least theoretically, to represent the views, interests and values of their electorate, and furthermore, be accountable to them by the vote. Yet for planners there is no such equivalent system. They are accountable in many ways only to themselves. If the public are to be safeguarded, participation is the only means by which this can come about. Not only is this the most ethically just, but is also probably the most efficient means by which to solve problems. "A policy is an hypothesis which is to be tested against reality and corrected in the light of experience. Detecting mistakes and inherent dangers by critical examination and discussion beforehand is an altogether more rational procedure, and one as a rule less wasteful of resources, people and time, than waiting till they reveal themselves in practice" (Magee, 1973).

To ask whether we live in the city we want, or deserve, one only has to look around. More important whose city is it? Are "cities the image of the society who has built them" (Borcherdt, 1967; Bockelmann, 1968)?

REFERENCES

Ball,C. & M. (1973) Education for a Change. Harmondsworth: Penguin.

Bernard,J. (1973) The Sociology of Community. Illinois: Scott, Foresman.

Bockelmann,W. (1968) The citizen and his town. In Building the Good City. German Federation for Housing, Town and Regional Planning. Koln-Mulheim.

Bonczck,W. (1968) Land economy and land regulations. In Building the Good City. German Federation for Housing, Town and Regional Planning. Koln-Mulheim.

Borcherdt,H. (1967) Uberlegungen zur stadt von morgen. In Polis und Regio, List Gesellschaft e.V., Vol.57. Basel-Tubingen.

Census 1971. (1974) Availability of Cars. London: H.M.S.O.

Eliot,T.S. (1934) The Rock. London: Faber and Faber.

Eyles,J. (1974) Increased spatial mobility: a minor social indicator. Area, 6, 305-308.

Garner,B.J. (1967) Models in urban geography and settlement location. Chorley,R.J., & Haggett,P. (eds.) Models in Geography. London: Methuen.

Garrison,W. et al. (1959) Studies in Highway Development and Geographic Change. Seattle.

Magee,B. (1973) Popper. London: Fontana.

Myers,S. (1969) Transit technology: the soft revolution. In Future Directions for Research in Urban Transportation. Paris: O.E.C.D.

Popper,K. (1968) The Open Society and its Enemies. London: Routledge & Kegan Paul.

Professional Institutions Council for Conservation (1975) The Urban Road in Relation to Conservation and Renewal of the Urban Environment. London.

Richardson,H.W. (1971) Urban Economics. Harmondsworth: Penguin.

Schaeffer,K.H. & Sclar,E. (1975) Access for All: Transportation and Urban Growth. Harmondsworth: Penguin.

Schofer,J.L. (1970) A systems approach to urban transportation. In Berry,B.J.L., & Horton,F.E. (eds.) Geographic Perspectives on Urban Systems. New Jersey: Prentice Hall.

Shepherd,J., Westaway,J., & Lee,T. (1974) A Social Atlas of London. Oxford: Clarendon Press.

Stegman,M.A. (1969) Accessibility models and residential location. Journal of the American Institute of Planners, 35, 22-29.

THE QUESTION OF LAND-USE IN THE PARTICIPATORY TRANSPORTATION STUDY

Ralph Gakenheimer

Department of Urban Studies and Planning

Massachusetts Institute of Technology

The role of land development analysis and policy in urban transportation planning is currently very unclear in the United States. The relationship between land use and access has been the basis of the formal models sustaining urban transportation planning. Over the last twenty years it has gotten considerable methodological development and application in public policy studies. During the last five years, however, it has come under serious criticism. In particular, increased participation in urban transportation studies by citizens and local public officials has put marked pragmatic pressures on the role of land-use studies. The result is a different picture of policy-making than was prevalent in the earlier period of U.S. transportation planning.

The applicability of the conclusions of this paper to other countries must be left for their respective professionals to consider. Transportation and urban development planning is an intensely culture-bound subject when considered in meaningful detail. Its context in the U.S. is one of conflict between classical land use-transportation planning on the one hand and, on the other, the less-formalized urban transportation study of the last few years, with its emphasis on intensive community participation and attention to a number of 'ad hoc' social and environmental issues. I will refer to the movement from the former to the latter stages as transition from the "classical" to the "open study".

What is the effect of community participation in urban transportation studies? It has had the clear effect of requiring serious policy-making attention to questions of community concern, no matter how difficult those questions may be to consider within the framework of formal approaches to the analysis of transport and land-use. At the same time, it results in omitting some topics. In an effort to be as responsive as possible to community concerns, studies are likely to give little attention to topics not pressed by participants.

This paper will argue that there different kinds of land-use planning, whose dissimilarities were unfortunately masked by the formal methodology applied to the topic in its earlier stages. Some of these kinds of land-use planning are thrown into bold relief by intensive participation in urban transportation planning. Others are shunted into neglect by participation, and may need to be defended by governmental planning policy. Otherwise they may be excluded to our grief in the long run.

THE CASE

The exposition here will be in the form of an essay rather than an empirical study; but it is based on observations of two major urban transportation studies in progress. They have constituted the major metropolitan transportation studies in the United States during the last five years. One was the Boston Transportation Planning Review (BTPR) 1972-73*. The other is the work of the San Francisco Bay Area Metropolitan Transportation Commission (MTC), a permanent agency observed by the writer in 1975. The studies were similar in interesting ways: (1) both were strongly oriented to community and local-official participation; (2) both were oriented insistently to project corridor problems (as differentiated from the studies which have directed analysis to the more generalized level in an effort to determine new project corridors); and (3) both tended to choose techniques for land-use and travel-demand analysis which were unambitious by the standards of technical progressiveness of the earlier land use-transportation studies. In both cases this was a conscious decision, intended primarily to maintain maximum flexibility for response to participant interests.

The differences between the two studies are also important. The most significant for present purposes are the three following. (1) The most prominent MTC participants are the cities and large, powerful counties of California's Bay Area, and the State Department of Transportation. In Boston the most prominent participants were regional pro- and anti-highway coalitions, the State Secretary of Transportation and the Governor. Only the City of Boston was very important as a local governmental participant. In the Western United States, local government takes a stronger part in

---

* For a fuller analysis of the open study in urban transportation in the light of the Boston Experience see, Ralph Gakenheimer (forthcoming) <u>Transportation Planning as Response to Controversy: the Boston Case</u>. Cambridge, Mass.: MIT Press.

public decision-making of this kind than in the East. The difference is important because local governments have a scale of interest in land-use which is pre-set by their size as jurisdictions. Regional interest coalitions have a scale of concern which is set by some combination of regional boundaries, project domains, and the locations of target groups (e.g. the inner-city poor). (2) The BTPR examined alternative corridor designs to a level of great detail, involving exact alignment and grade designs and the management of abutting land-uses. The MTC consigns small-scale matters to the local authority and concentrates almost exclusively on origin-destination zones and the larger scale. (3) During the observation of these studies there was an intense concern in Boston for social equity, particularly with respect to the inner-city poor -- including anticipated impacts of facilities on the amenity of neighborhoods and their specific access needs. Relatively little emphasis on these matters was observed in the Bay Area. This was largely a result of the particular projects under consideration by the two agencies. Those under current consideration in the Bay Area did not have large impacts of that kind. It was also a fact that the vigorous social commitment which characterized the nation at the end of the 1960's was not so visible by the time of observing the MTC. (4) Environmental concerns in Boston were mostly for particular natural resource sites, including especially a freshwater marsh and a forest reserve. In the Bay Area there was a concern for the "whole" natural environment. This included the overall impacts caused by the arrival of land-uses attracted by increased accessibility, and so forth. Although to some extent this also reflected the particular projects in question, it probably indicates stable characteristics of environmental perception by people on the two coasts.

THREE SCALES OF EFFECT ON LAND-USE

Whether considering policy for land development to accompany transportation planning, or just the examination of the impacts that would be caused by proposed projects, there are three quite different scales of attention. They are mutually distinguished by analytical and decision-making characteristics.

Metropolitan form is the largest scale. It is a level at which the gross impacts of transport facilities upon urban development can be considered - such as differences in aggregate travel-time for the entire urban population based on different overall density assumptions. Major options are considered such as the opportunity costs of occupying different peripheral unurbanized lands.

Planning of this kind was a popular aspect of transportation planning in the 1950's and early 1960's. It deals with very long term effects and with benefits which can be appreciated only by aggregating impacts over a very large area. It is virtually impossible to compare alternatives at this scale rigorously, since their consequences are so numerous and since they involve so many mutually incommensurate variables.

As a result of these characteristics, the planning of metropolitan form has fallen into a role of little concern in recent transportation studies. Such long-run, highly aggregated concerns are bound to be of interest only in times of stability - times when political pressures do not marshall attention to problems of the short run and small scale, such as social equity. In any case, the planning of metropolitan form never attracted much political interest in the U.S., because it did not correspond to the scale of anyone's vested interests. Its prospects for implementation are poor in the extreme, since the objectives of land development planning are likely to be different from those of the civil jurisdictions within whose responsibility it lies, and since their grip on land-use control is in most cases quite weak.

Activities allocation is the second scale of planning, the analytic back-bone of land-use projection for the classical urban transportation study. The scale of allocation is the origin-destination zone. By use of a quantitative model, or by manual means, new residential populations and employees are distributed by categories to zones, subject to rules defining development tendencies and constraints representing zonal absorbant capacities, etc. The rules used in allocating activities include the effects of increased access on land development conveyed by new transport facilities.

The most apparent reason for this form of projection in an applied study are to establish activities allocations as a basis for future transportation demand, and to gauge the land development consequences of enhanced accessibility. For the first of these two purposes, some form of activities allocation is clearly necessary in transportation planning. The second purpose, describing the land development impacts of transport facilities, involves greater complication.

The results of projection are numbers of residents and employees for each zone. This is different from dealing in detail with what will be on the ground as a result of the facility. For example, the density and locations at which anticipated residential population is absorbed would be important to an understanding of resulting environmental conditions. The impacts on land immediately abutting a proposed facility, averaged with those upon parcels some distance away within the same zone, constitutes a projection with specific limitations.

In particular, the use of these projections for the

estimation of future travel-demand is fairly straightforward. The relative accuracy of different projection methods may be of some interest; the assumptions about historical trends in land development are likely to create more interest*.

Their usefulness in connection with environmental conditions, on the other hand, depends on the nature of public concern for environment. Communities whose concern is for the "whole" environment of residential occupancy, in the fashion of the Bay Area MTC, find use for projections in this form. Debate can emerge in those cases over development forms implied by the zonal land-use impacts of different transport network design options. When the environmental interest is in the form of special resource sites, in the Boston manner described above, activities allocations are of considerably less interest. Only detailed design-level descriptions can foretell the fate of ecologically sensitive sites.

<u>Land-use design</u> is the final and smallest scale. The scale in this case is determined by the individual parcel-size or land-occupying establishment. This form of impact projection and planning is the forte of the open study. It did not receive much attention in the classical urban transportation study. In spite of the fact that considerable scientific effort has been given to the projection of impacts at this scale, the applicable skills in public policy studies are primarily the intuitive methods of urban design. Questions include the arrangement of designs which minimize disruption to neighborhoods by minimizing necessary dislocations and buffering sensitive land-uses. The design of corridor features to restore small-scale accessibility, replace displaced facilities, and enhance local amenity generally are elements of the job at this scale.

The scale of land-use analysis and planning is different from the two foregoing in various important ways. First, it is not much aided by methods developed for analysis at the larger scales, since they deal invariably only with aggregates. Secondly, however subjective the evaluation of alternatives may be at this scale, they are easily understood by study participants and tend to be clear in terms of whose interests are benefitted and disbenefitted.

---

* Changes in land development trends in certain major U.S. cities have been important issues in transportation debate, and have presented problems for the use of trend-based models. Central Boston is no longer loosing population and economic activities to decentralization, a trend upon which earlier land-use models were based for that city. Debate over the West Side Highway in New York found importance in the fact that office space was no longer rapidly increasing in Manhattan, as the land-use projections had assumed.

This is the scale at which a transportation facility, rather than being something which conveys mobility, is simply an incursion into the landscape. The balance of the impacts observed is invariably negative and inflicted upon low-income families. As a result, this scale is the most important for participants whose concern is social equity and whose sympathies are against the facility being proposed. (The first two scales of land-use tend to favor network planning, and to be a part of the calculus of those who assume that the project in question is desirable.)

In addition, this is the scale at which interest is in the short term (rather than depending on a long-term transport-land development equilibrium), and which is likely to be the most feasible for direct policy intervention. Capacity to alter land development characteristics is sometimes in the hands of the facility building agency. In any case, interest is concentrated in very particular, concrete situations.

As regards impacts on the natural environment, it is clear that small-scale land-planning is a basic element in the orientation to specific natural resource sites noted as a characteristic of the Boston study. The balance of nature in a marsh may be very much effected by the detailed location of a highway whether it is on piles or at-grade, and so forth. When the orientation is to the "whole" environment as mentioned in the Bay Area case, this scale is less important.

Across the country, this scale of concern has attracted a great deal of political attention, for the reasons mentioned above. It has substantially changed the characteristics of urban transportation planning in many areas, and is responsible for stalemating the construction of transportation facilities in many cities.

In conclusion, it is suggested that there are three different scales of land-use analysis and planning. The first two are traditional to the transportation planning field; the third is new. The first two have encountered great obstacles to effective planning; the third has become 'de facto' the most influential form in planning. This situation has been brought on, in good measure, as a consequence of intensive participation in urban transportation studies. The question is how to reinvigorate the land development concern at the first two scales in the context of intensive participation.

This will require some investigation of the peculiar incompatibilities between land-use and transportation planning.

LAND-USE AND TRANSPORTATION PLANNING: PROBLEMS OF COMPATIBILITY

The formal models relating access and land development were so intellectually compelling that they obscured some hard realities: that the two phenomena are very different from one another in practical terms, and are extremely difficult to plan together. Consider the following differences between transport planning and land-use planning of the first two types above.

1. Forms of Projection. Transport is projected merely in the form of person-trips per unit time. Land-use, to be useful for planning purposes, must be projected by location, density and forms of land-occupying activity. Thus, notwithstanding the interdependencies between the two, the fact that transport is projected as a single variable gives it the sense of a more stable projection.

2. Policy Sensitivity of Projected Values. Between the present and the target date, change in land development and trip-making are subject to very different kinds of governmental policy. The level of policy sensitivity of each is variable from one area to another. In the U.S. trip-making might be sensitive to increasing concerns in national government for energy conservation and pollution problems. In other countries it is affected far more, and by such things as high and changeable excise taxes on motor vehicles. Land-use in the U.S. is affected by local land development controls, typically rather weakly.

3. The Concept of Total Demand. Full demand for land-use in the U.S. is a defensible concept; that is, the need to supply land for consumption by growing urban activities is generally accepted. (There are local exceptions to this, but they are few). The concept of a full, irreducible demand for urban travel, on the other hand, has come under considerable criticism in the last years. Both as a result of the fragility of the methodology producing such estimates, and a sense by public decision participants that "full" demand for travel should not necessarily be served. This is a distinction with difficult implications for the effort to create some kind of equilibrium relationship between the two phenomena.

4. The Governments Responsible. In the U.S. urban transportation planning is usually the job of state authorities working within guidelines carefully prepared by the Federal Department of Transportation, and with considerable on-going influence from that source. Local governments are increasingly influential, but still limited in effect. They have a form of participation which primarily responds to planning rather than initiating it. Land-use planning, on the other hand, is done exclusively by local governments. That is, transportation is planned by large government, and land-use by small government. There are typically great differences between them as to ideological commitments, quality of

avaiblable professional services, nature of political pressures guiding decision-making, and so forth.

5. Objectives of Planning. While the wisdom of fully serving the demand for urban transportation has fallen upon criticism, the basic purpose for the provision of facilities remains fairly clear: to provide increased trip-making opportunity. Concern for the social distribution of that opportunity introduces an element of subtlety, but has so far not received enough attention to really complicate the objectives of public investment in mobility. Land-use planning, on the other hand, has objectives which are numerous, mutually incommensurate, likely to be ephemeral, and difficult to defend in public debate. The fact that transport planning objectives are strong and land-use planning objectives are weak has had the effect of centering the combined planning activity on transport, leaving the land-use component as the passive one.

6. Forms of Implementation. Transport projects are built directly by public action. The expenditure is from a special sector of the budget. The necessary land is acquired by public action for the purpose. Land-use, in contrast, is planned indirectly for the most part by the regulation of private initiative. Further, transport projects are lumpy investments; once started they represent a strong commitment to complete. Land development planning is incremental; the plan can always be changed to reflect more recent thinking, though a number of decisions to build may have been affected by the plan. As a result, the likelihood of implementation of a plan is much greater in transport than in land-use. This has had a considerable effect on the way the two sectors respectively are planned. Transportation planners, for example, are much more conservative in their assumptions about the extent to which future urban development is subject to policy that would change natural trends.

CONCLUSIONS

Several recommendations emerge in relation to the problems of land-use planning in the context of urban transportation planning.

1. It is necessary to monitor the allocation of effort among the three scales of land-use planning. As suggested by the discussion, the different scales of land-use planning favor different interest groups in transportation planning. Therefore, in a participatory study, the distribution of staff effort among them may gravitate to proportions by which the dominant participant group takes unfair advantage.

There are other distortive effects possible. To the extent

that the study in question is based on the classical tenets of land use-transportation studies, there will be a tendency for the effort to gravitate toward favoring the second (activities allocation) form of land-use planning. Commenting on the two cases introduced here, it is possible to allege that in Boston the larger scale of planning was slighted and that in the Bay Area projects too little attention is given to the smaller scale.

2. As a special case of the conclusion above, it is important to point out the possibility that strong social commitment in a transportation study may tend to orient all decisions to the small scale of land-use and virtually cancel attention to the larger scale. Tendencies of this sort were visible in the Boston case. (Symptomatically, after the Boston corridor decisions no one even took the trouble to tabulate the consequences of the decisions on large-scale land development patterns). The matter of how to handle the larger scales of land-use planning under such circumstances motivates many of the conclusions following below.

3. If larger scale land-use planning is to be taken seriously in transportation studies, there is an important need to define and stabilize the objectives of planning. The absence of strong objectives at the present time has been a major problem of influence. Classical transportation planning was sufficiently isolated from local political reality to operate on the basis of formalities which failed to be effective, but remained part of the methodology. The participatory study breaks down such formalities and demands more focused planning.

4. There is a need to consider more adequately the short-term consequences of transport network development. This is an obvious consequence of more active political concern, and of increased focus on the small-scale. At the small-scale, response follows impact very quickly. It includes especially the need to consider interim impacts on land development. These are impacts on land development which occur between the final decision about transport facilities and the beginning of construction, and also the interim of the construction process itself. The pre-construction interim may be a time of disinvestment in property which threatens its functions as well as those of surrounding, unaffected parcels. The construction interim may be a time of irreparable damage to the surrounding natural ecology and nuisance which badly damages local amenity. The two interims together may well be some twelve years or more in length, a period considerably longer than the average resident may expect to remain in the project locality. Thus the interim consequences of transportation planning on land development may be of considerably more importance to the surrounding population then the final consequences of the project.

5. There is need for a strong decision-maker to commit

himself to the final conclusion in participatory transportation planning. This need is suggested here by the fact that the different kinds of land-use planning tend to be within the concerns of different interest groups. That issue is typical of others in participatory transportation planning. There is really no basis for systematic compromise among interests. Each of them perceives the problem in a very different light from the others. It is not that they disagree in debate over the same issues; they rarely address the same issues. Therefore, if there is not an influential, respected decision-maker responsible for the final judgements, the participatory process is sure to disintegrate.

6. Small-scale land-use planning is a newcomer to urban transportation planning and needs somewhat different institutional arrangements than is normally provided by the studies. One aspect of this is the need to create contracts of trust between participants and government authority. For example, the land use component may be fully designed by the transport study, exposed to adequate participation, and included in the final decision. The implementing agencies, however, are apt not to consider the land-use aspect an inflexible commitment of the final decision. Consequently, after the study is over and the participatory effort disbanded, the implementing agency may vitiate the intent of decisions on this matter.

Indeed, the expectation of this is one thing that has tended to polarize opinion in controversial projects. The anti-highway groups may find that planners offer all kinds of designed amenities which reduce facility impacts or compensate for them - depressed alignment, air-rights building, special community facilities, right-of-way plantings, cross right-of-way access, baffles and so forth. The anti-highway group may suspect, however, that once it agrees to the facility on the strength of such compensations it enters a trap. Approval of the highway may be considered a firm commitment on their part, while compensations may be eaten away as the project afterwards passes through one cost-conscious agency after another.

There is a need for contracted agreements between transport planning agencies and higher authority which assures stable acceptance of all key features in a project design, thereby adding strength to promises. Such agreements would enhance the participatory process by making compromise more feasible.

CITIZEN PARTICIPATION IN TRANSPORTATION PLANNING

Louis E. Keefer

Bureau of Advance Planning

Pennsylvania Department of Transportation

At a 1972 national conference sponsored by the National Academy of Science Transportation Research Board, citizen participation in transportation planning in the United States was defined as:

> "An open process in which the rights of the community to be informed, to influence, and to get a response from government are reflected and, in which a representative cross section of affected citizens interact with appointed and elected officials on issues of transportation supply at all stages of planning and development. The participants in the process identify and examine all reasonable alternatives and their consequences to assist the appropriate decision makers in choosing the course that they believe is needed and that they feel will best serve the needs and objectives of the community".

My own first definition is less complicated. Citizen participation in transportation planning is any continuing process through which responsible transportation officials really pay attention to what people want in transportation. The process is conceptually no different than that used by any producer of goods and services: find out what the public wants, determine whether it can profitably be provided, and then market it.

The analogy breaks down, however, because transportation is basically a government monopoly, and what government may market depends more on federal and state law, federal and state budgets, regulatory rules and regulations, administrative interpretations, and plain, old-fashioned politics than on what the public may actually want and need. Thus, my amended definition is: Citizen participation in transportation planning is any continuing process through which responsible transportation officials really pay attention to what people want in transportation, and then compromise that want against that which government can actually provide.

As limited as this definition might at first seem, I think it would satisfy most critics. In my experience, what the rational layman wants most is the opportunity to voice his opinions, and to feel he has been given honest consideration in the final decision making process. So viewed, citizen participation in transportation planning is as simple as exchanging information with friends and neighbours about the homes they want, the cars they buy, or the colleges to which they send their children.

As a practitioner, rather than a theoretician, I say that citizen participation need not be complicated, and that it can and does work. Transportation planners should spend less time debating the merits of this or that technique, and spend more time with their sleeves rolled up <u>doing it</u>!

I see two major purposes for citizen participation:

1. <u>To produce plans that can be implemented</u>. This means that plans must be responsive to needs as perceived by the public and not just as perceived by planners. How often have plans, considered excellent by experts, failed for lack of public support? <u>Too</u> often. Planners must give up planning to please themselves and plan to please the public instead. I believe this can only be accomplished by involving the public directly in the planning process.

2. <u>To achieve credibility in the planning process</u>. All government planning is distrusted to some extent. The public is well aware that government has planned it into inflations, recessions, fuel shortages, higher taxes, environmental crises, and all manner of not-so-smart predicaments. Transportation departments seem especially distrusted (at least in the United States). Trust and credibility in the planning process can be rebuilt by really listening and acting upon what the public has to say.

In long-range urban systems planning in Pennsylvania, we have formally organized citizen advisory committees in every large urban area in the state. Membership in each committee ranges from twenty to seventy persons representing as many diverse interests and points of view as possible. The committees review work programmes, analytical reports, plans, and priority programmes, and report their positions to transportation study policy groups before the latter take any action on a significant transportation issue.

PennDOT exercises no restraints on the committees. They elect their own officers, write their own bylaws, keep their own minutes, elect their own membership, and make up their own minds on any issue. Since the committees are advisory to policy groups comprised principally of locally-elected officials, and not to PennDOT as such, their effectiveness depends largely on how much attention they are given by these officials. Unfortunately, such attention is sometimes lacking. Having once been elected by the public-at-large, some officials take the position that they will not be unduly swayed by small pressure groups, in whatever name. Even so, most committees are sufficiently persistent to seriously effect many actions taken by the transportation study policy groups.

PennDOT's prime role is providing understandable technical information to such committees and interpreting for them the maze of governmental policies and procedures that often determine what can or cannot be done. We attend committee meetings everywhere in the state. We help arrange meeting places, print newsletters, encourage news media coverage, and, short of providing travel expenses or honorariums for membership, do everything possible to make these committees successful. All meetings are announced in the news media and are open to the general public.

There are problems. Attendance at meetings is often disappointing. It has proved difficult to sustain membership interest in long-range planning because the principle issues tend toward the esoteric. For example, what is the elasticity between the cost of gasoline and the amount of automobile travel; and how can that be predicted in order to affect the size and shape of highway and transit plans needed over the next twenty years? The average citizen does not relate well to such issues. Moreover, committee membership can sometimes become dominated by unrepresentative factions. There is not always the sought-after balance of environmentalists, ecologists, businessmen, minority groups, aged, handicapped and so forth, necessary to constitute a valid community cross-section. PennDOT encourages that balance, but since it does not appoint members or impose higher restraints it cannot unilaterally correct imbalances.

At least annually, however, each entire community has an opportunity to be heard. The citizen advisory committees sponsor public meetings prior to federally-mandated annual plan and programme readoption by the transportation study policy groups.

Widely publicized throughout the metropolitan area, these meetings allow for open-ended presentation and discussion of existing plans of record. Transcripts are provided to the transportation study policy groups prior to official action on plan and programme re-adoption. In several instances, enough objections have been generated at these public meetings, over controversial portions of plans, that policy groups did subsequently amend them before readoption.

Having a community-oriented long-range plan, by the way, provides no guarantee that any given project can and will be built. In Pennsylvania, the long-range plan only provides the starting point for programming by the Pennsylvania Transportation Commission, budgeting by the Pennsylvania General Assembly, and subsequent project location and design studies. Recognition that plans are only starting points, and may be changed from year to year, no doubt detracts from the public's interest in debating the merits of alternative long-range futures, but that is unavoidable.

Although citizen groups are not directly consulted in programming and budgeting, they do become involved again in the location and design studies. A citizen advisory group is formed for each major study to follow its progress from first to last. Initial membership is drawn from the systems planning citizen advisory committee representatives residing in, or having special knowledge about, the specific corridor involved. Full membership is attained from other residents in the corridor.

It is relatively easy to develop membership in location and design study groups because the issues are more immediate: Where will the highway go? Whose homes and businessess will be taken? Whose properties will be adversely affected by air and noise pollution? Why build the highway at all? The fact that relatively more highway projects are stopped (and many more modified) at this stage than during systems planning points up another major purpose of gaining effective citizen reaction early on: the better the long-range plans, the more time and money saved during subsequent project development. Why wait until a location study is made to discover that a project is unacceptable to the community?

Project-related citizen advisory groups have drawn mixed reaction. Some critics consider them failures, first, because they have not helped overcome the flaws inherent in certain projects; and, second, because forming such groups at the very start of project development alledgedly only serves as an early-warning system for the prompt and stronger organization of opposition factions. Most citizens are pleased, obviously, since the concept represents a major departure from the old procedure of <u>not</u> involving citizens until the mandated public hearings which <u>follow</u> study completion.

The PennDOT citizen participation programme certainly is not limited to urbanized areas. We are concurrently developing statewide plans for highway, rail and aviation systems, and have various channels for involving citizen participation in each effort. I shall not recite the details; but these three efforts involve three statewide advisory committees and twenty regional advisory committees, numbering altogether more than 600 members. These are not citizen committees, per se, but they do include perhaps 100 persons with citizen viewpoints (i.e., persons with non-governmental, non-agency, and non-professional affiliations). Various public meetings and hearings will be held before statewide highway, rail and aviation plans are finally adopted.

It should be noted here that while PennDOT has legal responsibility for preparing long-range systems plans for all modes (highway, rail, rapid transit, bus, aviation, etc.) it has design, construction and operating authority for highways only. Thus, it is not directly involved in any citizen participation efforts associated with project planning for the non-highway modes.

We are, of course, constantly trying to improve our citizen participation programme generally. In our present Centre Region Area Transportation Study (CRATS), for example, we have contracted with the Transportation Institute of the Pennsylvania State University to conduct an "innovative and flexible citizen participation programme" embracing a broad range of involvement techniques. We have printed and distributed a 30-page "Citizen's Guide to CRATS" to some 25,000 households in the area. We mail regular newsletters to nearly 1,000 households. We have had over 50 citizen advisory committee meetings, almost as many general public meetings, and a special two-hour TV show where the public's questions submitted live by telephone were answered live on TV by a panel of CRATS planning officials. We are devoting considerable manpower to the effort, and consider it our most exemplary effort so far.

Those of you with special interest in this field may already be familiar with our Manual for Citizen Participation in Transportation Planning published last year. The result of a $100,000 research project funded in cooperation with the Federal Highway Administration, the Manual identifies some forty specific techniques to involve citizens in both systems and project planning, means for selecting the appropriate technique to match particular situations, and the 'how' of applying such techniques. A follow-on phase will test and more fully evaluate those techniques that go beyond our present practice.

I should like to sum up the main benefits I see coming from an effective citizen participation programme, and to try to answer the obvious question "What difference does it make?".

1. It can help us produce better plans. The best professional planners in the world cannot know everything and we can obtain good positive ideas from the public when we really try.

2. At the same time it can help us avoid long-range plans that include individual projects that will subsequently fail to gain public acceptance, thereby saving time, money and frustration for all concerned.

3. It can also assist us to regain a considerable measure of public confidence, and restore credibility in the whole transportation planning process.

4. It can keep before us the crucial question: Why? Trying to justify a new highway to a citizen group is a test of logic for which many transportation planners need a regular refresher course.

5. It can help focus news media attention on crucial transportation issues, which in turn can stimulate top transportation administrators and elected officials to react more promptly to what people really want.

6. It can also keep before us the certainty that societal values change over time, and that what the public perceives as today's transportation problems cannot be answered with yesterday's transportation solutions.

In Pennsylvania, I believe that such benefits are very real, and that our citizen participation programme has clearly affected what we have planned and built in recent years. I strongly recommend citizen participation as both necessary and rewarding in any public planning effort, and believe we will see more rather than less citizen participation in the future.

CITIZEN PARTICIPATION: AN ADMINISTRATIVE STRATEGY FOR TRANSPORTATION IMPROVEMENT

Gordon J. Fielding

School of Social Sciences

University of California, Irvine

Structuring effective citizen participation strategies is a difficult task which tends to be neglected in transportation improvement. Primary effort is devoted to elegant patronage models, innovative modes of transportation, and engineering design. However, if realistic citizen participation is neglected, a project will seldom be implemented, because the level of public support needed to adopt the plan and sustain criticism which arises during construction will not have been developed.

Realistic citizen participation is neither "tokenism" nor "cooption" (Burke, 1968). A strategy is needed which involves affected citizens from the outset of the planning programme and maintains their involvement throughout construction. A strategy of citizen participation which seeks to involve the affected community after the locational and engineering decisions have been made cannot be effective. Citizen participation must begin early in the process. It must seek to involve groups with differing interests and attitudes to the proposal and assist them to select the most beneficial, least detrimental, plan.

The pluralistic social structure and the preference for non-partisan systems of local government in American metropolitan areas makes realistic citizen participation more difficult. In each metropolitan area, there are numerous interest groups who will be affected by transportation improvement. Business interests have long advocated transportation improvement. They have now been joined by environmental interests requesting technology and locational changes as well as by senior citizens and the handicapped who desire special services. If there were partisan representation in local government, these interest groups could seek representation through an established political process. However, when there are no political parties willing to articulate the views of the different interest groups, they must compete with each other to influence the outcome. Competition frequently generates frictions and antagonisms which are difficult to reconcile. If mutually rewarding solutions are not outlined early in the planning process, groups may become so deeply divided that compromise on a final plan is impossible.

The urban administrator charged with developing a programme of transportation improvement in a pluralist society needs to adopt the role of a 'coordinator-catalyst' (Fielding, 1972). He must structure a range of alternative plans and allow the interest groups to evaluate the advantages and disadvantages of each plan in terms of their individual interests. Through the process of evaluation, a plan which is acceptable to the majority of interest groups may become apparent. If there is sufficient support to adopt the plan, then development is possible. Where there is not sufficient support for any one of the alternatives proposed, it is unlikely that the project can be completed. The urban administrator then knows, early in the planning process, whether progress will be possible.

TRANSPORTATION PLANNING

Changing approaches to transportation planning have been consistent with changes in the social and political environment in American urban areas. Particularly significant for the emphasis upon competitive citizen participation strategy has been the increase in citizen involvement expressed by the Civil Rights movement and anti-war organizations in the 1960's. The demonstrated effectiveness of ways in which citizens could make their desires known to government increased their willingness to object to prevailing approaches to transportation planning. They were no longer willing to accept technical descriptions of impacts and desired to evaluate them in terms of their own objectives for community development.

During the past two decades, America has been home to three actively competing and fashionable lines of thought in transportation planning (Webber, 1973).

1. Cost Benefit Analysis: Where the value of the transportation project was calculated in terms of the ratio or benefits which would be derived from the project to the costs of construction. Benefits were primarily economic assessments of the travel time saved and reduced fuel consumption and accidents.

2. Cost Effectiveness Analysis: Where the benefits were broadened to incorporate aspects of systems analysis. This was an attempt to introduce into the evaluation of alternative transportation plans an assessment of the impact on social and environmental factors. Although evaluation was often qualitative, it enabled a comparison between alternative plans.

3. Citizen Protest and Participation: Where citizen opposition to transportation programmes was legitimized as a mode of behaviour and citizen groups were given the opportunity to evaluate plans and to determine whether or not a transportation proposal would be an improvement. Groups who had been unskilled in the ways of politics were given courses in citizen participation, and transportation planners as either agency employees or private advocates were directed to assist citizen groups in articulating their support and objections. It is in this context that the transportation planner became a 'coordinator-catalyst', helping citizen groups select the most beneficial plan and articulate the reasons for their choice.

The aim of increased participation was to promote positive contributions to the transportation planning process. Affected communities were urged to articulate the consequences of a project in terms of their own interests and to encourage the acceptance or rejection of the project if adequate compensation for the local impact of regional-wide facilities was not provided (Wolpert et al., 1972, pp. 27-43).

Melvin Webber has shown that all three approaches to transportation planning are necessary in order to provide effective citizen participation. He indicates that the major questions in transportation planning are essentially "political in character. They can be equitably resolved only through bargaining - through debate and negotiation. But, of course, debates based on ignorance and negotiation without estimations of outcomes are not likely to serve any of the participants' own interests. It is here that analysis and systemic forecasting find their critical roles - as informers and sources of intelligence. Technics and politics are thus mutually interdependent in a true symbiotic relation, making the systems analyst and the participating citizen joint partners in the pursuit of social betterment". (Webber, 1973, p. 50)

DEFICIENCIES OF THE CITIZEN PARTICIPATION APPROACH

A participatory strategy for transportation planning has several deficiencies. Even if the transportation planner is wise enough or technically skilled enough to propose a range of transportation improvements which will satisfy all involved interest groups, there is no assurance that the transportation plan will be implemented. Some of the problems which have arisen through the use of this strategy are as follows:

1. Costs escalate beyond the capacity of the organization to provide funds, when attempts are made to incorporate improvements to satisfy a wide range of community interests. For example, undergrounding transit lines costs three or four times as much as designing facilities at grade. Yet communities insist facilities be undergrounded so as to avoid noise and visual impacts upon the immediate area.

2. Citizen involvement is more effective for short-term project plans than for system (regional) plans, which may be completed over twenty years. It is very difficult for citizens to realize the impact upon a community of a transportation corridor when the precise location has not been defined nor the technology selected. Citizens are much more adept at evaluating the impact of elevated or depressed facilities and one alignment as opposed to another. (Manheim, 1974, pp. 311-42)

3. Attitudes towards transportation proposals are not known in advance. The author demonstrated (1972) that attitudes towards a transportation proposal change during the evaluation of alternatives. Only after one plan has been selected, discussed and adopted at community meetings is a consistent attitude achieved.

4. Only a small minority of the community is involved in the evaluation of transportation proposals. Rosenau, (1974, pp. 6-16) distinguishes between the "Attentive Public" and the "Mass Public". He estimates that only between ten and fifteen percent of any community can be regarded as "attentive" and of this group only a proportion which has the time and interest will become involved in a transportation proposal. Successful involvement of the "Attentive Public" in the transportation evaluation process may ensure adoption of the plan, but it will not ensure construction if the citizens are required to vote for local fundings.

5. It is difficult to change the role of the transportation engineer from "project developer" to "project coordinator". The reward structure for the professional engineer is in terms of completed projects, whereas the coordinator-catalyst role requires the engineer to abandon the project if there is not sufficient community support to ensure adoption and completion.

These are some of the limitations of a citizen participation strategy for transportation planning. Adoption does not assure success. However, it is a viable strategy given the current desire for community participation.

MODEL FOR CITIZEN PARTICIPATION

An approach for citizen participation should include the following groups: elected or appointed officials charged with adopting the regional plan; the technical staff of the adopting agency; citizens whose professional training could be helpful in the clarification of impacts, and interested citizens from the "Attentive Public" (Figure 1).

The elected officials should determine the timing for the study and ensure that a wide spectrum of interest groups is represented or has an opportunity to participate. At the conclusion of the study, the officials will be required to make a decision on the evidence presented at the public hearings by competing interest groups. They should also insist that the staff provide adequate information and stimulate participation from those groups who would not normally participate in the transportation planning process.

Citizens appointed to study committees normally have either special technical competence or derive a benefit from the outcome. Because it is difficult to combine citizens with different viewpoints, different committees should be created to study the economic, social, and environmental impacts of alternative plans based upon demand projections and impact data supplied by the staff and consultants.

The elected decision-makers and the technical staff also have an obligation to inform the "Mass Public". This can be done through judicious use of press-releases and news-briefings.

Establishment of transportation goals and objectives which translate the goals into tangible achievements should be the first priority for the technical study committees. The technical staff should help the committees by providing information and demonstra-

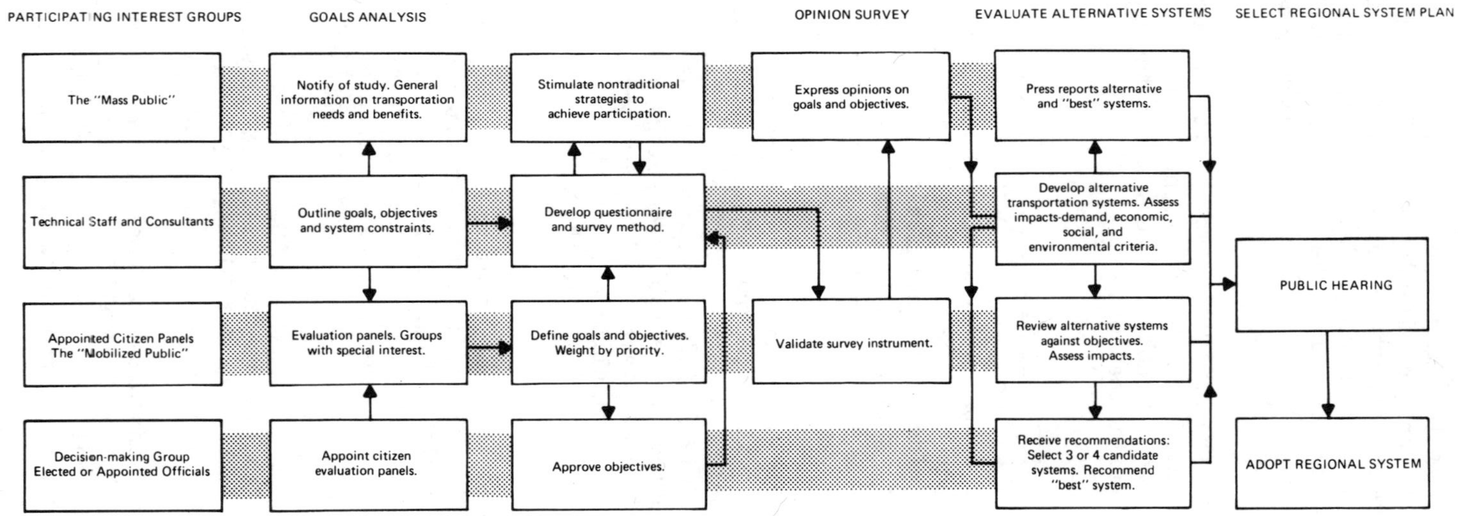

FIGURE 1. HYPOTHETICAL MODEL TO INCORPORATE CITIZEN PARTICIPATION IN THE METROPOLITAN REGION TRANSPORTATION PLANNING PROCESS

ting how the goals and objectives can be stated in a way in which they are useful for transportation planning.

Survey research can be used to check the validity of the goals and objectives. It is always possible that the views of the minority who choose to participate in the transportation planning process will differ substantially from those of the general public. By designing a survey instrument which summarizes the goals and objectives by interviewing a sample of the metropolitan population, the validity of the goals and objectives can be assessed. This is a time-consuming and expensive process. However, it does ensure that the goals and objectives fairly represent the overall attitudes towards transportation improvement which exist in the metropolitan area.

Using the trip-demand predictions and the social, economic, and environmental assessment information prepared by the technical staff and consultants, alternative system plans should be developed. Each plan should be assessed in terms of its advantages and disadvantages and weighted in respect to the community goals and objectives adopted by the decision-making authority. This is where the cost-benefit and cost-effectiveness approaches are amalgamated into the citizen participatory approach.

Three or four 'best' plans should be described in detail. Information should be provided to the press on these plans and they should be subjected to an in-depth review by the appointed citizen committees. A systems approach should be used by the committees so that the impact of each plan upon the criteria may be assessed and the criteria weighted for importance based upon community goals and objectives. By addition and deletion of route segments one plan may become more attractive from a technical viewpoint and derive substantial citizen support. If such a plan exists, the citizen committees should be encouraged to endorse this plan and organize their information to clarify the advantages and disadvantages.

The public hearing affords an opportunity for citizen committees to meet the "Mass Public". The advantages and disadvantages of the three or four best plans can be discussed and the superior quality of the selected plan can be outlined. It is then up to the elected officials to reconcile disagreements and, if there is sufficient level of public support, to adopt the final plan for the regional transportation system.

ORANGE COUNTY, CALIFORNIA

The development and implementation of the Master Plan for

Transit in Orange County, California, afforded the author an opportunity to test the above-described approach to citizen participation. Citizen groups were involved from the outset of the programme. They defined the goals and objectives; they assisted in the selection of the consultants and the writing of consultant contracts; they evaluated alternative plans and described them in community meetings. And, at the conclusion of the programme, a wide spectrum of citizen groups not only endorsed the plans as they were developed, but also endorsed the proposals for funding by local, state, and federal money.

Orange County is a suburban metropolitan area located between Los Angeles and San Diego on the Southern California coast. In 1975 it had a population of 1.7 million and anticipates 2.4 million by 1990.

Virtually no public transit was available when the planning and development programme was announced in 1972. The automobile was supreme. Within three years, 167 buses were operating county-wide and two Dial-A-Ride service modules were providing local services and connections to the conventional bus routes. Another six communities are scheduled to receive Dial-A-Ride services and another fifty conventional buses will be added before the short-range portion of the Master Plan is completed in 1976.

The long-range, rapid transit plan called for 137 miles of grade-separated rapid transit which was to be fully coordinated with the plans for Los Angeles and San Diego. Existing rail-lines were to provide the east-west corridor and freeway rights-of-way were proposed for a north-south service.

Whereas the backbone of the Master Plan was rapid transit, it was founded upon a saturation level of bus-service. Fifteen hundred buses were proposed to provide services along major highways and freeways on fifteen minute schedules, with DAR service for all communities.

Implementation of the entire Master Plan would have required fifteen years. It was estimated that the completed system would carry 600,000 passengers daily and would cost $1,350 million, plus annual operating subsidies of $111 million. All costs are in $1974.

Citizen Participation

Completion of the long-range portion of the Master Plan simultaneous with the introduction of bus services could not have been achieved without a broad base of public support. Such support was not pre-existing, but had to be developed by encouraging realistic participation for groups interested in transportation improvement. A commitment to transit was developed, and many individuals involved in the citizen panels have remained active to ensure implementation and periodic re-evaluation of the Plan.

Five panels were involved throughout the entire three-year development. They represented business, professional and civic associations who realized the long-range impact of transit upon the development of the County. The Technical Advisory Committee consisted of engineers and city managers from Orange County together with representatives from federal, state, regional, and private agencies. Business interests were represented by the Transportation Committee of the Orange County Chamber of Commerce. The Council of Aging created a panel to review proposals from the perspective of senior citizens, and the Human Relations Commission articulated the interest of the low income and minority populations.

A panel created by the League of Women Voters and the American Association of University Women was extremely influential. Members of this panel had a long-standing interest in transportation improvement. They were adamant that transit planning should be used as a means to promote corridors of development and redevelopment. They had the time and the skill to ensure that their point of view was heard and understood.

The five panels met separately to evaluate alternative plans. Public forums conducted by the Board of Directors toward the completion of the planning process served to bring the groups together. At these meetings, the groups articulated their particular interests in order to emphasize positions which might not otherwise have been incorporated by the staff and consultants. Perhaps the citizen's role is best demonstrated in their review and revision of the goals for the rapid transit corridor study.

Goals Analysis

Goal statements were prepared by the staff based upon the goals in other studies and desires expressed in community forums. Two meetings were held with each citizen panel to clarify the meaning of each goal. The goal statements were completely rewritten by the panels so that the meaning was clear to lay groups before any attempt was made to rank the goals in terms of priorities. Each panel ranked the goals separately. Panelists were given two dollars in pennies and asked to weigh their preference in monetary terms. All pennies were required to be used and no goal was to be given more than fifteen pennies (there were fifteen goal statements). Containers were labelled and pennies were allocated according to individual preference. Mean scores for each goal were then computed for each panel.

Considerable debate occurred between panelists over the relative merits of the fifteen goals, but by using money to indicate preference, panelists were placed in a choice situation with which they were familiar. Staff and consultants were always available to provide technical information on how the goals would influence the rapid transit plan.

Goal preferences were summarized for each interest group and presented to the Board of Directors. This comparison in ranking, together with oral presentations before the Board of Directors, assisted the Directors to adopt a set of goals.

The goals adopted by the Board of Directors became the objectives for the study and were used in the evaluation of alternative system plans. However, the assessments prepared by each interest group were retained by the consultants and used separately to weigh criteria representing the economic, social, and environmental impacts of each systems plan. By weighing the criteria separately for each interest group, a reliable index of group preferences for candidate plans was obtained.

An opinion survey was conducted to assess the correspondence between the attitudes of the panelists expressed by goal preference and the attitudes of the "Mass Public". A stratified random sample of households was selected and 1800 interviews were conducted. Although the survey indicated that there was a board basis of public support for the goals, this support was not uniform. The highest levels of support came from younger respondents and from those with upper levels of education and income and those in professional, technical and managerial occupations. These groups were less likely to believe that freeways and automobiles would remain the primary method of travel in the County. For this reason, they were willing to support the development of public transit. Middle-aged and middle-income people were less inclined to support long-range transit development. They saw the automobile as a satisfying means of transportation and supported goals which would improve transportation by automobiles. Because these preferences were known in advance, the consultants and staff were able to develop a hypothetical 'best' plan which was presumed to satisfy the objectives of the major interest groups.

A Hypothetical or 'Best' Systems Plan

Each panel was asked to take the hypothetical plan to the associations which they represented and to discuss the relative merits of the plan with the general public. Considerable discussion of the plan occurred which was led by influential community representatives. Staff and consultants were available as coordinators and information providers; their position was advisory. They served to coordinate the process and assist groups and individuals who desired to recommend changes.

Once the citizen groups had amended the hypothetical 'best' plan to reflect community desires, it was recommended to the Board of Directors. The Board held a series of public hearings on this plan. Considerable evidence of need was provided by cities who did not have direct access to one of the rapid transit corridors. The Board requested that the staff add two additional corridors to the plan. These had been considered as candidate plans, but had been omitted from the recommended plan because the additional ridership that these corridors generated was not sufficient to offset the increase in capital and operating cost. Because of the submissions made by the omitted cities as to future commercial and industrial developments in their areas, the Board of Directors added the corridors to the plan.

A philosophy that the plan was more flexible than the people underlay the 'coordinator-catalyst' approach. The willingness of the professional staff and consultants to accept the additional corridors was indicative of this approach. These corridors are of low priority, but if the developments occur which the city officials project, then the Plan can be re-evaluated. The additional corridors had some merit, and rejection would have created opposition from the cities which could have jeopardized the entire Plan.

CONCLUSION

Without the citizen panels, it is unlikely that the Master Plan could have been completed, and partly implemented, within three years. Only a small proportion of the public actually participated, but this was sufficient to direct the Plan in a manner which was acceptable to the "Attentive" or "Mobilized Public". It also secured adoption of the Plan by the Board of Directors. However, the public participation strategy was not effective in eliciting support from the "Mass Public".

The opinion survey detected a lack of support in the large, middle-income segment of the Orange County public. This group did not oppose adoption of the Plan. However, they voted against the proposal to accelerate development by adding one cent to the sales tax.

Only thirty-eight percent voted in favour in the referendum of November, 1974. Despite the overwhelming defeat, the Plan remains. The referendum was not a test of the Plan, but an indicator of the pace at which the community desired transit development. Considerable local support has been generated for transit improvements, and segments of the Plan will be implemented as local matching funds become available.

REFERENCES

Burke, M. (1968) Citizen participation strategies. Journal of the American Institute of Planners, 34, 287-94.

Fielding, J. (1972) Structuring citizen involvement in freeway planning. Highway Research Board Record, 380, 23-36

Manheim, M.L. et al. (1974) Transportation decision-making: a guide to social and environmental considerations. Urban Systems Laboratory. MIT, Cambridge.

Rosenau, J.N. (1974) Citizenship Between Elections: An Inquiry into the Mobilizable American. New York: Free Press.

Webber, M.M. (1973) On the technics and politics of transportation planning. In Citizen Participation in Transportation Planning, Highway Research Board Special Report, 142, 43-51.

Wolpert, J., Mumphrey, A., & Seley, J. (1972) Metropolitan Neighbourhoods: Participation and Conflict over Change. Research Paper No.16. Commission on College Geography. Association of American Geographers.

INDEX